P9-CCM-625

Second Edition

Applied Math for Technicians

Claude S. Moore

Bennie L. Griffin

Edward C. Polhamus, Jr.

Mathematics Department
Danville Community College
Danville, Virginia

Prentice-Hall, Inc. *Englewood Cliffs, New Jersey 07632*

Library of Congress Cataloging in Publication Data

Moore, Claude S., 1944–
Applied math for technicians.

Includes index.
1. Mathematics–1961– . I. Griffin, Bennie L., 1935– . II. Polhamus, Edward C., 1948– . III. Title.
QA39.2.M64 1982 512′.13 81-13916
ISBN 0-13-041178-7 AACR2

Editorial/production supervision: *Karen Wagstaff and Kathryn Gollin Marshak*
Interior design: *Karen Wagstaff*
Manufacturing buyer: *Gordon Osbourne*
Drawings: *George E. Morris*

Printed in the United States of America

10 9 8 7 6

ISBN 0-13-041178-7

Prentice-Hall International, Inc., *London*
Prentice-Hall of Australia Pty. Limited, *Sydney*
Prentice-Hall of Canada, Ltd., *Toronto*
Prentice-Hall of India Private Limited, *New Delhi*
Prentice-Hall of Japan, Inc., *Tokyo*
Prentice-Hall of Southeast Asia Pte. Ltd., *Singapore*
Whitehall Books Limited, *Wellington, New Zealand*

To

Evelyn, Cheryl, Sheila

and the *Children*

Contents

Preface

Applied Math for Technicians has been revised to better meet the mathematical needs of machinists, auto mechanics, printers, construction workers, and other technicians and trades people. This book may be used in technical and vocational schools as well as junior and community colleges. It is designed to be used for part or all of a full academic year, with proper selection of topics.

Emphasis is on practical applications. *Applied Math for Technicians* contains over 200 worked examples and more than 1,800 exercises. Many exercises were suggested by technical students and instructors. More than 500 of the exercises apply directly to one or more of the professions or vocations mentioned above. The last chapter contains five topics on special interest applications that can be used in conjunction with earlier chapters or used independently to conclude a study of technical mathematics. There is an instructor's manual available for this textbook.

The development of the book grew out of a recognized need for a suitable text designed for students in the technical areas mentioned above. Before beginning the manuscript, the authors conducted research relating to appropriate content and structure of such a text. This research included contact with shop workers, vocational-technical instructors, students, and employers.

In addition to improved layout and minor modifications, there are several major changes in this new edition. Chapter 4 is totally new, reflecting a desire to better meet student needs in special types of arithmetic problems. Chapter 5 on systems of measure (English and metric) has been completely revised to introduce the student more gradually to the arithmetic of measurement. The chapters on algebra have been rearranged and combined for more efficient teaching. The chapter on graphs now begins with a more general approach which deals with several types of graphs before concentrating on line graphs.

We would like to thank users of the first edition who offered suggestions for improvements in the new edition. All of the suggestions were helpful to us, and they were used as guides during the revision process.

We express our sincere appreciation to both Danville Community College students and instructors in the particular technical areas for which this book was designed. We appreciate the encouragement and support of the administration of Danville Community College during the preparation of this book. Our appreciation also goes to Prentice-Hall, Inc. and the individuals there who made this revision possible.

C.S.M.
B.L.G.
Danville, Virginia *E.C.P.*

Chapter 1

Whole Numbers

Since we use the Hindu-Arabic number system, we shall say a few words about what this system is and why we use it. The Hindu-Arabic number system is simply what is usually referred to as the *decimal system of numbers*.

A number is nothing more than an idea, an abstraction, or a property, something like honesty, bravery, or color. We have a feeling for what numbers are, but we need some symbols to represent them. For instance, if we say that we have five marbles, then in order to show this on paper we would use the symbol (or numeral) 5 to represent the number of marbles. The Romans would have used the symbol V to represent five marbles. The early Egyptians would have written I I I I I.

The Hindu-Arabic number system (or decimal system) is most widely used because it is the easiest and most economical. Consider writing the number *one thousand seven hundred seventy-six*. In the decimal system this would be 1,776. In the early Egyptian number system this would be 𓆼𓍢𓍢𓍢𓍢𓍢𓍢𓍢𓎆𓎆𓎆𓎆𓎆𓎆𓎆IIIIII, and in the Roman number system this would be MDCCLXXVI.

> *Definition:* A *whole number* is any number of the set $\{0, 1, 2, 3, \ldots\}$.

(The three dots mean that the numbers continue without stopping.)

1-1 PLACE VALUE

In order to understand the decimal system of whole numbers better, let us briefly review what is meant by place value. If we write a numeral such as 128, we really mean 100 + 20 + 8. That is, we have 1 hundred, we have 2 tens, and we have 8 ones. Listed below is a chart that should give us a quick review of place value for the numbers 128 (one hundred twenty-eight), 1,972 (one thousand nine hundred seventy-two), and 65,849,047 (sixty-five million eight hundred forty-nine thousand forty-seven).

TABLE 1-1 Place Value Chart

Ten Billions	Billions	Hundred Millions	Ten Millions	Millions	Hundred Thousands	Ten Thousands	Thousands	Hundreds	Tens	Ones
								1	2	8
							1	9	7	2
			6	5	8	4	9	0	4	7

Exercise Set 1-1: Write each number in Hindu-Arabic notation.

Example 1-1: One thousand nine hundred seventy-two: **1,972**

1. Twenty-five
2. Three thousand one hundred one
3. Five million four thousand four
4. Five million forty thousand four hundred four
5. Forty-seven thousand five hundred two
6. One hundred fifty million four thousand seventy-six
7. Thirteen thousand thirteen

Write each number as it should be read.

Example 1-2: 40,757: **Forty thousand seven hundred fifty-seven**

8. 278
9. 827
10. 414
11. 6,060
12. 77,777
13. 14,014
14. 7,077,700
15. 65,849,047

Answers

2. 3,101
4. 5,040,404

6. 150,004,076
8. Two hundred seventy-eight
10. Four hundred fourteen
12. Seventy-seven thousand seven hundred seventy-seven
14. Seven million seventy-seven thousand seven hundred

1-2 ADDITION

When adding numbers, we must add two numbers at a time. After adding two numbers, we may add other numbers to that sum and continue this process until we have added all the numbers. We must also be careful to position our numbers so that we add the ones digits together, the tens digits together, and so forth.

Example 1-3: Add 342 and 47.

$$\begin{array}{lr} \text{(addend)} & 342 \\ \text{(addend)} & \underline{+47} \\ \text{(sum)} & \mathbf{389} \end{array} \quad \text{or} \quad \begin{array}{r} 300 + 40 + 2 \\ \underline{+ 40 + 7} \\ 300 + 80 + 9 = \mathbf{389} \end{array}$$

Example 1-4: Add 297 and 165.

$$\begin{array}{r} 297 \\ \underline{+165} \\ \mathbf{462} \end{array} \quad \text{or} \quad \begin{array}{r} 200 + \ \ 90 + \ \ 7 \\ \underline{+100 + \ \ 60 + \ \ 5} \\ 300 + 150 + 12 \end{array}$$

Since 150 is more than 9 tens, we need to regroup (carry). This may be done as follows:

$$\begin{array}{l} 300 + 150 + 12 \\ \begin{array}{c}300\\100\end{array} + \ \ 50 + 12 \\ 400 + \begin{array}{c}50\\10\end{array} + \ \ 2 \\ 400 + \ \ 60 + \ \ 2 = \mathbf{462} \end{array}$$

Exercise Set 1-2: Find each of the following sums.

Part A

	(a)	(b)	(c)	(d)	(e)
1.	21 34	47 52	65 31	78 11	36 42

2.	38	54	65	91	87
	23	48	43	39	23
3.	301	257	634	399	489
	49	32	102	22	72
4.	114	432	780	279	365
	629	568	468	641	365
5.	3,267	6,902	5,073	6,623	4,321
	523	2,319	3,705	4,507	5,683

Part B

6.	342	561
	87	324
	102	132
7.	926	666
	32	333
	141	111
8.	8,201	5,460
	346	3,124
	5,213	503
	4,061	1,127

9. (a) 437 + 231 + 22 = (b) 506 + 422 + 101 =
10. (a) 924 + 613 + 210 = (b) 6,994 + 236 + 1,847 =

Part C

11. Find the sum of the first 10 whole numbers (0 + 1 + 2 + · · · + 9).
12. Sue March went to town and bought a skirt for $22. She bought a flower for $5, a pair of shoes for $42, and a hat for $17. How much money did Mrs. March spend?
13. During a basketball game, seven players scored the following points: 12, 9, 21, 16, 14, 17, and 35. If these were the only scores for this team, what was this team's final score?
14. There are three machines available to cut a certain metal. Machine A cuts 13 in. per minute, Machine B cuts 8 in. per minute, and Machine C cuts 11 in. per minute. How many inches of this metal can be cut in 1 min if all three machines are working?
15. An *odd whole number* may be represented as $2n + 1$ where n is any whole number. Find the sum of the first seven odd whole numbers.

Answers

2. (a) 61	(b) 102	(c) 108	(d) 130	(e) 110
4. (a) 743	(b) 1,000	(c) 1,248	(d) 920	(e) 730
6. (a) 531	(b) 1,017			
8. (a) 17,821	(b) 10,214			
10. (a) 1,747	(b) 9,077			
12. $86				
14. 32 in.				

1-3 SUBTRACTION

To subtract *one* number (A) *from another* number (B), we have

$$\begin{array}{r} \text{another} \\ -\text{one} \\ \hline \text{difference} \end{array} \quad \text{or} \quad \begin{array}{r} \text{B} \\ -\text{A} \\ \hline \text{difference} \end{array}$$

Example 1-5: Subtract 135 from 196.

$$\begin{array}{lr} \text{(minuend)} & 196 \\ \text{(subtrahend)} & -135 \\ \hline \text{(difference)} & \mathbf{61} \end{array} \quad \text{or} \quad \begin{array}{r} 100 + 90 + 6 \\ -100 - 30 - 5 \\ \hline +60 + 1 = \mathbf{61} \end{array}$$

Example 1-6: Suppose we wanted to subtract 238 from 353. Since 8 is larger than 3, we must borrow one 10 from the five 10's as follows:

$$\begin{array}{r} {\scriptstyle 4\ 13} \\ 3\,\not{5}\,\not{3} \\ -2\,3\,8 \\ \hline \mathbf{1\,1\,5} \end{array} \quad \text{or} \quad \begin{array}{r} 300 + 50 + 3 = \;\; 300 + 40 + 13 \\ -200 - 30 - 8 = -200 - 30 - \;\; 8 \\ \hline 100 + 10 + \;\; 5 = \mathbf{115} \end{array}$$

Exercise Set 1-3: Find the difference of each pair of numbers.

Part A

	(a)	(b)	(c)	(d)	(e)
1.	38 15	59 46	78 64	43 41	62 22
2.	96 47	34 28	83 64	77 58	94 52

3.

432	499	567	840	945
126	209	99	791	920

4.

5,834	7,401	9,043	6,431	4,109
5,625	4,893	6,020	5,555	343

5.

3,002	4,893	6,020	8,888	3,454
2,003	2,899	206	999	3,451

Part B

6.

43,021	67,682	10,910	14,163
3,242	6,768	7,893	13,159

7. (a) 432 - 16 = (b) 568 - 99 =
8. (a) 703 - 36 = (b) 421 - 43 =
9. (a) 3,021 - 119 = (b) 8,920 - 306 =
10. (a) 4,478 - 1,299 = (b) 6,403 - 4,235 =

Part C

11. If a board is 111 in. long, how much will be left if you cut a 79-in. piece off the board?
12. Joe weighs 236 lb now. Two years ago Joe weighed 294 lb. How many pounds did Joe lose during the past 2 years?
13. How much money does Jane have now if she had $146.58 and spent $79.69?
14. The cooling system of a car will hold 24 qt of water and antifreeze. If the cooling system is full and contains 8 qt of antifreeze, how many quarts of water are in the cooling system?
15. Jack sold 38 bolts from a box containing 150 bolts. How many bolts should be in the box?
16. If you printed 4,000 leaflets and distributed 2,461 of them, could you fill an order for 1,540 more?

Answers

2. (a) 49 (b) 6 (c) 19 (d) 19 (e) 42
4. (a) 209 (b) 2,508 (c) 3,023 (d) 876 (e) 3,766
6. (a) 39,779 (b) 60,914 (c) 3,017 (d) 1,004
8. (a) 667 (b) 378
10. (a) 3,179 (b) 2,168
12. 58 lb
14. 16 qt
16. No, you have 1,539 left.

1-4 MULTIPLICATION

Multiplication may be performed as repeated addition.

Example 1-7: The product of 4 times 147 is the same as

$$\begin{array}{r} 147 \\ 147 \\ 147 \\ +147 \\ \hline 588 \end{array}$$

Since repeated addition may be very time-consuming or infeasible, we choose to use multiplication instead of repeated addition whenever possible. The result of multiplication is called the *product*.

Example 1-8: Multiply 147 by 4.

$$\begin{array}{lrcr} \text{(multiplicand)} & 147 & & 100 + \ 40 + \ 7 \\ \text{(multiplier)} & \underline{\times 4} & \text{or} & \underline{\qquad\qquad\qquad \times 4} \\ \text{(product)} & \mathbf{588} & & 400 + 160 + 28 = \mathbf{588} \end{array}$$

The multiplicand and the multiplier are called the *factors* of the product.

Example 1-9: Multiply 28 by 12.

$$\begin{array}{rcc} 28 & & 28 \quad 28 \\ \underline{\times 12} & \text{or} & \underline{\times 10 + \times 2} \\ 56 & & 280 + \ 56 = \mathbf{336} \\ \underline{28\ \ } & & \\ \mathbf{336} & & \end{array}$$

Since $12 = 10 + 2$, the product of 28 and 12 is the same as 28 times 10 plus 28 times 2. It should be pointed out that (28)(12), $28 \cdot 12$, and 28×12 all indicate the multiplication of 28 and 12.

Exercise Set 1-4: Find the products.

Part A

	(a)	(b)	(c)	(d)	(e)
1.	$\begin{array}{r} 31 \\ \underline{9} \end{array}$	$\begin{array}{r} 42 \\ \underline{4} \end{array}$	$\begin{array}{r} 15 \\ \underline{7} \end{array}$	$\begin{array}{r} 63 \\ \underline{5} \end{array}$	$\begin{array}{r} 92 \\ \underline{6} \end{array}$

2.	14	21	43	67	73
	12	11	22	32	41
3.	402	514	938	604	372
	6	3	5	7	8
4.	382	927	476	925	653
	13	31	24	25	43
5.	792	105	814	222	584
	67	45	52	22	69

Part B

6.	9,243	3,065
	701	223
7.	4,488	3,334
	804	333

8. (a) (47)(50) = (b) (834)(20) =
9. (a) (7,620)(40) = (b) (615)(201) =
10. (a) (64)(20)(15) = (b) (55)(9) =

Part C

11. Six men weigh 213 lb each. Together, how much do the six men weigh?
12. If you drove your car for 12 hr at an average speed of 55 mph, how many miles did you go?
13. A certain machine can cut a 2-ft piece of metal in 13 sec. If 13 of these machines are used, how many feet of this metal can be cut in 13 sec?
14. In order to build a certain bookshelf, 13 pieces of 1 in. by 12 in. board 14 in. long are needed. How many inches of this 1 in. by 12 in. board are needed?
15. An electrician wants to buy 17 rolls of a type of wire that costs $97.43 per roll. How much will the wire cost? (You may omit tax.)
16. A ream of paper contains 500 sheets. How many sheets are contained in 23 reams of paper?

Answers

2.	(a) 168	(b) 231	(c) 946	(d) 2,144	(e) 2,993
4.	(a) 4,966	(b) 28,737	(c) 11,424	(d) 23,125	(e) 28,079
6.	(a) 6,479,343	(b) 683,495			
8.	(a) 2,350	(b) 16,680			
10.	(a) 19,200	(b) 495			
12.	660 mi				
14.	182″				
16.	11,500 sheets				

1-5 DIVISION

Division is the inverse operation of multiplication. In order to work division problems, we must know the multiplication facts up to 10 times 10.

Example 1-10: We say that 10 divided by 5 is equal to 2 because 5 times 2 is equal to 10. That is, $10 \div 5 = 2$ because $5 \cdot 2 = 10$.

Example 1-11: Divide 128 by 4.

$$\begin{array}{lrl} & 32 & \text{(quotient)} \\ \text{(divisor)} & 4\overline{)128} & \text{(dividend)} \\ & \underline{12} & \\ & 8 & \\ & \underline{8} & \\ & 0 & \text{(remainder)} \end{array}$$

so $128 \div 4 = 32$ because $4 \cdot 32 = 128$.

Example 1-12: Divide 327 by 12.

$$\begin{array}{lrl} & 27 & \text{(quotient)} \\ \text{(divisor)} & 12\overline{)327} & \text{(dividend)} \\ & \underline{24} & \\ & 87 & \\ & \underline{84} & \\ & 3 & \text{(remainder)} \end{array}$$

Check: $12 \cdot 27 = 324$, and then adding the remainder 3 gives $324 + 3 = 327$.

Example 1-13: Divide 3,124 by 31.

$$\begin{array}{r} 100 \\ 31\overline{)3{,}124} \\ \underline{3\,1} \\ 02 \\ \underline{0} \\ 24 \\ \underline{0} \\ 24 \end{array}$$

Check: $31 \cdot 100 = 3100$
$3100 + 24 = 3124$

Exercise Set 1-5: Find the quotient and remainder for each problem.

Part A

	(a)	(b)	(c)	(d)
1.	$3\overline{)225}$	$8\overline{)4{,}324}$	$6\overline{)3{,}201}$	$7\overline{)54{,}210}$
2.	$5\overline{)12{,}025}$	$4\overline{)36{,}021}$	$9\overline{)87{,}300}$	$7\overline{)30{,}426}$

3. $12\overline{)4{,}824}$ $15\overline{)75{,}030}$ $13\overline{)1{,}001}$ $36\overline{)18{,}972}$
4. $23\overline{)4{,}260}$ $21\overline{)42{,}021}$ $87\overline{)49{,}122}$ $56\overline{)32{,}450}$
5. $193\overline{)42{,}306}$ $102\overline{)36{,}018}$ $240\overline{)48{,}240}$ $347\overline{)92{,}451}$

Part B

6. (a) $432\overline{)65{,}056}$ (b) $304\overline{)42{,}100}$ (c) $542\overline{)65{,}204}$
7. (a) $563\overline{)74{,}312}$ (b) $248\overline{)147{,}026}$ (c) $472\overline{)35{,}603}$
8. (a) $3{,}402\overline{)617{,}428}$ (b) $5{,}421\overline{)524{,}076}$
9. (a) $432 \div 21 =$ (b) $7{,}440 \div 12 =$
10. (a) $63{,}200 \div 31 =$ (b) $754{,}500 \div 1{,}500 =$

Part C

11. The product of two numbers is 476. Find the second of the two numbers if the first one is 14.
12. A machinist bought 27 hacksaw blades, and gave the salesperson $26.19 for the blades. How much did each blade cost?
13. How many miles per gallon of gasoline can you travel if you can go 128 mi on 16 qt of gasoline?
14. If 28 times 14 is 392, what number must be divided into the product of 28 and 14 to give a quotient of 56?
15. A woman bought a car and made monthly payments for 3 years and 6 months. If the total cost (cost of car plus the insurance and carrying cost) was $8,427.30, how much did she pay per month?
16. The local cab company has 17 cars with six-cylinder engines that need the spark plugs replaced. If the plugs come in packages of eight, how many packages do you need to handle this job?
17. You are drilling holes in a beam to attach a sheet metal wall, and the weight of the metal suggests spacing the bolts no more than 9 in. apart. If the beam is 24 ft long, what is the minimum number of bolts needed for this job?
18. A printer tells you that your 5 in. by 8 in. paper stock is about 30 picas by 48 picas. When he suggests substituting some stock that is 54 picas by 72 picas, what size is he talking about in inches?

Answers

2. (a) 2,405 (b) 9,005, r of 1 (c) 9,700 (d) 4,346, r of 4
4. (a) 185, r of 5 (b) 2,001 (c) 564, r of 54 (d) 579, r of 26
6. (a) 150, r of 256 (b) 138, r of 148 (c) 120, r of 164
8. (a) 181, r of 1,666 (b) 96, r of 3,660
10. (a) 2,038, r of 22 (b) 503
12. 97¢ each

14. 7
16. 13
18. 9″ by 12″

1-6 POSITIVE POWERS OF WHOLE NUMBERS

Powers of whole numbers is a short form for expressing successive multiplication with the same number. A power is indicated by using a superscript (or exponent).

Example 1-14: $3 \cdot 3$ can be written as 3^2, read *the second power of 3*. The superscript 2 indicates that 3 is used twice in the product.

Example 1-15: $5^3 = 5 \cdot 5 \cdot 5 = 125$
(5 is used three times.)

Example 1-16: $10^4 = 10 \cdot 10 \cdot 10 \cdot 10 = 10{,}000$
(10 is used four times; four zeros are used.)

We shall discuss zero and negative powers of whole numbers in Chapter 3.

1-7 ORDER OF OPERATIONS

Up to this point, we have dealt with one operation at a time. Suppose you were asked to evaluate $\{12 - [(6 \cdot 7) \div 14] + 3^2\}$. What would your answer be? The parentheses ( ), brackets [ ], and braces { } are called *grouping symbols* and usually appear in this order: { [( )] }. Before stating the rules for order of operations, we need to place the operations in three different classes:

Rules for Order of Operations

Class A. Powers of numbers.

Class B. Multiplication and division.

Class C. Addition and subtraction.

Rule 1. Operations in parentheses, brackets, or braces are performed first. (The operations inside the innermost pair of grouping symbols should be performed first.) Apply Rules 1 and 2 as needed inside grouping symbols.

Rule 2. If operations are not in the same class, perform all operations in Class A, all in Class B, and all in Class C, in that order.

Rule 3. If operations are in the same class, perform operations from left to right as they appear.

Example 1-17

$$
\begin{aligned}
12 - [(6 \cdot 7) \div 14] + 3^2 &= 12 - (42 \div 14) + 3^2 \\
&= 12 - 3 + 3^2 \\
&= 12 - 3 + 9 \\
&= 9 + 9 \\
&= 18
\end{aligned}
$$

Note that we started with the innermost set of grouping symbols $(6 \cdot 7)$ and worked our way outward. Also, notice the 3^2 was written as 9 before any additions or subtractions because 3^2 is a Class A operation, while the addition is a Class C operation.

Example 1-18

$$
\begin{aligned}
4 \div 2 \cdot 5 \cdot 9 \div 3 &= 2 \cdot 5 \cdot 9 \div 3 \\
&= 10 \cdot 9 \div 3 \\
&= 90 \div 3 \\
&= 30
\end{aligned}
$$

> *Note:* When all operations are in the same class, Rule 3 tells us to work from left to right.

Example 1-19

$$
\begin{aligned}
86 - [5 \cdot (16 + 7 - 2) \div 3] &= 86 - (5 \cdot 21 \div 3) \\
&= 86 - (105 \div 3) \\
&= 86 - 35 = 51
\end{aligned}
$$

Exercise Set 1-7: Evaluate each expression below.

Part A

1. $12 + 4 \div 2$
2. $26 - 2 \cdot 8$
3. $5 \cdot 3 + 6^2 - 20 \div 2$
4. $6 \cdot 5 + 4 - 2 \cdot 3$
5. $10 \div 5 + 10 \div 2 \cdot 4$

Part B

6. $96 \div (3 \cdot 2) + 4 \cdot (3 + 5)$
7. $5^2 + 3 + 7 \cdot 4^2$
8. $16^2 - (9 + 3)^2 \div 6$
9. $[6 + (3 - 1) \cdot 2]$
10. $[(8 \div 2) - 3] + 7^2$

Part C

11. What is the combined length of 11 cinder blocks each of which is 8 in. short of 2 ft?
12. In wiring five houses, each house needs 250 ft of 110-volt wire and 75 ft of 220-volt wire. How many total feet of wire do you need for the job? (Work this problem in two different ways.)

Answers

2. 10
4. 28
6. 48
8. 232
10. 50
12. 1,625′

1-8 FACTORS AND MULTIPLES

The multiplicand and multiplier (Sec. 1-4) are called *factors*. For instance, since $3 \cdot 4 = 12$, we call 3 and 4 factors or divisors of 12.

> *Definition:* A *factor* or *divisor* of a given number is a number that will divide into the given number and leave a remainder of zero.

Since 3 is a factor of 12, we call 12 a multiple of 3.

> *Definition:* A number (A) is a *multiple* of another number (B) if B is a factor or divisor of A.

Divisors of 36 are 2, 3, 4, 6, 9, 12, and 18 because each of these numbers divides into 36 with a remainder of zero. The number one (1) and the number itself are always divisors of the number so 1 and 36 are also divisors of 36. [You should recall that multiplying or dividing a number by one (1) does not change the number.

Also, dividing a nonzero number by itself equals one (1).] We see that 36 is a multiple of 1, 2, 3, 4, 6, 9, 12, 18, and 36.

The process of determining whether or not a number is a divisor of another number often reduces to a matter of inspection. The following are five rules for the numbers 2, 3, 5, 9, and 10, respectively.

> Rule for 2: Two is a divisor of a number if and only if the last digit of the number is even (a multiple of 2, that is, 0, 2, 4, 6, 8).

Example 1-20: Is 2 a divisor of 3,478? Yes, because the last digit 8 is even (a multiple of 2).

> Rule for 3: Three is a divisor of a number if and only if the sum of the digits of the number is a multiple of 3.

Example 1-21: Is 3 a divisor of 1,731? Yes, because $1 + 7 + 3 + 1$ is equal to 12, and 12 is a multiple of 3.

> Rule for 5: Five is a divisor of a number if and only if the last digit of the number is 0 (zero) or 5.

Example 1-22: Is 5 a divisor of 78,401? No, because the last digit is not 0 or 5.

> Rule for 9: Nine is a divisor of a number if and only if the sum of the digits of the number is a multiple of 9.

Example 1-23: Is 9 a divisor of 387? Yes, because $3 + 8 + 7 = 18$, and 18 is a multiple of 9.

> Rule for 10: Ten is a divisor of a number if and only if the last digit of the number is 0.

Example 1-24: Is 10 a divisor of 3,475? No, because the last digit is not 0.

> *Definition:* A *prime number* is a number greater than one (1) that has only one and itself as factors or divisors.

Example 1-25: The number 2 is prime because its only factors are 1 and 2.

Example 1-26: The number 4 is not prime because 2 is a factor of 4 ($2 \cdot 2 = 4$).

The first 10 primes are 2, 3, 5, 7, 11, 13, 17, 19, 23, and 29.

> *Definition:* A *prime factor* of a given number is a factor of the given number that is also prime.

Example 1-27: The number 2 is a prime factor of 6. All prime factors of 6 are 2 and 3.

Example 1-28: Write 9, 12, 25, and 36 as products of prime factors.

$$9 = 3 \cdot 3 = 3^2 \qquad 12 = 3 \cdot 4 = 3 \cdot 2 \cdot 2 = 2^2 \cdot 3$$

$$25 = 5 \cdot 5 = 5^2 \qquad 36 = 4 \cdot 9 = 2 \cdot 2 \cdot 3 \cdot 3 = 2^2 \cdot 3^2$$

Example 1-29: Write 420 as the product of prime factors. Since 420 is much larger than the previous numbers, we shall work this out using two different methods.

Method 1	Method 2
$420 = 42 \cdot 10$	**2** | 420
$= 6 \cdot 7 \cdot 2 \cdot 5$	**2** | 210
$= 2 \cdot 3 \cdot 7 \cdot 2 \cdot 5$	**3** | 105
$= 2 \cdot 2 \cdot 3 \cdot 5 \cdot 7$	**5** | 35
$= 2^2 \cdot 3 \cdot 5 \cdot 7$	**7** | 7
	1

> *Explanation of Method 2:* Use the Rules for 2, 3, and 5 to determine the prime factor. Divide the smallest prime factor into the number ($420 \div 2$). Divide the quotient by its smallest prime factor ($210 \div 2$). Repeat this process until a quotient of 1 is reached. The prime factors of the number (420) will be the boldface numbers in the left column. So we write $420 = 2 \cdot 2 \cdot 3 \cdot 5 \cdot 7 = 2^2 \cdot 3 \cdot 5 \cdot 7$.

Exercise Set 1-8

Part A

1. Write the following as a power of a whole number.

   Example 1-30: $2 \cdot 2 \cdot 2 \cdot 2 = 2^4$

   (a) $6 \cdot 6 \cdot 6 =$
   (b) $4 \cdot 4 \cdot 4 \cdot 4 \cdot 4 =$
   (c) $12 \cdot 12 \cdot 12 \cdot 12 =$
   (d) $73 \cdot 73 =$
   (e) $17 \cdot 17 \cdot 17 \cdot 17 \cdot 17 \cdot 17 =$
   (f) $100 \cdot 100 \cdot 100 =$

2. Write the following as indicated products.

   Example 1-31: $14^3 = 14 \cdot 14 \cdot 14$

   (a) $17^2 =$ (b) $5^4 =$
   (c) $19^3 =$ (d) $13^5 =$
   (e) $193^4 =$ (f) $100^6 =$

3. List all possible factors of each of the following numbers.

   Example 1-32: 12: 1, 2, 3, 4, 6, 12

   (a) 5: (b) 8:
   (c) 10: (d) 22:
   (e) 45: (f) 100:

4. Find five multiples of each of the following numbers.

   Example 1-33: 7: 7, 14, 21, 28, 35

   (a) 3: (b) 4:
   (c) 6: (d) 10:
   (e) 13: (f) 23:

5. List the next five prime numbers:
   23, ___, ___, ___, ___, ___

Part B

6. Without actually dividing, test each of the following numbers to see if **2** is a factor, if **3** is a factor, if **5** is a factor, if **9** is a factor, and if **10** is a factor.
   (a) 224 (b) 333 (c) 3,475
   (d) 6,944 (e) 62,553 (f) 83,430
7. Using Method 1, write each number as the product of prime factors.
   (a) 56 = (b) 45 = (c) 96 =
   (d) 105 = (e) 250 = (f) 700 =
8. Using Method 2, write each number as the product of prime factors.
   (a) 150 = (b) 135 = (c) 500 =
   (d) 256 = (e) 351 = (f) 506 =

Using either Method 1 or Method 2, write each number as the product of prime factors.

9. (a) 402 = (b) 850 = (c) 900 =
10. (a) 1,000 = (b) 1,300 = (c) 1,480 =

Part C

11. Show that 59,706 is a multiple of 62.
12. Verify that 17 is a factor of 544.
13. If you traveled 543 mi in 9 hr, can your average speed in miles per hour be expressed using a whole number? Explain your answer.
14. Write the product of 12 and 22 in terms of prime factors.
15. Write five numbers so that 2,310 will be a multiple of each of them.

Answers

2. (a) $17 \cdot 17$
   (b) $5 \cdot 5 \cdot 5 \cdot 5$
   (c) $19 \cdot 19 \cdot 19$
   (d) $13 \cdot 13 \cdot 13 \cdot 13 \cdot 13$
   (e) $193 \cdot 193 \cdot 193 \cdot 193$
   (f) $100 \cdot 100 \cdot 100 \cdot 100 \cdot 100 \cdot 100$
4. (a) 3, 6, 9, 12, 15
   (b) 4, 8, 12, 16, 20
   (c) 6, 12, 18, 24, 30
   (d) 10, 20, 30, 40, 50
   (e) 13, 26, 39, 52, 65
   (f) 23, 46, 69, 92, 115
6. (a) 2 (b) 3, 9 (c) 5 (d) 2 (e) 3
   (f) 2, 3, 5, 9, 10
8. (a) $2 \cdot 3 \cdot 5^2$ (b) $3^3 \cdot 5$ (c) $2^2 \cdot 5^3$ (d) 2^8
   (e) $3^3 \cdot 13$ (f) $2 \cdot 11 \cdot 23$
10. (a) $2^3 \cdot 5^3$ (b) $2^2 \cdot 5^2 \cdot 13$ (c) $2^3 \cdot 5 \cdot 37$
12. $544 = 17 \times 32$
14. $2^3 \cdot 3 \cdot 11$

1-9 HIGHEST COMMON FACTOR

Now we are going to work with a special factor—the highest common factor.

> *Definition:* The *highest common factor* or *greatest common divisor* of two or more numbers is the highest (or largest) number that is a factor or divisor of each of the given numbers.

Example 1-34: The highest common factor of 6 and 4, denoted by HCF (6, 4), is 2 because 2 is the largest factor of 6 and 4.

In order to find the highest common factor of two numbers we have a three-step procedure.

> *Steps for Finding the Highest Common Factor*
>
> Step 1. Write each number in terms of prime factors so that like factors are in the same column.
>
> Step 2. Copy the factors that appear in every row.
>
> Step 3. Find the product of these factors. This product is the highest common factor.

Example 1-35: Find the HCF (60, 42).

Step 1. $60 = 2 \cdot 2 \cdot 3 \cdot 5$
$\quad\quad 42 = 2 \cdot \quad 3 \cdot \quad 7$

Step 2. $\quad 2 \quad 3$

Step 3. HCF $(60, 42) = 2 \cdot 3 = 6$

To find the greatest common divisor of more than two numbers, we can follow the same procedure as that for two numbers.

Example 1-36: Find the greatest common divisor of 75, 15, and 150, denoted by GCD (75, 15, 150).

Step 1. $75 = \quad 3 \cdot 5 \cdot 5$
$\quad\quad 15 = \quad 3 \cdot 5$
$\quad\quad 150 = 2 \cdot 3 \cdot 5 \cdot 5$

Step 2. $\quad 3 \quad 5$

Step 3. GCD $(75, 15, 150) = 3 \cdot 5 = 15$

1-10 LEAST COMMON MULTIPLE

In Sec. 1-8 we defined the word *multiple*. Now we shall define the least common multiple.

> *Definition:* The *least common multiple* of two or more numbers is the least (or smallest) number that is a multiple of each of the given numbers.

Example 1-37: Find the least common multiple of 6 and 15, denoted by LCM (6, 15). Some multiples of 6 are 6, 12, 18, 24, 30, 36. Some multiples of 15 are 15, 30, 45, 60. By looking at the multiples of 6 and those of 15, we see that the LCM (6, 15) = 30.

One way to find the least common multiple of two numbers is outlined below.

Steps for Finding the Least Common Multiple

Step 1. Write each number in terms of prime factors so that like factors are in the same column.

Step 2. Copy one factor from each column.

Step 3. Find the product of these factors. This product is the least common multiple.

Example 1-38: Find the LCM (45, 25).

Step 1.
$$\begin{aligned} 45 &= 3 \cdot 3 \cdot 5 \\ 25 &= \phantom{3 \cdot 3 \cdot{}} 5 \cdot 5 \end{aligned}$$

Step 2. 3 3 5 5

Step 3. LCM (45, 25) = $3 \cdot 3 \cdot 5 \cdot 5 = 225$

To find the least common multiple of more than two numbers, we can follow the same procedure as that for two numbers.

Example 1-39: Find the least common multiple of 9, 30, and 24, denoted by LCM (9, 30, 24).

Step 1.
$$\begin{aligned} 9 &= \phantom{2 \cdot 2 \cdot 2 \cdot{}} 3 \cdot 3 \\ 30 &= \phantom{2 \cdot 2 \cdot{}} 2 \cdot 3 \cdot \phantom{3 \cdot{}} 5 \\ 24 &= 2 \cdot 2 \cdot 2 \cdot 3 \end{aligned}$$

Step 2. 2 2 2 3 3 5

Step 3. LCM (9, 30, 24) = $2 \cdot 2 \cdot 2 \cdot 3 \cdot 3 \cdot 5 = 360$

Exercise Set 1-10

Part A

Find the highest common factor for each pair of numbers.

1. (a) 26, 12 (b) 39, 52
2. (a) 16, 36 (b) 102, 75
3. (a) 105, 70 (b) 68, 85

Find the least common multiple for each pair of numbers.

4. (a) 3, 7 (b) 8, 14 (c) 9, 21
5. (a) 16, 48 (b) 18, 8 (c) 15, 50

Part B

Find the highest common factor for each group of numbers.

6. (a) 16, 24, 72 (b) 15, 35, 70
7. (a) 48, 96, 128 (b) 4, 12, 16, 36
8. (a) 21, 35, 777 (b) 36, 72, 900

Find the least common multiple for each group of numbers.

9. (a) 6, 30, 14 (b) 2, 3, 7, 9 (c) 5, 12, 6
10. (a) 12, 18, 15 (b) 8, 12, 16 (c) 10, 105, 21

Part C

11. Jane has three pieces of old angle iron. She wants to cut each piece into shorter pieces that are as long as possible, but all the shorter pieces must have the same length. If her three pieces are 30 in., 72 in., and 108 in. long, how long should each of the shorter pieces be?
12. Two numbers (a and b) have an LCM of 35. If two other numbers (c and d) have LCM of 6, what is the LCM of all four numbers (a, b, c, and d)?
13. What is the HCF of the first 10 prime numbers?
14. A carpenter is building a kennel. He has determined that he will need some 2-ft, 3-ft, 4-ft, and 6-ft pieces of 2 in. by 4 in. lumber. If he can buy the 2 in. by 4 in. lumber in 8-ft, 10-ft, 12-ft, or 16-ft pieces, which length should he buy so that he can cut all 2-ft, 3-ft, 4-ft, or 6-ft lengths from one piece with no waste? (Assume he can buy only one length.)
15. What is the HCF of the LCM (3, 8) and the LCM (6, 12)?
16. Your machine shop produces special bolts that customers may order in multiples of 8, 12, or 20. How many should you package per box so that you will never have to send a partially filled box?
17. The newspaper that you print is delivered by paper carriers who have 45, 60, or 90 customers. How many papers should you bundle together to avoid splitting bundles for the paper carriers?
18. Over a period of several months, you have noticed that orders for a special item almost always call for 12, 18, 30, or 48 of them. How should you stack these items for easy filling of orders?

Answers

2. (a) 4 (b) 3
4. (a) 21 (b) 56 (c) 63
6. (a) 8 (b) 5
8. (a) 7 (b) 36
10. (a) 180 (b) 48 (c) 210
12. 210
14. 12′
16. 4
18. 6 per stack

Review Exercises

Part A

1. Find the sum.

   (a) $\begin{array}{r} 346 \\ 47 \\ \underline{204} \end{array}$ (b) $\begin{array}{r} 9{,}831 \\ 642 \\ \underline{509} \end{array}$ (c) $22 + 43 + 89 =$ (d) $75 + 105 + 20 =$

2. Find the difference.

   (a) $\begin{array}{r} 45{,}702 \\ \underline{36{,}813} \end{array}$ (b) $\begin{array}{r} 62{,}004 \\ \underline{3{,}005} \end{array}$ (c) $4{,}204 - 296 =$ (d) $5{,}862 - 5{,}814 =$

3. Find the product.

   (a) $\begin{array}{r} 4{,}721 \\ \underline{360} \end{array}$ (b) $\begin{array}{r} 6{,}298 \\ \underline{404} \end{array}$ (c) $(58)(35) =$ (d) $(3{,}264)(25) =$

4. Find the quotient and remainder.

   (a) $375\overline{)42{,}601}$ (b) $506\overline{)43{,}956}$ (c) $5{,}427 \div 9 =$ (d) $702 \div 32 =$

5. What is the maximum number of 12¢ stamps you can buy for $5.68?

Part B

6. A woman earned $4,355 in 13 weeks. How much did she earn per week?
7. Jean and Tim had $541.92 together. If Jean had $339.84 and Tim spent $27.43, how much does Tim have now?
8. A certain machine runs at an average of 720 rpm. How many revolutions per hour does this machine make?
9. The weight of 90 ft of a particular type of wire is 2 lb. What is the weight of 3 mi of this wire?

10. How long a piece of carpet should you buy for a runner for the steps shown in Fig. 1-1?

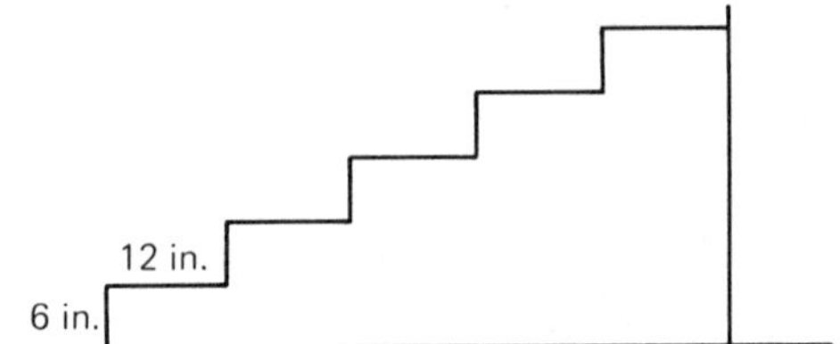

Figure 1-1

Part C

11. A service station owner buys spark plugs for $11.76 a dozen. If they are sold for $1.52 each, how much money is made on a dozen spark plugs?
12. Machine A will cut a piece of metal three times as fast as Machine B. If Machine A can cut 20 ft in 5 min, how long would it take Machine B to cut 20 ft?
13. A company sells regular gas for $1.54 per gallon and premium for $1.58 per gallon. You know that your car will average 24 mi per gallon on regular and 28 mi per gallon on premium. Which gas should you buy in order to save money? How much money would you save on gas for a trip of 126 mi?
14. You are asked to cut some 2 in. by 3 in. cards. You have cards that are 5 in. by 8 in., some 4 in. by 6 in., and some 8 in. by 10 in. Which size card should you use to cut in order to have the least amount of waste each time a larger card is cut? How many small pieces can be cut from each larger piece?
15. A garage has five mechanics, each of whom is paid $6.80 per hour for 40 hr and $10.20 per hour over 40 hr. It takes each of the five mechanics 7 hr overtime per week to do all the work. If the manager hires one more mechanic, the six mechanics can do all the work in 40 hr per week. Should the manager hire the sixth mechanic? How much money would he lose by making the wrong decision?

Answers

2. (a) 8,889 (b) 58,999 (c) 3,908 (d) 48
4. (a) 113, r of 226 (b) 86, r of 440 (c) 603 (d) 21, r of 30
6. $335
8. 43,200
10. 90″ or 7 1/2′
12. 15 min
14. 4″ by 6″

Chapter 2

Fractional Numbers

Suppose you had a 7-in. piece of lumber, and you wanted to cut it into two equal pieces. How long should each piece be? You are probably thinking that each piece would be three and one-half inches (written 3 1/2 in.) long. Now we are using a number that is not a whole number. The number 3 1/2 is called a *fractional number*. Some other fractional numbers are 1/3, 2/5, 7/4, 4 1/8. These numbers may also be written as

$$\frac{1}{3} \quad \frac{2}{5} \quad \frac{7}{4} \quad 4\frac{1}{8}$$

2-1 FRACTIONAL NUMBER

A fraction of a quantity may be thought of as part (or a portion) of a quantity.

> *Definition:* A *fractional number* is any number that can be written as one whole number divided by another nonzero whole number (i.e., $\frac{a}{b}$ or a/b, where a and b are whole numbers and b is not zero).

In the fraction above, the top number a is called the *numerator* and the bottom number b is called the *denominator*. From the definition, we see that the denominator cannot be zero.

The denominator tells us into how many pieces (or fractional parts) the quantity has been divided. The numerator tells us how many of these pieces are being considered. For example, suppose you have 2/3 of a pie. The 3 means that

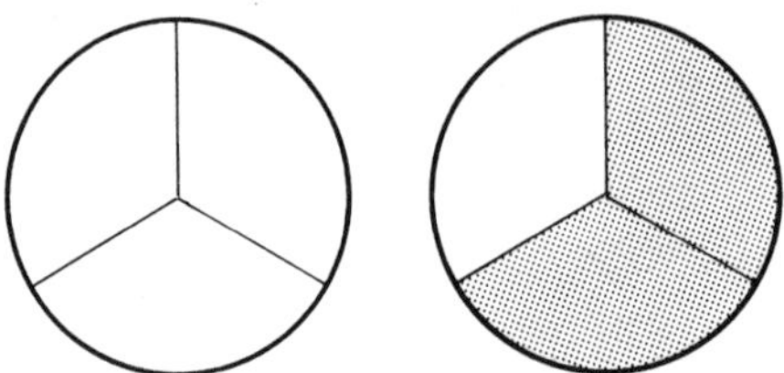

Figure 2-1

the pie has been divided into 3 equal pieces. The 2 means that you have (or are considering) 2 of these pieces. In Fig. 2-1, the sketch on the left shows the pie cut into 3 equal pieces. In the sketch on the right, the shaded region represents the 2 pieces referred to by the numerator, that is, 2/3 of the pie.

If the numerator of a fraction is smaller than the denominator, then the fraction is called a *proper fraction*. Some proper fractions are 1/3, 2/5, 7/9, and 5/10. When the numerator is equal to or larger than the denominator, the fraction is called an *improper fraction*. (The adjective *improper* does not suggest avoidance; it is just a technical classification.) Some improper fractions are 3/2, 12/7, 9/4, and 11/5. Each improper fraction can be expressed, and better understood, as a *mixed number*, that is, a whole number and a proper fraction combined. Some mixed numbers are 1 1/2, 1 5/7, 2 1/4, and 2 1/5. All of these numbers are fractional numbers. Notice that a whole number can be written as a fraction, such as 2 = 2/1 = 6/3. In order to change a mixed number to an improper fraction, we may follow the steps below.

Steps for Changing a Mixed Number to an Improper Fraction

Step 1. Multiply the whole number by the denominator of the fraction.

Step 2. Add the numerator to this product.

Step 3. Place the sum over the original denominator.

The result obtained from these three steps is the improper fraction desired.

Example 2-1: Change 5 7/8 to an improper fraction. Since $5 \cdot 8 = 40$ and $40 + 7 = 47$, we can write 5 7/8 = 47/8.

In order to change an improper fraction to a mixed number, we may follow the steps below.

Steps for Changing Improper Fractions to Mixed Numbers

Step 1. Divide the numerator by the denominator.

Step 2. Use the quotient as the whole number part of the mixed number.

Step 3. Form the fractional part using the original denominator and the remainder as the numerator.

Example 2-2: Convert 47/9 to a mixed number.

$$\begin{array}{r} 5 \\ 9\overline{)47} \\ \underline{45} \\ 2 \end{array} \qquad \frac{47}{9} = 5\frac{2}{9}$$

You may check this answer by converting 5 2/9 to an improper fraction.

2-2 EQUIVALENT FRACTIONAL NUMBERS

If John has a piece of paper 10/16 in. long and Sue has a piece of paper 5/8 in. long, are both pieces of paper the same length? By looking closely at Fig. 2-2, we can see that 10/16 and 5/8 represent the same portion of an inch. Therefore, both pieces of paper are the same length.

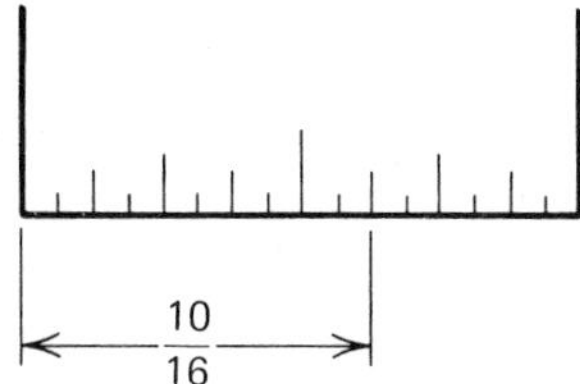

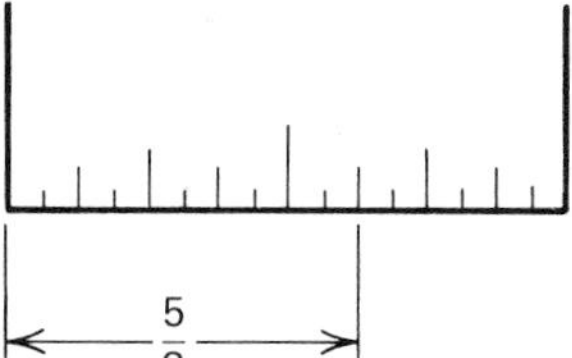

Figure 2-2

> *Definition:* Two fractional numbers or fractions are said to be *equivalent* (or equal) if they represent the same fractional amount of a quantity.

We may also say that two fractions are equivalent if both fractions can be written with the same numerator and the same denominator.

In Sec. 2-6 we shall show that multiplying the numerator and denominator of a fraction by the same number does not change the value of the fraction. The use of this fact to check the equivalence of two fractions is illustrated by the following examples.

> *Fundamental Principle of Fractions:* The numerator and the denominator of a fraction can be multiplied or divided by the same nonzero number without changing the value of the fraction.

Example 2-3: Is 2/3 equivalent to 4/6? We can write

$$\frac{2}{3} = \frac{2 \cdot 2}{3 \cdot 2} = \frac{4}{6}$$

Therefore, 2/3 and 4/6 are equivalent fractions.

Example 2-4

$$\frac{7}{8} = \frac{\#}{64}$$

Replace the # with the number that will make a true statement.

$$\frac{7}{8} = \frac{7 \cdot ?}{8 \cdot ?} = \frac{\#}{64}$$

Since $8 \cdot 8 = 64$, we can see that ? should be replaced by the number 8, giving us

$$\frac{7}{8} = \frac{7 \cdot 8}{8 \cdot 8} = \frac{56}{64}$$

Therefore, the # should be replaced by 56.

Example 2-5: What would we do if we had to find a replacement for # where #/5 = 4/10? We could write

$$\frac{4}{10} = \frac{2 \cdot 2}{5 \cdot 2} = \frac{2}{5}$$

Therefore, # should be 2.

A more mathematical definition is stated below.

> *Definition:* For any two fractions a/b and c/d, a/b = c/d if and only if (a)(d) = (b)(c).

Example 2-6: Is 7/16 equal to 20/47? By definition 7/16 = 20/47 if and only if (7)(47) = (16)(20). Since (7)(47) = 329 ≠ 320 = (16)(20), we say that 7/16 ≠ 20/47.

If two fractions are not equivalent, then one must be smaller than the other. If two or more fractions have the same denominator, then the fractions are related in the same way that their numerators are related.

> *Definition:* For any two fractions whose denominators are the same
>
> 1. a/b < c/b if a < c,
> 2. a/b > c/b if a > c,
> 3. a/b = c/b if a = c.
>
> *Note:* a < c is read *a is less than c* and a > c is read *a is greater than c*.

Example 2-7: Which sign, $<$, $>$, or $=$, shows the correct relationship between 9/11 and 10/11? From the previous definition, $9/11 < 10/11$ because $9 < 10$.

Example 2-8: Which sign, $<$, $>$, or $=$, shows the correct relationship between 3/7 and 3/8? Before we can compare these fractions, we must write each original fraction as an equivalent fraction so that *both new fractions* have the same denominator.

$$\frac{3}{7} = \frac{3 \cdot 8}{7 \cdot 8} = \frac{24}{56} \quad \text{and} \quad \frac{3}{8} = \frac{3 \cdot 7}{8 \cdot 7} = \frac{21}{56}$$

It is correct to write $21/56 < 24/56$ because $21 < 24$. So $3/8 < 3/7$.

Example 8 illustrates the fact that if two fractions have the same numerators, then the fraction with the larger denominator is the smaller fraction.

> *Definition:* A fraction is said to be in *lowest terms* or *reduced form* if the numerator and denominator have no common factor other than 1 (highest common factor).

Example 2-9: The fraction 2/5 is in lowest terms because HCF $(2, 5) = 1$.

The usual method for reducing fractions is to start by factoring both numerator and denominator into prime factors. Any factor that appears in both numerator and denominator should be marked out of both.

Example 2-10

$$\frac{15}{35} = \frac{3 \cdot \not{5}}{\not{5} \cdot 7} = \frac{3}{7}$$

Example 2-11

$$\frac{120}{32} = \frac{3 \cdot \not{2} \cdot \not{2} \cdot \not{2} \cdot 5}{2 \cdot \not{2} \cdot \not{2} \cdot \not{2} \cdot 2} = \frac{3 \cdot 5}{2 \cdot 2} = \frac{15}{4}$$

Exercise Set 2-2

Part A

1. Which of the following are fractional numbers?

   (a) 14 (b) 7 1/3 (c) 2/0 (d) 0/4 (e) 0/0

2. Identify each fractional number as a proper fraction, an improper fraction, or a mixed number.

   (a) 3/4 (b) 2/1 (c) 7/3 (d) 2 1/3 (e) 3/19
   (f) 2 5/8 (g) 14/5 (h) 3/7 (i) 104 2/9 (j) 14/1

3. Change each mixed number to an improper fraction.

(a) 7 2/3 (b) 11 5/8 (c) 3 14/15 (d) 5 2/5 (e) 4 1/8
(f) 9 3/4 (g) 100 2/7 (h) 99 1/2 (i) 33 4/5 (j) 18 9/10

Find the number that will make each statement true.

4. (a) 6/8 = ?/4 (b) 16/24 = ?/12
(c) 12/48 = ?/16 (d) 14/84 = ?/12

5. (a) ?/32 = 12/64 (b) 7/? = 56/64
(c) 2/3 = ?/12 (d) 12/33 = 4/?

Part B

Write the correct sign (<, >, =) between each pair of numbers.

6. (a) 4/5, 4/6 (b) 7/8, 5/8 (c) 11/16, 11/16 (d) 1/3, 1/4

7. (a) 25/2, 37/3 (b) 4/3, 5/2 (c) 17/4, 17/3 (d) 114/25, 23/5

Reduce each fraction to lowest terms.

8. (a) 12/14 (b) 15/20 (c) 36/8 (d) 13/39

9. (a) 114/7 (b) 93/102 (c) 75/35 (d) 453/33

10. Arrange in order with the smallest number first.

(a) 13/16, 7/8, 3/4 (b) 2/5, 2/3, 1/2 (c) 1/4, 1/5, 1/3
(d) 3/4, 11/16, 23/32 (e) 19/64, 19/32, 5/8

Part C

11. In Fig. 2-3(a)-(d), each number given represents the length of one side. Assign the proper value to each side of the figures.

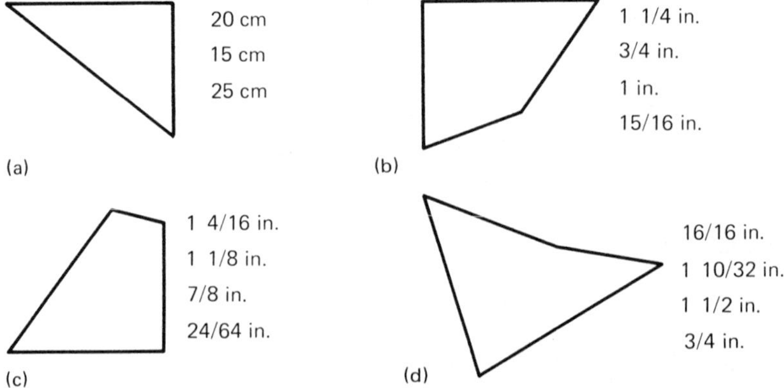

Figure 2-3

12. John wants to nail two boards together. The boards have a combined thickness of 2 7/8 in. John has three nails that are the same size except in length. The nails are 2 13/16 in., 2 15/16 in., and 2 3/4 in. long. If he wants to use

the longest nail that will not come out on the opposite side, what length nail should John use?

13. A piece of steel rod 5 13/64 in. long is needed. Jane cut a piece 5 3/16 in. long. Joe cut a piece 5 7/32 in. long. If the rod is too short, it will not work; however, if it is too long, it may work. Which rod is more likely to work?

14. Mrs. Smith has a piece of string that is 45/4 in. long. Mrs. Jones said that she needs a string at least 11 5/16 in. long. Is Mrs. Smith's string long enough for Mrs. Jones?

15. If you made a measurement that was halfway between 3/4 in. and 7/8 in., how could you express the exact measurement as a fraction?

16. An ad for your newspaper was canceled at the last minute, leaving a space 1 7/8 in. by 3 1/2 in. An AP wirephoto just came in with dimensions of 1 13/16 in. by 3 3/16 in. Will the photo need cropping in order to fit into the ad space?

Answers

2. (a) Proper (b) Improper (c) Improper (d) Mixed (e) Proper
   (f) Mixed (g) Improper (h) Proper (i) Mixed (j) Improper
4. (a) 3 (b) 8 (c) 4 (d) 2
6. (a) > (b) > (c) = (d) >
8. (a) 6/7 (b) 3/4 (c) 9/2 (d) 1/3
10. (a) 3/4, 13/16, 7/8 (b) 2/5, 1/2, 2/3 (c) 1/5, 1/4, 1/3
    (d) 11/16, 23/32, 3/4 (e) 19/64, 19/32, 5/8
12. 2 13/16″
14. No, 1/16″ short
16. No

2-3 ADDITION BY DEFINITION

In order to add two fractions, we use a rule that consists of four steps.

Steps for Adding Two Fractions

Step 1. Find the product of the first numerator and the second denominator.

Step 2. Find the product of the first denominator and the second numerator.

Step 3. Use the sum of the two products as the numerator of the answer.

Step 4. Use the product of the two denominators as the denominator of the answer.

These four steps are usually written in shorter form as the following definition.

Definition: If a/b and c/d are two fractions, then

$$\frac{a}{b} + \frac{c}{d} = \frac{(a \cdot d) + (b \cdot c)}{b \cdot d}$$

Example 2-12: Add 3/8 and 1/5. Express answer in lowest terms.

$$\frac{3}{8} + \frac{1}{5} = \frac{(3 \cdot 5) + (8 \cdot 1)}{8 \cdot 5} = \frac{15 + 8}{40} = \frac{23}{40}$$

and 23/40 is in lowest terms because HCF (23, 40) = 1.

Example 2-13: Find the sum of 3/4 and 7/12. Express sum in lowest terms.

$$\frac{3}{4} + \frac{7}{12} = \frac{(3 \cdot 12) + (4 \cdot 7)}{4 \cdot 12} = \frac{36 + 28}{48} = \frac{64}{48} = \frac{4}{3}$$

and 4/3 is in lowest terms because HCF (4, 3) = 1.

Example 2-14: Add 3 1/2 and 2/3. First convert the mixed number 3 1/2 to an improper fraction. Since 3 1/2 = 3 + 1/2, we have

$$3\ 1/2 = 3 + 1/2 = \frac{3}{1} + \frac{1}{2} = \frac{3 \cdot 2 + 1 \cdot 1}{1 \cdot 2} = \frac{6 + 1}{2} = \frac{7}{2}$$

so

$$3\ 1/2 + 2/3 = \frac{7}{2} + \frac{2}{3} = \frac{7 \cdot 3 + 2 \cdot 2}{2 \cdot 3} = \frac{21 + 4}{6} = \frac{25}{6} \quad \text{or} \quad 4\ 1/6$$

Example 2-15: Find the sum of 12 2/3 and 3 1/6. Converting each mixed number to an improper fraction gives 12 2/3 = 38/3 and 3 1/6 = 19/6, so

$$12\ 2/3 + 3\ 1/6 = 38/3 + 19/6 = \frac{38 \cdot 6 + 3 \cdot 19}{3 \cdot 6} = \frac{228 + 57}{18} = \frac{285}{18}$$

Therefore 285/18 = 95/6 or **15 5/6.**

Another approach to this problem is as follows:

$$12\frac{2}{3} + 3\frac{1}{6} = \left(12 + \frac{2}{3}\right) + \left(3 + \frac{1}{6}\right) = (12 + 3) + \left(\frac{2}{3} + \frac{1}{6}\right)$$

$$= 15 + \left(\frac{4}{6} + \frac{1}{6}\right)$$

$$= 15 + \frac{5}{6} \quad \text{or} \quad \mathbf{15\ 5/6}$$

2-4 ADDITION USING LEAST COMMON DENOMINATOR

In Sec. 1-10, we discussed the least common multiple, LCM, of two or more numbers. Now we wish to discuss the least common denominator of two or more fractions.

> *Definition:* The *least common denominator* of two or more fractions is the smallest number which is a multiple of each denominator (least common multiple).

In many problems it is easier to add fractions by using the least common denominator, denoted LCD, than by using the definition given in Sec. 2-3.

Example 2-16: Add 3/16 and 5/24 by definition.

$$\frac{3}{16} + \frac{5}{24} = \frac{3 \cdot 24 + 16 \cdot 5}{16 \cdot 24} = \frac{72 + 80}{384} = \frac{152}{384} = \frac{19}{48}$$

Example 2-17: Add 3/16 and 5/24 using LCD.

$$\text{LCD}\left(\frac{3}{16}, \frac{5}{24}\right) = \text{LCM}\ (16, 24) = 48$$

$$\frac{3}{16} + \frac{5}{24} = \frac{9}{48} + \frac{10}{48} = \frac{19}{48}$$

In Example 2-17, we changed each fraction to an equivalent one with the least common denominator 48 as the denominator. That is,

$$\frac{3}{16} = \frac{3 \cdot 3}{16 \cdot 3} = \frac{9}{48} \quad \text{and} \quad \frac{5}{24} = \frac{5 \cdot 2}{24 \cdot 2} = \frac{10}{48}$$

Once this is done, we add the numerators and put that sum over the LCD. So the answer is 19/48.

Example 2-18: Find the sum of 18 1/24 and 2 5/48. The LCD is 48. Therefore, we have

$$18\frac{1}{24} = 18 + \frac{1}{24} = 18 + \frac{2}{48} \quad \text{and} \quad 2\frac{5}{48} = 2 + \frac{5}{48}$$

Thus

$$18\frac{1}{24} + 2\frac{5}{48} = \left(18 + \frac{2}{48}\right) + \left(2 + \frac{5}{48}\right) = (18 + 2) + \left(\frac{2}{48} + \frac{5}{48}\right)$$

$$= 20 + \frac{7}{48} \quad \text{or} \quad 20\frac{7}{48}$$

Exercise Set 2-4

Part A

Add the following fractional numbers and express your answers in reduced form.

1. (a) 2/3 + 1/3 (b) 3/4 + 1/2 (c) 7/8 + 1/4 (d) 5/9 + 1/3
2. (a) 2/7 + 2/5 (b) 4/5 + 2/9 (c) 1/6 + 1/7 (d) 3/10 + 1/3
3. (a) 2 1/3 + 1/3 (b) 4 2/5 + 3 (c) 1 6/7 + 2/7 (d) 3 4/5 + 13 5/8
4. Find the least common denominator for the following pairs of fractions.
   (a) 2/3, 1/2 (b) 3/4, 3/8 (c) 7/8, 1/3 (d) 1/8, 1/9
   (e) 3/5, 2/7 (f) 15/2, 4/9 (g) 1/6, 1/24 (h) 5/13, 4/39
   (i) 14/56, 7/42 (j) 5/6, 2/5 (k) 3/2, 7/8 (l) 4/3, 3/4
5. Find the sum of each pair of fractions in Exercise 4.

Part B

Find the missing dimensions in Exercises 6–8.

6\.

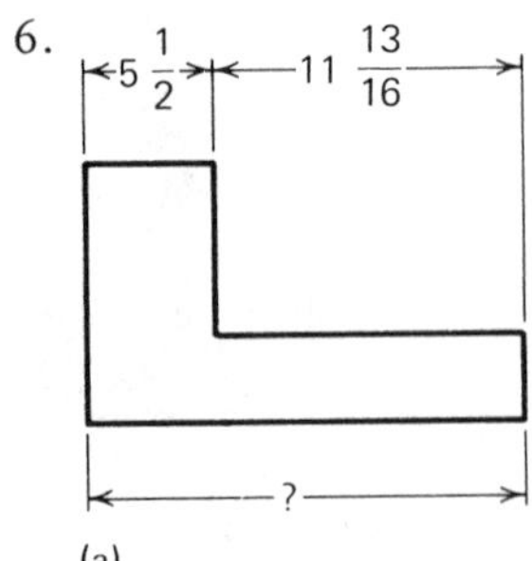

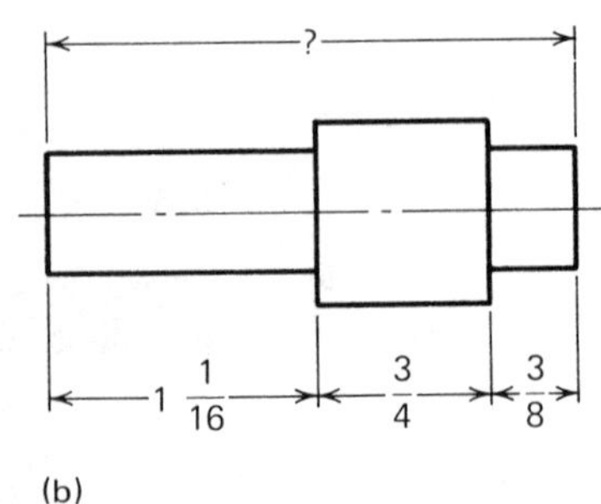

Figure 2-4

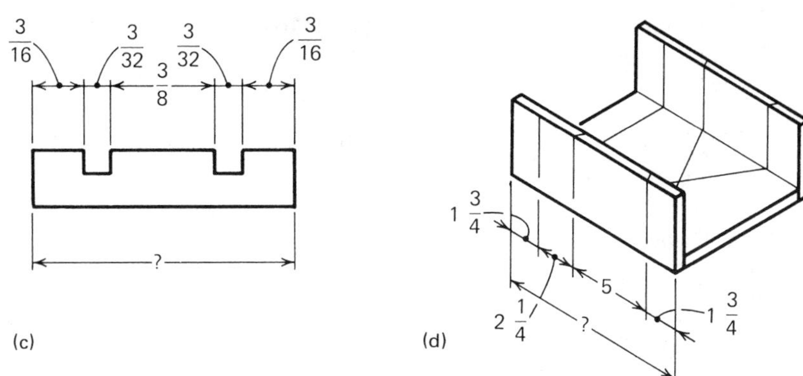

Figure 2-4. continued.

7.

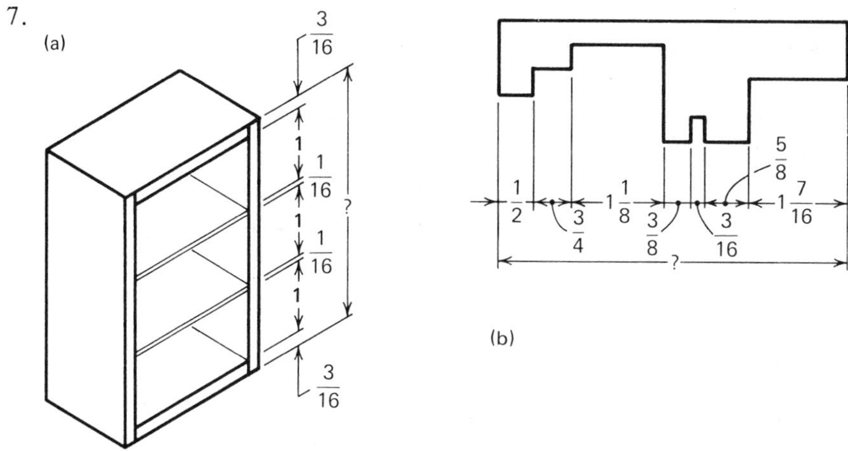

Figure 2-5

8.

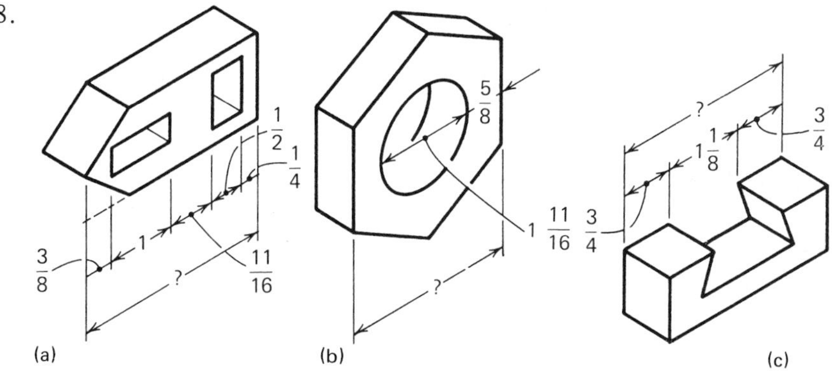

Figure 2-6

9. The inside diameter of a steel tube is 2 5/16 in. Its wall thickness is 7/32 in. What is its outside diameter?
10. How many 64ths of an inch are there in 5/8 of an inch?
11. You are asked to print calling cards for your boss. Her name section is 8 1/2 picas wide and the address section is 6 3/4 picas wide. You must leave a width of 2 3/4 picas between the name and the address and a 1 1/2 pica margin on each side. How wide should you make the card?

Part C

12. A man does 1/3 of a job before lunch and 3/5 of it after lunch. How much of the job has he done? Which part of the day did he do the most work?
13. Four boards of thicknesses 5/8 in., 15/32 in., 3/4 in., and 3/8 in. are to be glued together. Allowing 1/16 in. for glue between each two boards, how thick will the final product be?
14. A stock that closed Monday at 42 5/8 gained 1 3/4 points on the following day. At what price did it close on Tuesday?
15. Which of the following fractions is not equivalent to the other three fractions: 14/21, 38/57, 18/28, 24/36?
16. Jennifer has two bottles the same size. Both bottles contain the same type of developing solution. If one bottle is 2/3 full and the other is 1/4 full, can she put all the solution in one bottle? Explain.

Answers

2. (a) 24/35 (b) 46/45 or 1 1/45 (c) 13/42 (d) 19/30
4. (a) 6 (b) 8 (c) 24 (d) 72 (e) 35 (f) 18
(g) 24 (h) 39 (i) 168 (j) 30 (k) 8 (l) 12
6. (a) 17 5/16 (b) 2 3/16 (c) 15/16 (d) 10 3/4
8. (a) 2 13/16 (b) 2 15/16 (c) 2 5/8
10. 40
12. 14/15, afternoon
14. 44 3/8
16. Yes, because 2/3 + 1/4 = 11/12, which is less than one bottle full.

2-5 SUBTRACTION

To subtract *one* fraction *from another* fraction, we have

$$\begin{array}{r} \text{another} \\ -\text{one} \\ \hline \text{difference} \end{array}$$

Steps for Subtracting Two Fractions

Step 1. Find the product of the first numerator and the second demoninator.

Step 2. Find the product of the first denominator and the second numerator.

Step 3. Use the difference of the two products as the numerator of the answer.

Step 4. Use the product of the two denominators as the denominator of the answer.

The four steps above are written in shorter form in the following definition.

Definition: If a/b and c/d are two fractions, then

$$\frac{a}{b} - \frac{c}{d} = \frac{a \cdot d - b \cdot c}{b \cdot d}$$

Example 2-19

$$\frac{5}{8} - \frac{1}{4} = \frac{5 \cdot 4 - 8 \cdot 1}{8 \cdot 4} = \frac{20 - 8}{32} = \frac{12}{32} = \frac{3}{8}$$

Example 2-20

$$\frac{23}{25} - \frac{1}{2} = \frac{23 \cdot 2 - 25 \cdot 1}{25 \cdot 2} = \frac{46 - 25}{50} = \frac{21}{50}$$

Just as in the addition of fractions, it is usually easier to subtract fractions by using the least common denominator.

Example 2-21: 23/28 – 1/12 (using LCD). The LCD is 84.

$$\frac{23}{28} = \frac{23 \cdot 3}{28 \cdot 3} = \frac{69}{84} \quad \text{and} \quad \frac{1}{12} = \frac{1 \cdot 7}{12 \cdot 7} = \frac{7}{84}$$

So

$$\frac{23}{28} - \frac{1}{12} = \frac{69}{84} - \frac{7}{84} = \frac{62}{84} = \mathbf{\frac{31}{42}}$$

Example 2-22: 23/28 – 1/12 (by definition)

$$\frac{23}{28} - \frac{1}{12} = \frac{23 \cdot 12 - 28 \cdot 1}{28 \cdot 12} = \frac{276 - 28}{336} = \frac{248}{336} = \mathbf{\frac{31}{42}}$$

The example by definition looks easier; however, when you consider finding the products 23 · 12 and 28 · 12 and reducing 248/336 to lowest terms, the example using LCD is easier.

Example 2-23

$$5\frac{3}{8} - 1\frac{1}{8} = \frac{43}{8} - \frac{9}{8} = \frac{34}{8} = \frac{17}{4} = 4\frac{1}{4}$$

Exercise Set 2-5

Part A

1. Find each of the following sums.
   (a) 3 5/6 + 7 2/5 (b) 9 1/2 + 12 2/3
   (c) 5 3/8 + 2 1/5 + 12 3/10 (d) 15 3/4 + 17 4/5
   (e) 4 1/4 + 9 5/32 (f) 19 + 16 3/5 + 12 4/7 + 1/2
2. Subtract the first fraction from the sum of the fractions for each part of Exercise 1.
3. Find the difference for each pair of fractional numbers.
   (a) 7/16 - 1/4 (b) 13 1/2 - 4/15 (c) 14 1/10 - 13 1/8
   (d) 1/8 - 1/32 (e) 7/8 - 4/5 (f) 107 15/16 - 99 29/32

Part B

4. If 3/7 and 1/14 are subtracted, what would be the denominator of the answer in reduced form?

Find the missing dimension(s) in each figure.

5.

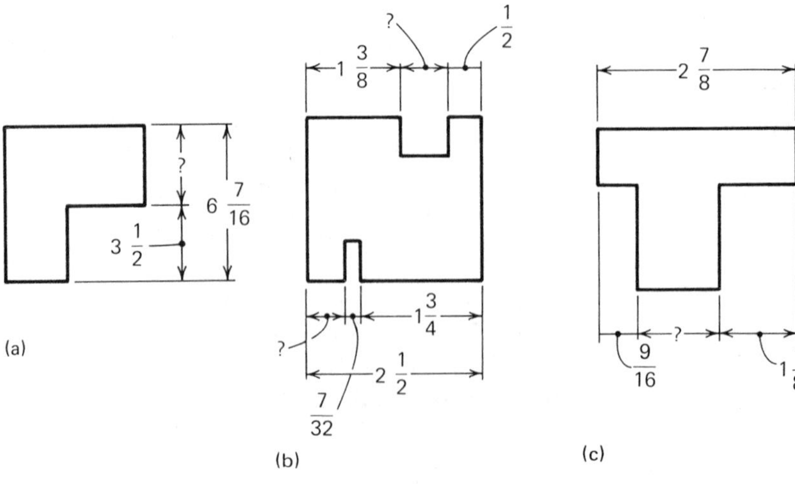

Figure 2-7

6.

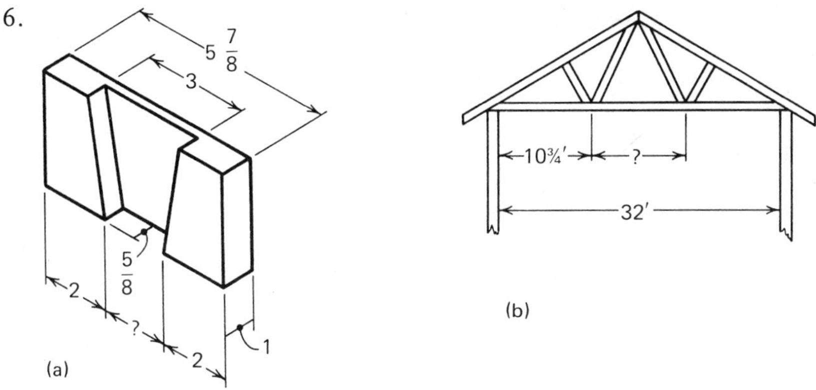

Figure 2-8

7.

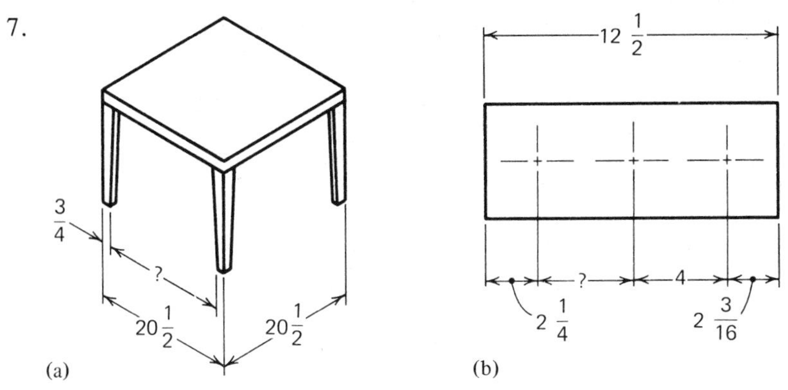

Figure 2-9

8. The difference between two numbers is 5/8. The smaller number is 2 1/3. What is the larger number?

Part C

9. From a steel strip 59 1/2 in. long, pieces 8 1/2, 11 1/4, 21 1/2, 3 9/32, 6 3/16, and 2 1/8 in. long were sheared. What length piece was left?
10. Find the difference between the sum of 3/4 and 7/9 and the sum of 2/5 and 3/4.
11. Pam has a board that is 27 3/16 in. long. If she cuts 5 8/32 in. off the board, will the remaining piece of board fit a length of 21 15/16 in.?
12. Jerry has 20 3/8 lb of paper. Together Tom and Jerry have 89 5/12 lb of paper. How many pounds of paper has Tom?

Answers

2. (a) 7 2/5 (b) 12 2/3 (c) 14 1/2
   (d) 17 4/5 (e) 9 5/32 (f) 29 47/70
4. 14
6. (a) 1 7/8 (b) 10 1/2
8. 2 23/24
10. 17/45
12. 69 1/24

2-6 MULTIPLICATION

Suppose you had a sheet of paper that measures 8 1/2 in. wide and 11 in. long. How many square inches of paper do you have? To find this we multiply the width by the length. (Areas are discussed in Chapter 12.) Our problem is to find the product of 8 1/2 and 11. There are different ways of working this problem, but we want to use a simple way that works for all multiplication problems.

Steps for Multiplying Fractions

Step 1. Multiply the two numerators and use the product as the numerator of the answer.

Step 2. Multiply the two denominators and use the product as the denominator of the answer.

These two steps for multiplying fractions are expressed in a shorter form in the following definition.

Definition: If a/b and c/d are fractions, then

$$\frac{a}{b} \cdot \frac{c}{d} = \frac{a \cdot c}{b \cdot d} \quad \text{or} \quad \frac{ac}{bd}$$

Example 2-24: Find the product of 1/2 and 2/3.

$$\frac{1}{2} \cdot \frac{2}{3} = \frac{1 \cdot 2}{2 \cdot 3} = \frac{2}{6} = \frac{1}{3}$$

or

$$\frac{1}{2} \cdot \frac{2}{3} = \frac{1 \cdot 2}{2 \cdot 3} = \frac{2 \cdot 1}{2 \cdot 3} = \frac{1}{3}$$

Example 2-25: Multiply 8 1/2 by 11. Since 8 1/2 = 17/2 and 11 = 11/1,

$$8\frac{1}{2} \cdot 11 = \frac{17}{2} \cdot \frac{11}{1} = \frac{17 \cdot 11}{2 \cdot 1} = \frac{187}{2} \quad \text{or} \quad 93\frac{1}{2}$$

Example 2-26: Multiply 8 1/2 by 11. Since 8 1/2 = 8 + 1/2,

$$8\frac{1}{2} \cdot 11 = \left(8 + \frac{1}{2}\right) \cdot 11 = 8 \cdot 11 + \frac{1}{2} \cdot 11 = 88 + 5\frac{1}{2} = 93\frac{1}{2}$$

Therefore, an 8 1/2 in. by 11 in. sheet of paper contains 93 1/2 in.^{2} (square inches) of paper.

Note

$$\frac{3}{5} \cdot \frac{2}{2} = \frac{3 \cdot 2}{5 \cdot 2} = \frac{6}{10}$$

Since 2/2 = 1, we are really looking at 3/5 · 1 = 3/5. We have shown that 3/5 = 6/10, and this justifies the statement in Sec. 2-2 that multiplying the numerator and denominator of a fraction by the same number does not change the value of the fraction. For instance when you write

$$\frac{7}{8} = \frac{7 \cdot 4}{8 \cdot 4} = \frac{28}{32}$$

you are really doing the following:

$$\frac{7}{8} = \frac{7}{8} \cdot 1 = \frac{7}{8} \cdot \frac{4}{4} = \frac{7 \cdot 4}{8 \cdot 4} = \frac{28}{32}$$

Exercise Set 2-6

Find each of the indicated products and express the answer in reduced form.

Part A

	(a)	(b)	(c)	(d)
1.	2/3 · 1/2	3/5 · 1/3	3/4 · 2/9	5/7 · 2
2.	3/4 · 2/5	3/5 · 4/5	3/5 · 1/3	7/8 · 1/7
3.	3/2 · 1/4	4/7 · 1/8	5/9 · 3/7	4/5 · 2/3
4.	1/12 · 2/3	4/3 · 5/2	9/4 · 1/12	6/5 · 5/6
5.	7/9 · 3/5	12/9 · 10/9	8/3 · 4/6	21/7 · 3/6

Part B

	(a)	(b)	(c)	(d)
6.	3 1/5 · 4 1/2	3 1/3 · 4	2 2/5 · 7/3	5 · 12/5
7.	8 1/2 · 5 1/2	9 · 7 2/3	10 1/4 · 4	6 3/5 · 3 5/6
8.	4 2/9 · 3 2/9	11 1/3 · 3 1/11	2 4/5 · 5	12 1/3 · 3 1/3

9. If John has 5/8 of a pint of Scotch and Tom has 1 1/2 glasses of Scotch, who has more Scotch? How much more does he have? (*Hint:* There are 16 oz in a pint and 8 oz in a glass.)
10. You have a sheet of paper that measures 8 1/2 in. by 11 in. What are three possible dimensions of one-fourth of the sheet of paper?

Part C

11. A piece of flat metal measures 13 1/2 in. long and 6 3/4 in. wide. If the metal is cut into two equal pieces, how many square inches are there in each piece?
12. Find the product of 17 2/7, 3 1/2, and 4 2/3.
13. Seven people earn $379.84 each per week. If each person saves 3/16 of his earnings, how much will all seven people save in 1 year? (Consider 52 paychecks per person per year.)
14. A certain machine can cut 168 3/8 in. of metal in 1 min. How many inches of the metal can this machine cut in 15 sec?
15. A cinder block wall is 23 1/2 blocks long and 14 blocks high. With the concrete joint a block measures 8 1/4 in. by 16 1/4 in. If you are going to paint this wall, how large an area (in square inches) should you plan to paint?
16. A V-8 engine has cylinders with a 7 1/16-in.^{2} cross-sectional area and a stroke length of 4 1/4 in. What is the total displacement in cubic inches?

Answers

2.	(a) 3/10	(b) 12/25	(c) 1/5	(d) 1/8
4.	(a) 1/18	(b) 10/3	(c) 3/16	(d) 1
6.	(a) 14 2/5	(b) 13 1/3	(c) 5 3/5	(d) 12
8.	(a) 13 49/81	(b) 35 1/33	(c) 14	(d) 41 1/9

10. 4 1/4 by 5 1/2; 8 1/2 by 2 3/4; 2 1/8 by 11
12. 282 1/3
14. 42 3/32″
16. 240 1/8 in.^{3}

2-7 DIVISION

Recall that in Chapter 1 we discussed division of whole numbers in terms of multiplication of whole numbers. For instance, we said that $128 \div 4 = 32$ because

$4 \cdot 32 = 128$. Now consider 4 divided by 1/2. Using the same approach as above, we can write $4 \div 1/2 = \#$ because $1/2 \cdot \# = 4$. It should be fairly easy to see that the # should be replaced by 8. So we have $4 \div 1/2 = 8$ because $1/2 \cdot 8 = 4$.

Thinking of the problem above we could ask, how many times will 1/2 a dollar divide into 4 dollars? That is, how many 50¢ pieces does it take to make $4? Since there are two 50¢ pieces in $1, $4 would have $4 \cdot 2$ or 8 of the 50¢ pieces.

The *reciprocal* of a fractional number is formed by inverting it, that is, by turning the fraction upside down. For example, the reciprocal of 3/7 is 7/3; similarly, the reciprocal of 5 1/3 = 16/3 is 3/16.

Rule for Dividing Fractions: Take the reciprocal of the divisor and multiply.

Definition: If a/b and c/d are two fractions, then

$$\frac{a}{b} \div \frac{c}{d} = \frac{a}{b} \cdot \frac{d}{c} = \frac{a \cdot d}{b \cdot c} = \frac{ad}{bc}$$

Example 2-27

$$\frac{3}{4} \div \frac{7}{8} = \frac{3}{4} \cdot \frac{8}{7} = \frac{3 \cdot 8}{4 \cdot 7} = \frac{24}{28} = \frac{6}{7}$$

Check: $3/4 \div 7/8 = 6/7$ because $7/8 \cdot 6/7 = 42/56 = 3/4$.

Example 2-28

$$4\frac{2}{3} \div 3\frac{1}{2} = \#$$

First change the mixed numbers to improper fractions. 4 2/3 = 14/3 and 3 1/2 = 7/2. So

$$4\frac{2}{3} \div 3\frac{1}{2} = \frac{14}{3} \div \frac{7}{2} = \frac{14}{3} \cdot \frac{2}{7} = \frac{28}{21} = \frac{4}{3}$$

You should verify that 3 1/2 times 4/3 is equal to 4 2/3.

Exercise Set 2-7

Part A

Divide and express the answer in reduced form.

	(a)	(b)	(c)	(d)
1.	$3/4 \div 2/5$	$3/5 \div 4/5$	$3/4 \div 1/3$	$7/22 \div 2/11$

2. 2/3 ÷ 6	8 ÷ 3/4	7/36 ÷ 14/27	24/7 ÷ 2/3
3. 17 ÷ 34/35	5/3 ÷ 20/3	7 ÷ 1/3	6 ÷ 2/3
4. 7/8 ÷ 1/2	8/21 ÷ 4/35	6/7 ÷ 6	5/6 ÷ 2/5
5. 4/5 ÷ 16/25	6/7 ÷ 9/14	4 ÷ 1/2	1/2 ÷ 1/2

Part B

Change each mixed number to an improper fraction, divide, and express your answer in simplest form.

	(a)	(b)	(c)	(d)
6.	6 2/3 ÷ 1 2/3	6 3/4 ÷ 9 1/3	4 8/19 ÷ 5 1/4	9 1/3 ÷ 4 2/3
7.	6 3/5 ÷ 11	51 ÷ 3 2/5	10 2/5 ÷ 2/5	12 1/3 ÷ 3

8. Find the quotient of 32 2/5 and 2 1/5.
9. Fill in the blanks with the numbers from Exercise 8 so that the answer will be the same for Exercises 8 and 9. How much is _____ divided by _____?
10. A certain farm has 217 1/2 acres of land. If it is divided into five equal tracts of land, how many acres will each tract contain?

Part C

11. How many pieces of candy can you buy with 18¢ if each piece costs 1 1/2 cents? (Neglect tax.)
12. Nine boys together bought a 12-ft metal rod. How many whole pieces of rod 3 9/16 in. long can each boy get, and how much of the rod will be left over? Each boy must get the same amount. (Neglect the waste of cutting the rod.)
13. A certain piece of cardboard weighs 2 1/8 oz. If a bundle of these pieces of cardboard weighs 21 1/4 lb, how many pieces are in a bundle?
14. A service station manager bought 15 qt of oil for $22.47. If she sells it for $1.30 a quart, how much is her gross profit per quart? (Express your answer in a fraction of a cent.)
15. A milling machine turning at 300 rpm will cut 3 1/4 in. of a certain metal in a minute. How much time will be required to make a 9 1/2-in. cut?
16. How many bricks are needed for the bottom row of a 42-ft brick wall if the bricks are 7 5/8 in. long with 1/4 in. of mortar mix between them?
17. The sports editor has softball standings for several leagues. Each league's listing requires 7/8 of a column inch. If you leave 3/16 in. between listings, how many listings can you list in 6 3/8 column inches?

Answers

2. (a) 1/9	(b) 32/3	(c) 3/8	(d) 36/7
4. (a) 14/8 = 7/4	(b) 10/3	(c) 1/7	(d) 25/12

6. (a) 4	(b) 81/112	(c) 16/19	(d) 2

8. 14 8/11
10. 43 1/2
12. 4 pieces; 15 3/4″ left
14. 19 4/5 cents
16. 64 bricks

Review Exercises

Perform each of the indicated operations.

Part A

	(a)	(b)	(c)	(d)
1.	1/2 + 3/8	1/3 + 3/5	1/2 + 2/3	2/5 + 5/6
2.	3/4 + 3/7 + 3/8	15 3/4 + 17 4/5	4 1/4 + 9 5/32	5/6 + 5/8 + 5/12
3.	4/5 - 3/4	7/8 - 1/3	15 - 10 9/32	10 1/5 - 3/4
4.	3 1/8 - 3/5	11 3/10 - 10 7/8	11 1/4 - 8 3/5	20 1/2 - 10 3/4
5.	2/5 · 3/4	1/2 · 1/5	14/15 · 10/21	75/100 · 2/3

Part B

	(a)	(b)	(c)	(d)
6.	3 1/2 · 3 1/7	2 1/2 · 1 3/10	13 1/2 · 2 7/8	5 3/4 · 3 9/16
7.	3/4 ÷ 2/3	3/4 · 3 1/4 ÷ 8	1/4 ÷ 6 · 120	1/8 ÷ 1/32 · 2 3/8

8. When 12 holes are drilled in a straight line and the same distance apart, the distance from the center of the first hole to the center of the last hole is 33 in. What will be the center-to-center distance between any two adjacent holes?

9. How many inches of stock is used to cut 33 bolts when each is to be 2 1/4 in. long and 3/16 in. of waste is to be allowed for each cut?

10. What is the weight of 16 pieces of drill rod each 6 1/2 ft long if drill rod of this kind and size weighs 5/8 lb per foot?

Part C

11. A certain alloy is to be made of 3/8 copper, 1/4 lead, 1/16 zinc, and 5/16 tin, by weight. We need 480 lb of this alloy. How many pounds of each metal must we use?

12. The net weight of the nails in a keg is 240 lb. The weight of each nail is 3/4 oz. How many nails are in the keg?

13. How many whole cards can be cut from 5 ft of stock if each card is to be 1 5/8 in. long? Allow 5/16 in. waste.

14. Find the total weight of three castings weighing 76 1/2 lb each, seven castings weighing 23 1/4 lb each, and two castings weighing 23 1/8 lb each.

15. A piece of stock in a lathe is 45 in. long. The tool feed is 5/64 in. How many revolutions will be required to turn the stock?

16. The two latches on your convertible top are 4 ft apart. If you want to add two additional latches (equally spaced), how far apart should they be?

Answers

2. (a) 1 31/56 (b) 33 11/20 (c) 13 13/32 (d) 1 7/8
4. (a) 2 21/40 (b) 17/40 (c) 2 13/20 (d) 9 3/4
6. (a) 11 (b) 3 1/4 (c) 38 13/16 (d) 20 31/64
8. 3″
10. 65 lb
12. 5,120 nails
14. 438 1/2 lb
16. 1 1/3′

Chapter 3

Decimal Numbers

Suppose you ordered a box of steel bolts 2 1/16 in. long with a 7/32-in. diameter. When you received the shipment, the size of the bolts was given as 2.0625 in. long and 0.21875-in. diameter. Should you return the bolts or are they the correct size? This problem will be discussed in Sec. 3-6.

3-1 PLACE VALUE IN DECIMAL NUMBERS

In Chapter 1, we discussed place value for whole numbers. For instance, the 5 in 58 represents 5 tens, and the 8 represents 8 ones. From this and the place value chart in Sec. 1-1, we see that each number in the decimal system of whole numbers is 1/10 of the number to the left of it. Following this procedure, the number to the right of 1 would be 1/10 of 1 or 1/10. The next number to the right would be 1/10 of 1/10 or 1/100. In a shorter form, 1/10 can be represented as 0.1 and 1/100 as 0.01. (It is customary to place a zero to the left of the decimal point in numbers less than one.)

Place Value Chart

Hundred Thousands	Ten Thousands	Thousands	Hundreds	Tens	Ones	.	Tenths	Hundredths	Thousandths	Ten Thousandths	Hundred Thousandths	Millionths
4	0	2	5	1	6	.	3	4	1	2	7	9

In the chart, we see that 341 thousandths would be written in decimal form as 0.341. So 16.34 would be read *16 and 34 hundredths*. The decimal point (.) in a number is read *and*.

A proper fraction that has as its denominator a power of ten (10; 100; 1,000; etc.) and is written in decimal form rather than fractional form is called a *decimal fraction*. A whole number and decimal fraction written together is called a *mixed decimal*. For example, 0.78 is a decimal fraction and 3.65 is a mixed decimal.

3-2 EXPANDED FORM AND ROUNDING OFF

Positive powers of whole numbers was discussed in Sec. 1-6. Now we wish to discuss powers of 10 where the exponents may be negative or positive whole numbers or zero.

> *Definition:* If X and k are positive whole numbers, then
>
> $$X^0 = 1 \quad \text{and} \quad X^{-k} = \frac{1}{X^k}$$

At present we are primarily concerned about the values when X = 10. For instance,

$$10^0 = 1 \qquad 10^{-1} = \frac{1}{10} \qquad 10^{-3} = \frac{1}{10^3} = \frac{1}{1{,}000}$$

Changing these to decimal form, we see that $10^{-1} = 0.1$ and $10^{-3} = 0.001$.

With -1 as an exponent, we have one digit to the right of the decimal point. If the exponent were -2, we would have two digits to the right of the decimal point. For exponents of $-3, -4, -5, \ldots$, we would have three, four, five, . . . digits to the right of the decimal point.

Example 3-1: Write the decimal fraction 0.453 in expanded form. The 4 represents 4 tenths or 4/10, the 5 represents 5 hundredths or 5/100, and the 3 represents 3 thousandths or 3/1,000. Using negative exponents of 10, we can write

$$\frac{4}{10} = 4 \cdot \frac{1}{10} = 4 \cdot 10^{-1}$$

$$\frac{5}{100} = 5 \cdot \frac{1}{100} = 5 \cdot 10^{-2}$$

$$\frac{3}{1{,}000} = 3 \cdot \frac{1}{1{,}000} = 3 \cdot 10^{-3}$$

so

$$0.453 = (4 \cdot 10^{-1}) + (5 \cdot 10^{-2}) + (3 \cdot 10^{-3})$$

Example 3-2: Express 17.42 in expanded form.

$$17.42 = 1 \cdot 10^1 + 7 \cdot 10^0 + 4 \cdot 10^{-1} + 2 \cdot 10^{-2}$$

Check: $10^1 = 10, 10^0 = 1, 10^{-1} = 1/10, 10^{-2} = 1/100$

so

$$\begin{aligned} 1 \cdot 10^1 + 7 \cdot 10^0 + 4 \cdot 10^{-1} + 2 \cdot 10^{-2} &= 10 + 7 + \frac{4}{10} + \frac{2}{100} \\ &= 17 + \frac{40}{100} + \frac{2}{100} \\ &= 17\frac{42}{100} \quad \text{or} \quad 17.42 \end{aligned}$$

When rounding off decimal numbers, we can follow the method given by the following steps.

Steps for Rounding Off Decimal Numbers

Step 1. Choose the position to which the number is to be rounded.

Step 2. If the digit in the position just to the right of that in Step 1 is less than 5, leave the digit of Step 1 unchanged and drop all digits to the right of it.

Step 3. If the digit in the position just to the right of that in Step 1 is 5 or more, add 1 to the digit of Step 1 and drop all digits to the right of it.

Example 3-3: Round 47.032 off to the nearest hundredth.

Step 1. The chosen position is the position of the 3.

Step 2. Since 2 is less than 5, our answer is 47.03.

Example 3-4: To the nearest tenth, round 5.764.

Step 1. The 7 is the digit to be considered.

Step 2. Does not apply.

Step 3. Since 6 is more than 5, increase 7 to 8 and our answer is 5.8.

Example 3-5: Round 586.349731 to the nearest thousandth. Since 7 is more than 5, 1 is added to the 9 in the thousandths position, changing the digits 349 to 350. So the number 586.349731 rounded to the nearest thousandth is 586.350. The zero must be kept because it is in the thousandths place. Without the zero, the number 586.35 would be to the nearest hundredth.

Exercise Set 3-2

1. Write the decimal number that represents each of the following.
   (a) Seven tenths ______
   (b) Thirty-five thousandths ______
   (c) Four hundred twenty-two and fifty-five hundredths ______
   (d) Nine and sixteen ten thousandths ______
   (e) Twenty-two thousand and twenty-two thousandths ______

2. Write each of the following in expanded form.
   (a) 2.74 ______
   (b) 14.031 ______
   (c) 0.1402 ______
   (d) 542.606 ______
   (e) 99.9192 ______

3. Round off each of the following to tenths, hundredths, and thousandths.
   (a) 0.78315 ______
   (b) 43.60542 ______
   (c) 0.98603 ______
   (d) 99.99750 ______
   (e) 5.5056 ______

Answers

2. (a) $2 \times 10^0 + 7 \times 10^{-1} + 4 \times 10^{-2}$
   (b) $1 \times 10^1 + 4 \times 10^0 + 3 \times 10^{-2} + 1 \times 10^{-3}$
   (c) $1 \times 10^{-1} + 4 \times 10^{-2} + 2 \times 10^{-4}$
   (d) $5 \times 10^2 + 4 \times 10^1 + 2 \times 10^0 + 6 \times 10^{-1} + 6 \times 10^{-3}$
   (e) $9 \times 10^1 + 9 \times 10^0 + 9 \times 10^{-1} + 1 \times 10^{-2} + 9 \times 10^{-3} + 2 \times 10^{-4}$

3-3 ADDITION AND SUBTRACTION

Recall that when you add two fractions, the most direct approach is to represent both with the same denominator before adding. When adding two decimal numbers, we must follow a similar policy.

Suppose we wanted to add 0.47 and 1.3. Since $0.47 = 4 \cdot 10^{-1} + 7 \cdot 10^{-2}$ and $1.3 = 1 \cdot 10^{0} + 3 \cdot 10^{-1}$, we could write $0.47 + 1.3 = (4 \cdot 10^{-1} + 7 \cdot 10^{-2}) + (1 \cdot 10^{0} + 3 \cdot 10^{-1})$. Now combine all terms that possess the same power of 10.

$$\begin{aligned} 0.47 + 1.3 &= 1 \cdot 10^{0} + (4 \cdot 10^{-1} + 3 \cdot 10^{-1}) + 7 \cdot 10^{-2} \\ &= 1 \cdot 10^{0} + 7 \cdot 10^{-1} + 7 \cdot 10^{-2} \\ &= 1.77 \end{aligned}$$

This may be written in shorter form as

$$\begin{array}{r} 0.47 \\ +1.3 \\ \hline 1.77 \end{array}$$

> *Steps for Adding (or Subtracting) Decimal Numbers*
>
> Step 1. Arrange the numbers such that all decimal points are in a vertical line.
>
> Step 2. Add (or subtract) the numbers as if they were whole numbers.
>
> Step 3. Place the decimal point in the answer in a vertical line with the other decimals.

Example 3-6: Add 0.702, 30.4, and 823.06.

$$\begin{array}{r} 0.702 \\ 30.4\mathbf{00} \\ +823.06\mathbf{0} \\ \hline 854.162 \end{array}$$

You may supply zeros where necessary to have the same number of digits to the **right** of the decimal in each number.

Example 3-7: Subtract 3.098 from 15.37.

$$\begin{array}{r} 15.37\mathbf{0} \\ -3.098 \\ \hline 12.272 \end{array}$$

Here the zero at the end of the first number is necessary so that we shall have a digit from which to subtract 8.

Exercise Set 3-3

Part A

1. Express each of the following in decimal form:

(a) 1/10	(b) 4/100	(c) 1/1,000	(d) 22/1,000
(e) 1/10,000	(f) 231/1,000	(g) 78/100	(h) 56/10,000

2. Add:

(a)	(b)	(c)	(d)
0.367	7.20	22.04	8.032
5.01	1.003	6.3335	17.9

3. Subtract:

(a)	(b)	(c)	(d)
0.907	5.820	12.01	147.0
0.32	4.69	3.782	21.75

4. Perform the indicated operations.

   (a) 7.50 + 2.314 + 0.09 = (b) 58.06 + 6.1 + 147.19 =

   (c) 14.7803 - 2.4619 = (d) 449.701 - 448.6998 =

5. Perform the indicated operations.

   (a) (47.04 - 16.99) + 0.786 = (b) 750.1 - (14.82 - 10.9) =

   (c) 562.9 + 9.07 - 15.15 = (d) (58.72 - 3.8) - (7.163 + 0.98) =

Part B

6. What is the wall thickness of a pipe having an outside diameter of 4.785 in. and an inside diameter of 3.95 in.?

7. If 50 ft of wire weighs 2.50 lb and 73 ft of the same type wire weighs 3.65 lb, how much does 23 ft of this wire weigh?

8. Mr. Smith has five different sizes of metal rods–call them A, B, C, D, and E. Size B is 0.38 in. larger than A, C is 0.034 in. larger than B, D is 0.12 in. larger than C, and D is 0.513 in. smaller than E. If size A has a 1.387-in. diameter, what is the diameter of the other four sizes?

9. Billy has a piece of metal which is 2.315 in. thick. How much must be milled off to make the piece 2.025 in. thick?

10. Mrs. Jones bought a one-owner used car. She drove it for 2,378.2 mi. The odometer now shows 65,776.4 mi. After making inquiries about the car, she found that the original owner had turned the odometer back from 59,381.0 miles to 46,298.2 miles while he owned it.

    (a) How many actual miles does Mrs. Jones' car have?

    (b) What was the odometer reading when she bought the car?

Part C

11. Find the missing dimensions in Fig. 3-1 (a)-(d).

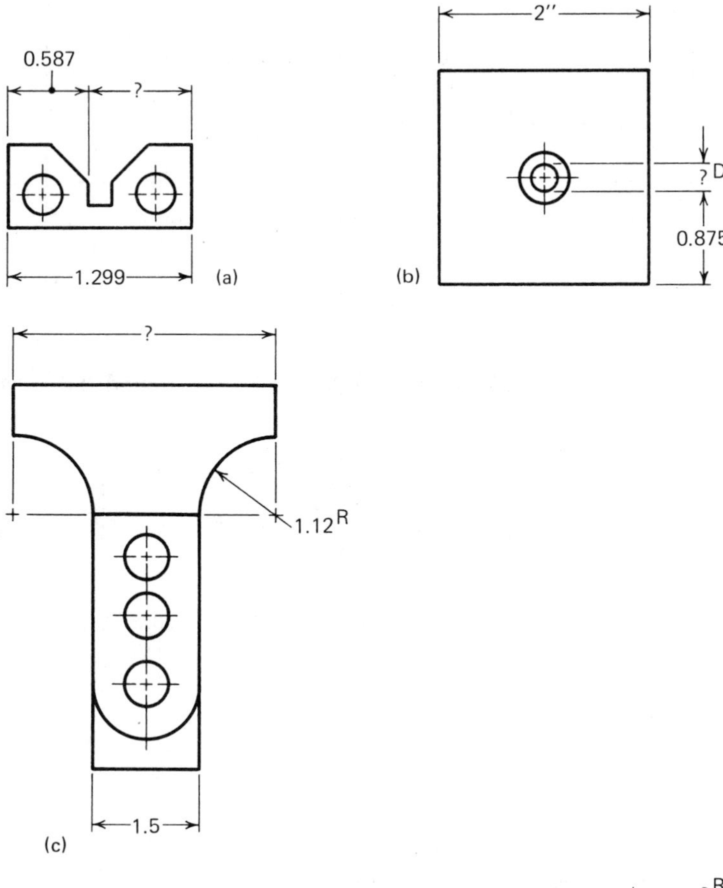

Figure 3-1

12. Find the values of X and Y on the template in Fig. 3-2. Round answers to hundredths.

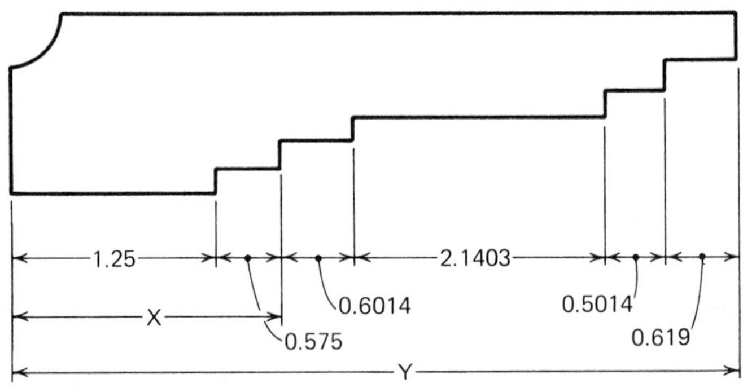

Figure 3-2

13. Determine values for A, B, X, and Y in Fig. 3-3. Round the answers to three decimal places.

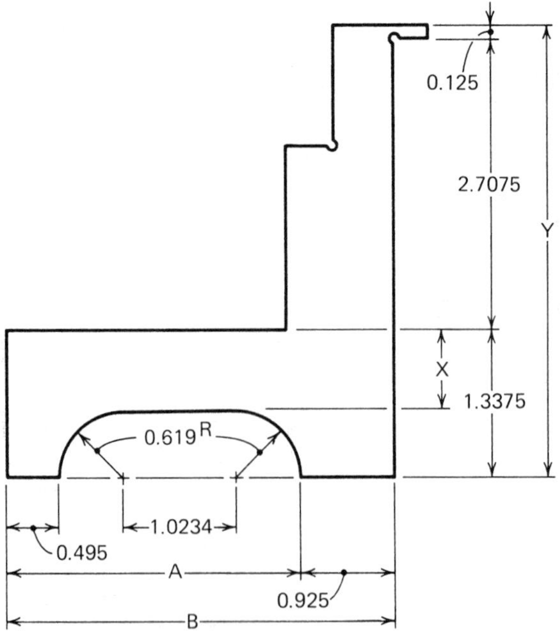

Figure 3-3

14. Suppose you have a piece of aluminum rod 37.25 in. long and you want to cut three shorter pieces from it. The first piece is to be 5.375 in. long, the second is to be 8.459 in. long, and the third is to be 12.625 in. long. Allow 0.125 in. for waste at each cut. How long a piece of aluminum rod will you have left?

15. Determine the values for A, B, C, D, and E (Fig. 3-4) correct to the nearest thousandth of an inch.

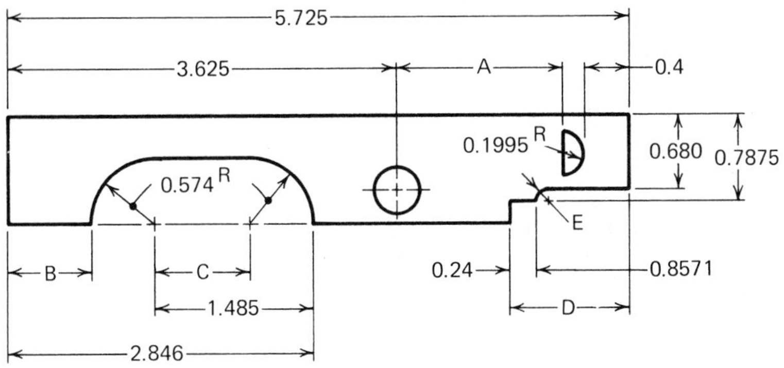

Figure 3-4

Answers

2. (a) 5.377 (b) 8.203 (c) 28.3735 (d) 25.932
4. (a) 9.904 (b) 211.35 (c) 12.3184 (d) 1.0012
6. 0.4175″
8. B = 1.767; C = 1.801; D = 1.921; E = 2.434.
10. (a) 78,859.2 (b) 63,398.2
12. X = 1.83; Y = 5.69
14. 10.416″

3-4 MULTIPLICATION AND DIVISION

A positive power of 10 is the number formed by using 1 and as many zeros as the exponent of 10. For instance, 10^5 is the same as $10 \cdot 10 \cdot 10 \cdot 10 \cdot 10$ or 100,000. (See Sec. 1-6.) Similarly, for negative exponents of 10 the value will be a number with as many digits to the right of the decimal as the positive value of the exponent of 10 and with 1 as the rightmost digit. For example, $10^{-1} = 0.1$ and $10^{-3} = 0.001$. (See Sec. 3-2.) With this in mind, we can state three steps for decimal multiplication.

> *Steps for Multiplying Two Decimal Numbers*
>
> Step 1. Multiply the two decimal numbers as if they were whole numbers.
>
> Step 2. Count the total number of digits to the right of the decimal point in the multiplicand and the multiplier.
>
> Step 3. Using the number obtained in Step 2, place the decimal point that number of digits from the right side of the answer.

Example 3-8: Multiply 4.701 by 0.3.

Step 1.
```
  4,701   or   4.701
      3          0.3
 ------       ------
 14,103       1.4103
```

Step 2. There are three digits to the right in the multiplicand. There is one digit to the right in the multiplier. The total is four, and the product should have four digits to the right.

Step 3. The answer is 1.4103.

Example 3-9: Find the product of 704.1 and 0.002 correct to the nearest hundredth.

```
 704.1       one digit to the right
 0.002      +three digits to the right
------      -------------------------
1.4082       four digits to the right
```

After rounding off to the nearest hundredth, the answer is 1.41.

In order to divide decimal numbers, we may follow a three-step procedure.

Steps for Dividing Decimal Numbers

Step 1. Move the decimal point in the divisor to the right of the last digit.

Step 2. Move the decimal point in the dividend the same number of places to the right.

Step 3. Divide as whole numbers and carry the decimal point straight up from the dividend to the quotient.

Example 3-10: Divide 174.3 by 12.45.

Step 1. Change 12.45 to 1245 (two places).

Step 2. Change 174.3 to 17430 (two places).

Step 3.

```
          14.
1245 )17430.
      1245
      ----
       4980
       4980
       ----
```

Note: It was necessary to add a zero in Step 2 in order to move the decimal two places.

Example 3-11: Find the quotient of 9,834.09 and 47.1 to two-decimal place accuracy.

Solution: Two-decimal place accuracy means that our answer will be rounded to the nearest hundredth. Thus we must divide to three digits to the right of the decimal point.

```
          208.791
   47.1)9834.0900
        942
         4140
         3768
          3729
          3297
           4320
           4239
             810
             471
             339
```

The answer to two-decimal place accuracy is 208.79

Exercise Set 3-4

Part A

1. Find the product of each of the following:

(a) 30.24 × 1.2 (b) 16.031 × 40.8 (c) 9.472 × 0.803 (d) 740.61 × 5.20

(e) 428.6 × 15.9 (f) 66.47 × 0.59 (g) 2.4103 × 6.94 (h) 81.096 × 43.01

(i) (48.93)(61.4) = (j) (7.98)(48.91) =
(k) (586.2)(2.81) = (l) (55.61)(2.02) =

2. Find the quotient to the nearest tenth.

(a) 4.6)58.92 (b) 36.1)821.4 (c) 7.82)59.876

(d) 42.015)2486.72 (e) 593.789)345.497106

3. Find the quotient to three-decimal place accuracy.

(a) 3.8)7.42 (b) 91.4)83.706 (c) 875.42)698.1

(d) 56.890)621.5 (e) 148.341)90,473.201

4. Divide and round off each answer to hundredths.

   (a) $147.9\overline{)12.45}$ (b) $23.65\overline{)14.5}$ (c) $5.896\overline{)629.41}$

   (d) $702.14\overline{)58.9032}$ (e) $7{,}403.219\overline{)590{,}352.601}$

5. Divide the product of 52.6 and 4.8 by the quotient of 45.75 and 6.7. Round off all quotients to two decimal places.

Part B

6. How long will it take you to drive 395.25 mi if you average 42.5 mph?
7. Jim can run the 50-yd dash in 4.7 sec. How many yards per second can Jim run at this speed? Express your answer to the nearest hundredth of a yard.
8. The Navy uses a certain metal that has the following composition: For each 100 lb of the metal, there are 3.8 lb of antimony, 3.0 lb of lead, 83.3 lb of zinc, 7.6 lb of tin, and 2.3 lb of copper. How many pounds of each additive would be needed to make 2,475 lb of this metal?
9. You were given a piece of metal with two holes drilled (see Fig. 3-5). You were asked to drill three more holes equally spaced between the existing two holes. What is the center-to-center distance?

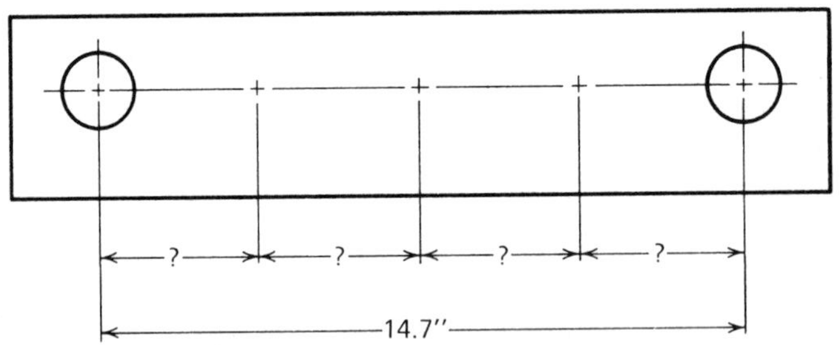

Figure 3-5

10. If 3 gal of transmission fluid weigh 25.08 lb, how many quarts of fluid are there in 501.6 lb of the fluid? How much will 501.6 lb of fluid cost if 1 qt costs $1.35?

Part C

11. An iron bar sells for 31.5¢ per pound. If the bar weighs 1.963 lb per foot, how much will a piece 17.45 ft long cost?
12. Find the average height per story of a 14-story building that is 153.75 ft high. Express your answer accurate to two decimal places.
13. If 35 sheets of a certain paper is 0.140 in. thick, how thick is 1 sheet? To the nearest whole sheet, how many sheets would be in a bundle that is 7.8 in. thick?

14. Scrap metal sells for \$27.58 per ton. Your truck loaded with scrap metal weighs 3,479 lb. If your truck (empty) weighs 2,047 lb, how many pounds of metal do you have? How much money should you receive when you sell it?

15. A piece of wire with a diameter of 0.035 in. is wound around a tube 15.5 in. long. How many whole turns does the wire make around the tube? Assume that there is no space between consecutive turns.

Answers

2. (a) 12.8 (b) 22.8 (c) 7.7 (d) 59.2 (e) 0.6
4. (a) 0.08 (b) 0.61 (c) 106.75 (d) 0.08 (e) 79.74
6. 9.3 hr
8. 94.05
   74.25
   2,061.675
   188.1
   56.925
10. 240 qt; \$324
12. 10.98′
14. 1,432 lb; \$19.75

3-5 USING POWERS OF 10 IN DECIMAL MULTIPLICATION

Since our number system is based on 10, the place values discussed in Sec. 1-1 can each be written as powers of 10. You may want to review Secs. 1-1, 1-6, and 3-2 before finishing this section.

$$10^0 = 1$$

$$10^1 = 10 \qquad 10^{-1} = 1/10 = 0.1$$

$$10^2 = 100 \qquad 10^{-2} = 1/100 = 0.01$$

$$10^3 = 1000 \qquad 10^{-3} = 1/1000 = 0.001$$

$$10^4 = 10,000 \qquad 10^{-4} = 1/10,000 = 0.0001$$

$$10^5 = 100,000 \qquad 10^{-5} = 1/100,000 = 0.00001$$

Now, consider the following examples which involve multiplication by powers of 10.

Example 3-12

(a)
```
  302
  X10
 ----
 3020
```

(b)
```
   302
  X100
------
30,200
```

(c)
```
       302
  X100,000
----------
30,200,000
```

(d) $\begin{array}{r} 3.02 \\ \times 10 \\ \hline 30.20 \end{array}$ (e) $\begin{array}{r} 3.02 \\ \times 100 \\ \hline 302.00 \end{array}$ (f) $\begin{array}{r} 3.02 \\ \times 100{,}000 \\ \hline 302{,}000.00 \end{array}$

In each case, the digits 3, 0, and 2 appear in the answer in their original order. The only change is their *place value*.

In Example 3-12(a) notice that the single zero in the multiplier 10 produces a single zero at the end of the answer. In Example 3-12(b) the two zeros in 100 produce two zeros at the end of the answer. Likewise, in Example 3-12(c) the five zeros in 100,000 produce five zeros at the end of the answer.

In Example 3-12(d)–(f), the same pattern works in the answer, except that the location of the decimal point must be more carefully determined. There are two ways to think through a problem like any of these. The first way is to ignore the decimal at first, multiply by the power of 10, and then insert a decimal in the answer using the rules for multiplication of decimal numbers (see Sec. 3-4). For instance, to multiply 76.5 × 100,

ignore the decimal and begin with	765×10^2
which is	76500
and then insert the decimal point (1 place)	7650.0

The other way to think through this problem is to concentrate on place value of the first digit. The first digit of 76.5 is 7, but this 7 actually represents 70, or 7×10, or 7×10^1. Now, multiplying that part of 76.5 by 100 would give us

$$(7 \times 10) \times (100)$$
$$7 \times (10 \times 100)$$
$$7 \times 1000$$
$$7000$$

Finally, bringing the other digits back into the problem, we get 7650.

> Rule 3-1. When *multiplying* a decimal number by a positive power of 10, move the decimal to the *right*, providing extra zeros as needed.
>
> Rule 3-2. When *dividing* a decimal number by a positive power of 10, move the decimal to the *left*, providing extra zeros as needed.

Example 3-13

(a)
$$\begin{array}{r} \mathbf{302.00} \\ \times 10 \\ \hline \mathbf{3020.00} \end{array} \quad (10^1)$$

or, by Rule 3-1

302 0.0

(b)
$$\begin{array}{r} \mathbf{302.00} \\ \times 100 \\ \hline \mathbf{30200}.00 \end{array} \quad (10^2)$$

or, by Rule 3-1

302 00.

(c)
$$\begin{array}{r} \mathbf{302.00} \\ \times 100{,}000 \\ \hline \mathbf{30{,}200}{,}000.00 \end{array} \quad (10^5)$$

or, by Rule 3-1

302 00**000**.

(d)
$$\begin{array}{r} 3.02 \\ \times 10 \\ \hline 3\,0.2 \end{array} \quad (10^1)$$

30.2

(e)
$$\begin{array}{r} 3.02 \\ \times 100 \\ \hline 3\,02. \end{array} \quad (10^2)$$

302

(f)
$$\begin{array}{r} 3.02 \\ \times 100{,}000 \\ \hline 3\,02\mathbf{000}. \end{array} \quad (10^5)$$

302,000

Example 3-14 (using Rule 3-2)

(a) 302.00 ÷ 10 = 30.2 00 = 30.20
(b) 302.00 ÷ 100 = 3.02 00 = 3.02
(c) 3.02 ÷ 10 = .3 02 = 0.302
(d) 3.02 ÷ 100 = .03 02 = 0.0302

Example 3-15

(a) 467.25 × 10 = 4672.5
(b) 3.14159 ÷ 10 = 0.314159
(c) 6.725 × 100 = 672.5
(d) 1350 ÷ 1000 = 1.350

Exercise Set 3-5

Part A

1. Find each product.
   (a) 5 × 100 (b) 57 × 10 (c) 357 × 100
   (d) 36.2 × 100 (e) 6.85 × 10 (f) 0.25 × 1000
2. Find each quotient.
   (a) 600 ÷ 10 (b) 57 ÷ 10 (c) 637,000 ÷ 100
   (d) 0.9 ÷ 10 (e) 89.765 ÷ 10 (f) 685.2 ÷ 1000
3. Perform the indicated operations.
   (a) 37.9 × 100 (b) 0.648 ÷ 10 (c) 0.085 × 1000
   (d) 96.8 × 10 (e) 387.2 ÷ 10 (f) 9.876 ÷ 1000

Part B

4. If a box of 1,000 nails costs $4.68, what is the cost of one nail?
5. If one quart of oil costs $1.50, what is the cost of 1,000 quarts of that oil?
6. Suppose your company has 100 employees on Purple X hospitalization, at a cost of $32.50 per month for each employee. How much must your company send to Purple X each month?
7. If the average number of children a woman has is 2.07, how many children would you expect 1,000 women to have?
8. If one box of nails has approximately 1,000 nails in it, how many nails are in 100 boxes?

Part C

9. Find the total value of this inventory list.

Item	Value Each	No. on Hand	Total Value
Desk	$425.50	10	
Chair	$152.25	100	
Stapler	$13.55	1000	
Paper Clip	$0.01	10,000	
		Grand Total	

10. Find the profit which should be distributed to each of 1,000 stockholders if a company had net profits of $86,255.00 last year.
11. The cost of a 3-day convention in Hampton has been set at $11,290 for rooms, $985 for meals, and $674 for other expenses. If 100 people are attending, how much should each pay for rooms, meals, and other expenses?
12. Find the total value of this inventory list.

Item	Value Each	No. on Hand	Total Value
Paint, gal.	$18.95	1000	
Thinner, pt.	$2.95	1000	
Solder, roll	$3.52	100	
Bolts, 2 in.	$0.60	10,000	
Nuts	$0.35	10,000	
		Grand Total	

Answers

2. (a) 60 (b) 5.7 (c) 6,370 (d) 0.09 (e) 8.9765 (f) 0.6852
4. $0.00468 (about half a cent)
6. $3,250.00
8. Approximately 100,000 nails
10. $86.26
12. Paint: $18,950.00; thinner: $2,950.00; solder: $352.00; bolts: $6,000.00; nuts: $3,500.00; grand total: $31,752.00

3-6 CONVERTING FRACTIONAL NUMBERS AND TERMINATING DECIMAL NUMBERS

Suppose you ordered a box of steel bolts 2 1/16 in. long with a 7/32-in. diameter. When you received the shipment, the size of the bolts was given as 2.0625 in. long and 0.21875-in. diameter. Should you return the bolts or are they the correct size? In order to answer this question, let us discuss converting a fractional number to a decimal number. The following steps may be used.

Steps for Converting a Fraction to a Decimal

Step 1. Change a mixed fractional number to an improper fraction.

Step 2. Divide the denominator into the numerator of the fraction. Add a decimal point and zeros to the dividend as necessary.

Step 3. Round off the quotient to the desired number of digits.

Example 3-16: Convert 2 1/16 to a decimal number to four decimal places.

Step 1. Change 2 1/16 to 33/16.

Step 2.

```
      2.0625
 16)33.0000
    32
     1 00
       96
        40
        32
         80
         80
          0
```

Step 3. Rounded to four decimal places, 2 1/16 = 2.0625

Example 3-17: Express 7/32 as a decimal number to the nearest hundred thousandth.

```
     0.21875
 32)7.00000
    6 4
     60
     32
     280
     256
      240
      224
       160
       160
         0
```

The nearest hundred thousandth means the same as five-decimal place accuracy; so 7/32 = 0.21875.

Now we see that you should keep the box of bolts that you ordered. The shipment you received is the same size as you ordered because 2 1/16 in. = 2.0625 in. and 7/32 in. = 0.21875 in.

In each of the examples above, a remainder of zero appeared. This leads us to state the following: A decimal is called a *terminating decimal* if it is obtained by dividing a fraction until a remainder of zero appears.

The decimal numbers 2.0625 and 0.21875 could be changed to fractional numbers and compared with 2 1/16 and 7/32. Then you would know whether to return the bolts. The following steps may be used to convert a terminating decimal to a fractional number.

Steps for Converting a Terminating Decimal to a Fraction

Step 1. Use the digits to the right of the decimal point as the numerator of the fraction.

Step 2. Use a power of 10(10, 100, 1,000, etc.) corresponding to the place value of the rightmost digit as the denominator of the fraction.

Step 3. Reduce the fraction to lowest terms.

Step 4. Affix the whole number (digits to the left of the decimal point) to the fraction to form a mixed fraction.

Example 3-18: Convert 2.0625 to a mixed number.

Step 1. 0625 is numerator.

Step 2. $\dfrac{0625}{10{,}000}$ (*Note:* One zero in the denominator for each digit to the right of the decimal point.)

Step 3. $\dfrac{625}{10{,}000} = \dfrac{125}{2{,}000} = \dfrac{25}{400} = \dfrac{5}{80} = \dfrac{1}{16}$

Step 4. Affix the 2. So, 2.0625 = 2 1/16.

Example 3-19: Express 0.21875 as a fraction.

Solution

$$0.21875 = \frac{21875}{100{,}000} = \frac{4375}{20{,}000} = \frac{875}{4{,}000} = \frac{175}{800} = \frac{35}{160} = \frac{7}{32}$$

So, 0.21875 = 7/32.

Now we see that you should keep the box of bolts that you ordered. The shipment you received is the same size as you ordered because 2.0625 in. = 2 1/16 in. and 0.21875 in. = 7/32 in.

Exercise Set 3-6

Part A

1. Convert each fraction to a decimal number.

   (a) 2/5 (b) 7/8 (c) 3/6 (d) 5/32
   (e) 1 3/16 (f) 2 3/4 (g) 9/16 (h) 9 1/4

2. Express each decimal as a fractional number in reduced form.

   (a) 0.038 (b) 0.71875 (c) 1.5625 (d) 0.034
   (e) 3.1875 (f) 0.555 (g) 2.0625 (h) 5.675

3. Change the following decimals of an inch to the nearest 32nd of an inch.

   (a) 0.394 (b) 1.416 (c) 0.708 (d) 1.89

4. Convert the following measures to the nearest fraction that can be read on a carpenter's square:

   (a) 2.6972 in. (b) 1.4295 in. (c) 3.6258 in.

5. Using U.S. standard, the gauge and thickness for sheet steel are as follows: No. 00, 0.34375 in.; No. 2, 0.265625 in.; No. 4, 0.234375 in.; No. 28, 0.015625 in. Find the approximate thickness of each to the nearest 64th of an inch.

Part B

In the following exercises, convert each fraction to the nearest hundredth in decimal form.

6.

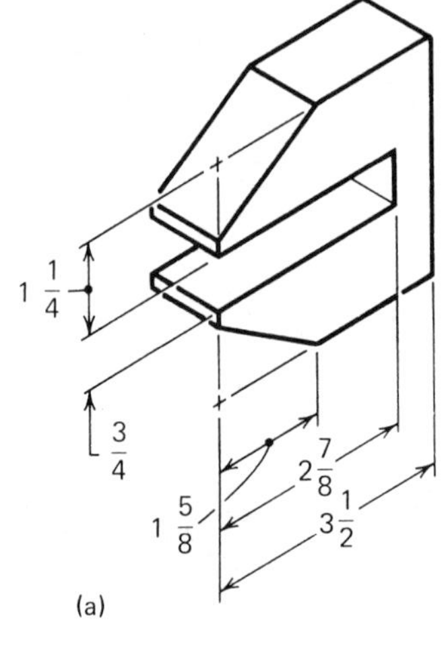

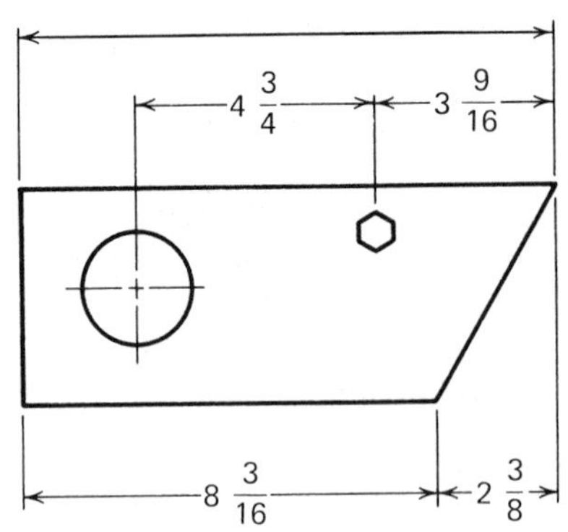

Figure 3-6

7.

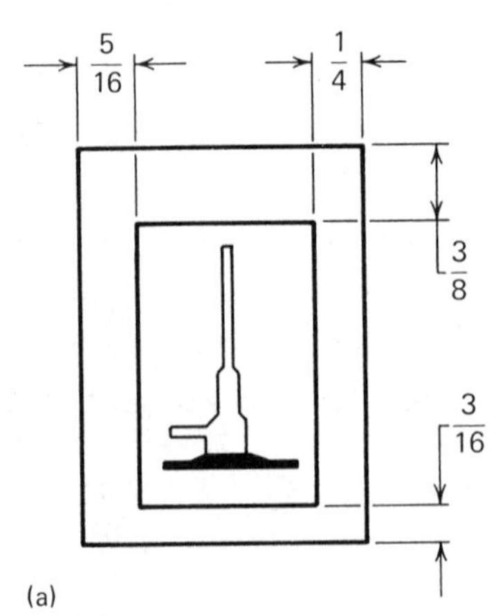

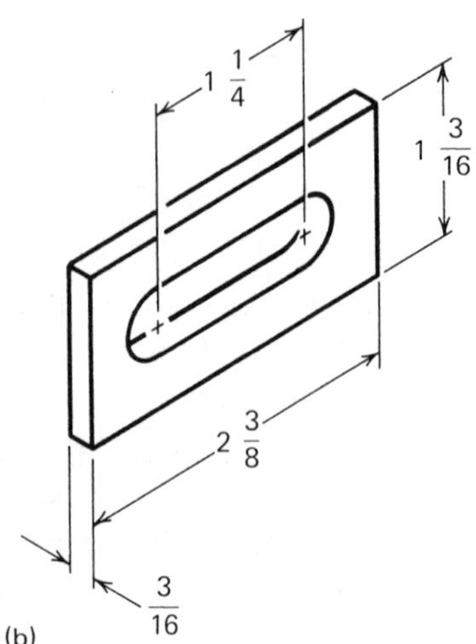

Figure 3-7

8.

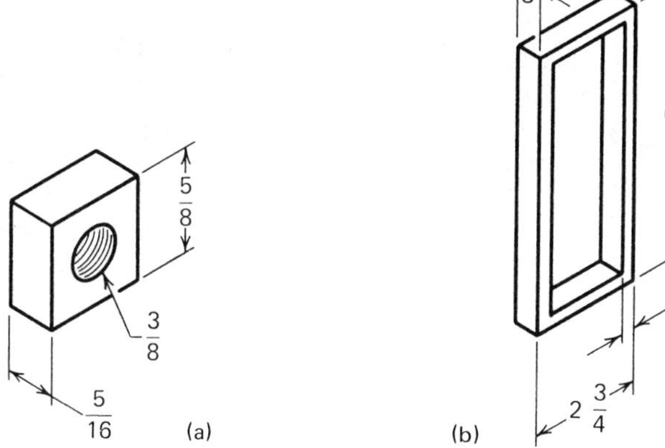

Figure 3-8

9. The blueprint for a 3/4-in. shaft that you are machining calls for 0.001-in. tolerance. What are the largest and smallest acceptable diameters?

10. You have received some boxes in which to pack 3 in. by 5 in. cards. Each box has an inside dimension of 3 1/8 in. by 5 1/8 in. by 3 1/2 in. high. If each card is 0.02 in. thick, how many cards would you expect to stack into each box?

11. Add the following numbers and express your answer to the nearest hundredth: 2.67, 31.8, 4 3/4, 56 4/5, and 8 1/16.

12. What is the displacement of a piston that has a 14.24-in.^{2} area and a 2 1/2-in. stroke length?

13. Subtract the product of 0.4343 and 3 1/3 from (22/7)(1.732), divide the difference by 21/25, and express the result as a decimal to the nearest thousandth.

14. Complete the following table.

Fractional Value	1/2	15/32	7/16	13/32	3/8	11/32	1/4	7/32	3/16
Decimal Value	0.5								

15. On the crankshaft shown in Fig. 3-9, change all measurements to decimal numbers to the nearest thousandth.

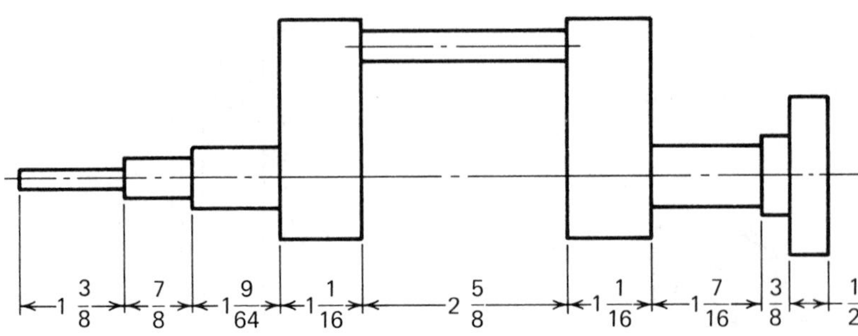

Figure 3-9

Answers

2. (a) 19/500 (b) 23/32 (c) 1 9/16 (d) 17/500
   (e) 3 3/16 (f) 111/200 (g) 2 1/16 (h) 5 27/40
4. (a) 2 11/16 (b) 1 7/16 (c) 3 5/8
6. (a) 2.875
   3.5
   1.625
   0.75
   1.25

   (b) 4.75
   3.5625
   2.375
   8.1875
8. (a) 0.375
   0.3125
   0.625

   (b) 0.375
   2.75
   0.625
   6.75
10. 175 cards
12. 35.6 in.^{3}
14. 0.46875
    0.4375
    0.40625
    0.375
    0.34375
    0.25
    0.21875
    0.1875

3-7 REPEATING DECIMAL NUMBERS

In Examples 3-16 and 3-17, when converting a fraction to a decimal, we divided far enough to obtain a remainder of zero. How many decimal places would we need to reach a remainder of zero when changing 1/6 to a decimal number? If we change

1/6 to decimal form, we find 0.1666 . . . , where the three dots indicate that the string of 6's continues indefinitely.

$$
\begin{array}{r}
0.1 \\
6\overline{)1.0} \\
\underline{6} \\
4
\end{array}
\qquad
\begin{array}{r}
0.16 \\
6\overline{)1.00} \\
\underline{6} \\
40 \\
\underline{36} \\
4
\end{array}
\qquad
\begin{array}{r}
0.166 \\
6\overline{)1.000} \\
\underline{6} \\
40 \\
\underline{36} \\
40 \\
\underline{36} \\
4
\end{array}
\qquad
\begin{array}{r}
0.1666 \\
6\overline{)1.0000} \\
\underline{6} \\
40 \\
\underline{36} \\
40 \\
\underline{36} \\
40 \\
\underline{36} \\
4
\end{array}
$$

Note that the remainder of 4 repeats, and in the quotient the 6 repeats and would do so endlessly if the division were continued.

If we change 1 3/11 or 14/11 to a decimal, we do not obtain an exact value.

$$
\begin{array}{r}
1.27 \\
11\overline{)14.00} \\
\underline{11} \\
30 \\
\underline{22} \\
80 \\
\underline{77} \\
3
\end{array}
\qquad
\begin{array}{r}
1.2727 \\
11\overline{)14.0000} \\
\underline{11} \\
30 \\
\underline{22} \\
80 \\
\underline{77} \\
30 \\
\underline{22} \\
80 \\
\underline{77} \\
3
\end{array}
\qquad
\begin{array}{r}
1.272727 \\
11\overline{)14.000000} \\
\underline{11} \\
30 \\
\underline{22} \\
80 \\
\underline{77} \\
30 \\
\underline{22} \\
80 \\
\underline{77} \\
30 \\
\underline{22} \\
80 \\
\underline{77} \\
3
\end{array}
$$

We find 1.272727 . . . . Here the three dots indicate that the string of 27's continues indefinitely.

Note that the remainder of 3 repeats. Observation of the quotient shows that the digits 27 repeat. They would do so endlessly if we continued the division.

Decimals of this type that have a digit or group of digits repeating endlessly are called *repeating decimals*. As a shorter form of writing these decimals, we write

0.1666 . . . as $0.1\overline{6}$ and 1.272727 . . . as $1.\overline{27}$. The bar over a single digit (or group of digits) indicates that that digit (or group of digits) repeats endlessly.

Now we shall discuss the process of changing a repeating decimal to a fraction. The following steps may be used.

Steps for Converting a Repeating Decimal to a Fraction

Step 1. Let N equal the repeating decimal number.

Step 2. Find the number of digits in the repeating string.

Step 3. Express a power of 10 using the number obtained in Step 2 as the exponent.

Step 4. Multiply both sides of the equation of Step 1 by the power of 10 found in Step 3.

Step 5. Subtract the result of Step 1 from that of Step 4.

Step 6. Divide both sides of the resulting equation by the coefficient of N (the number to the left of N).

Step 7. Reduce the fraction to lowest terms. This is the fractional equivalent of the decimal number.

Example 3-20: Express 0.272727 . . . as a fraction

Step 1. 1N = 0.272727 . . .

Step 2. There are two digits in the string.

Step 3. $10^2 = 100$

Step 4. 100N = 27.272727 . . .

Step 5.

$$\begin{array}{r} 100N = 27.272727\ldots \\ \underline{-1N = \ \ 0.272727\ldots} \\ 99N = 27. \end{array}$$

Step 6. $N = \dfrac{27}{99}$

Step 7. $N = \dfrac{27}{99} = \dfrac{\overset{3}{\cancel{27}}}{\underset{11}{\cancel{99}}} = \dfrac{3}{11}$

Example 13-21: Express 0.0333 . . . as a fraction.

Step 1. N = 0.0333 . . .

Step 2. There is one digit in the string.

Step 3. $10^1 = 10$

Step 4. $10N = 0.333 \ldots$

Step 5.
$$\begin{array}{r} 10N = 0.333 \ldots \\ -1N = 0.0333 \ldots \\ \hline 9N = 0.3000 \ldots = 0.3 \end{array}$$

Step 6. $N = \frac{0.3}{9}$

Step 7. $N = \frac{0.3 \times 10}{9 \times 10} = \frac{3}{90} = \frac{1}{30}$

Example 3-22: Convert 23.53535 . . . to a fraction.

Step 1. $N = 23.53535 \ldots$

Step 2. There are two digits repeating (notice the repetition of the digits 35).

Step 3. $10^2 = 100$

Step 4. $100N = 2353.535 \ldots$

Step 5.
$$\begin{array}{r} 100N = 2353.535 \ldots \\ -N = 23.53535 \ldots \\ \hline 99N = 2330.00000 \ldots = 2330 \end{array}$$

Step 6. $N = \frac{2330}{99}$

Step 7. $N = 23\frac{53}{99}$

Exercise Set 3-7

1. Express each of the following fractional numbers as a decimal number. Indicate whether each decimal is either a terminating decimal or a repeating decimal.

   (a) 1/3 (b) 2/7 (c) 2 3/8 (d) 11/15
   (e) 19/64 (f) 4 7/13 (g) 2/5 (h) 6 9/16

2. Express each of the following decimal numbers as a fractional number in reduced form.

   (a) 0.222 . . . (b) 0.0666 . . . (c) 5.555 . . .
   (d) 3.2727 . . . (e) 0.1666 . . . (f) 0.123123 . . .

3. If all terminating and repeating decimal numbers can be expressed as fractional numbers, is the number 1 equal to 0.999 . . .? Explain your answer.

Answers

2. (a) 2/9 (b) 1/15 (c) 5 5/9 (d) 3 3/11 (e) 1/6 (f) 41/333

Review Exercises

Part A

1. How many leads 15 picas long (2.5 in.) can be cut from a 30-in. strip? Assume there is no waste in each cut.
2. If 3/16 in. is allowed for each cut, how many pieces 8.75 in. long can be cut from a 2-in. X 4-in. board that is 10 ft long?
3. What is the weight of 655.5 ft of wire when 50 ft weighs 2.5 lb?
4. Round steel bars 7/8 in. in diameter weigh 1.785 lb per foot. What is the weight of 11 such bars, each 4.5 ft long?
5. A certain carpet sells for \$15.62 per yd^2. How much would a 3 yd by 4 yd piece of carpet cost?

Part B

6. A wooden mock-up of an engine weighs 75.5 lb. Wood weighs 0.04 lb per in.^{3} and cast iron weighs 0.26 lb per in.^{3} What would the iron engine weigh?
7. In Fig. 3-10, determine the center-to-center dimensions A and B to the nearest hundredth.

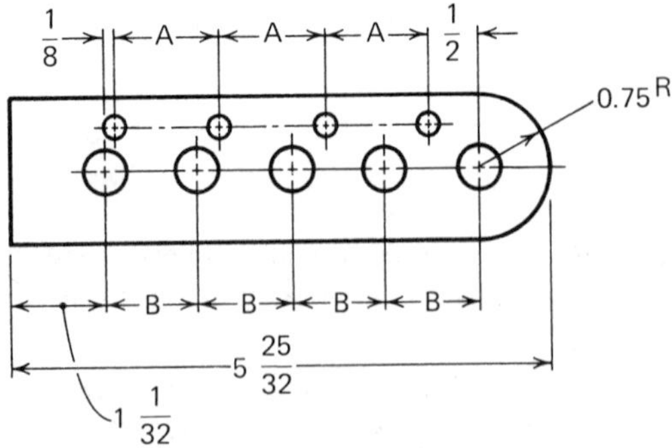

Figure 3-10

8. Compute dimensions A, B, C, D, and E in Fig. 3-11. Express answers to the nearest 32nd.

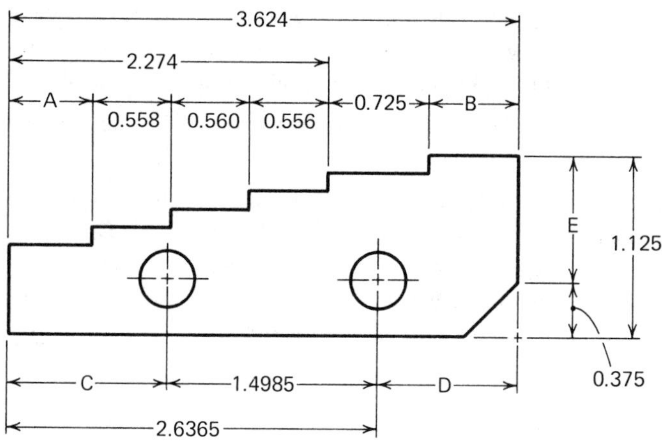

Figure 3-11

9. Jim paid 95¢ for 0.5 lb of nails. Half the nails came from a box marked 80¢ per lb and the rest came from a box marked $1.10 per lb. Give an explanation as to how the error was made.

10. The college print shop ordered 36 binders valued at $1.50 each and 48 packs of paper valued at $2.10 each. They wish to return 45 notebooks valued at $3.60 each. How much should the print shop owe or receive after the transaction is completed? Neglect tax.

11. A contractor agreed to do a job for $3,250. He hired three skilled workers and five helpers to do the job. The skilled workers earn $8.00 per hour and the helpers receive $4.50 per hour. In order for the contractor to make a profit of at least $800, the men must complete the job in how many days?

12. Determine the distance in inches that a tool travels for each of the five cuts A, B, C, D, and E. Each distance should be rounded off to one decimal place. Also determine the total distance.

Cuts	Feed per Revolutions	rpm	Time (min)
A	0.005	300	1.5
B	0.008	212	2.25
C	0.015	184	6.75
D	0.062	168	5.75
E	0.062	64	25.25

13. While the paymaster is on vacation, you have the job of figuring the weekly earnings for the workers. There are seven mechanics who each earn $5.70 per

hour. There are three other workers who each earn \$4.10 per hour. How much is the total weekly payroll if the mechanics work 6 days and the others work 6 1/2 days for 8 hr per day? They all receive time-and-a-half after 40 hr.

14. If a pump delivers 16.16 gal of water per stroke, how many pounds of water will it deliver in 55 strokes? (One gallon of water weighs 8.355 lb.)

15. You earn \$234.90 a week without commission. Your weekly commission on metal rods is 5¢ for each pound you sell over 3,000. One week you sold 125 ft of size A, 342.5 ft of size B, and 3,435 ft of size C. Size A weighs 4.125 lb per foot, size B weighs 2.25 lb per foot, and size C weighs 1.025 lb per foot. What were you total earnings that week?

Answers

2. 13 pieces
4. 88.3575 lb
6. 490.75 lb
8. 0.6 or 19/32
   0.625 or 5/8
   1.138 or 1 1/8
   0.9875 or 1
   0.75 or 3/4
10. Receive \$7.20
12. 2.3
    3.8
    18.6
    59.9
    100.2
    184.8
14. 7,425.9 lb

Chapter 4

Special Topics in Arithmetic

Now that we have studied whole numbers, common fractions, and decimal fractions, the basic methods of arithmetic are all available. However, there are special techniques and "tricks" which can be very helpful. Also, the low cost of modern hand-held calculators has made them widely available, so we need to learn something about calculator arithmetic.

4-1 CALCULATORS

There are two types of calculators: *Algebraic* and *RPN*. Most calculators are Algebraic and they can be identified by having an "=" button. RPN calculators do *not* have an "=" button, but they do have an "ENTER" button in most cases. RPN stands for Reverse Polish Notation, and an RPN calculator can make certain types of problems quite a bit easier once you understand how to use it. The following example will illustrate the basic difference between Algebraic and RPN logic.

Example 4-1: How should you evaluate $12 \div 3 + 5$ on a calculator?

First, we must decide which operation comes first. From the rules in Sec. 1-7, we know division comes before addition.

Algebraic		RPN	
Punch	Display	Punch	Display
12	12	12	12
÷	12	ENTER	12
3	3	3	3
+	4	÷	4
5	5	5	5
=	9	+	9

Input to the Algebraic calculator is done exactly like the written order in the expression, so it is easy to learn to use. The RPN calculator, on the other hand, always enters both numbers before specifying the operation. This may seem strange, but it has advantages in evaluating complex expressions. *On an Algebraic calculator with store capability, complex expressions often require storing preliminary results before continuing with another part of the problem.* But on an RPN calculator, careful planning can usually avoid any need to store partial results.

Example 4-2: Use a calculator to evaluate $(5 + 9) \div 2 + 6 \times 5$.

Algebraic		RPN	
Punch	Display	Punch	Display
5	5	5	5
+	5	ENTER	5
9	9	9	9
÷	14	+	14
2	2	2	2
=	7	÷	7
Store	7	6	6
6	6	ENTER	6
X	6	5	5
5	5	X	30
+	30	+	37
Recall	7		
=	37		

Since most of you will probably use Algebraic calculators, from now on we will only work with Algebraic logic. This does not mean that we prefer Algebraic logic to RPN, but we realize that most student calculators are built with Algebraic logic.

In addition to the four operations of arithmetic and a storage register for preliminary answers, a good calculator for a technician should also have trigonometric (trig) functions (see Chapter 15) and an exponent key. The exponent key, commonly denoted by Y^X, takes powers of numbers (see Chapter 8).

There are other keys found on most calculators which are particularly helpful.

[+/-] or [CHG] This key lets you change any entry from positive to negative or negative to positive. (See Chapter 7 for positive and negative numbers.)

Example 4-3: 3 X (-8)

Punch	Display
3	3
X	3
8	8
[+/-]	-8
=	-24

[1/x] This key simply gives you the reciprocal of the number in display. It would be helpful in evaluating fractions whose denominators involved computations, or in converting mixed numbers to decimals.

Example 4-4: $\frac{1}{5 \times 8}$

Punch	Display
5	5
X	5
8	8
=	40
[1/x]	0.025

Example 4-5: Convert 5 1/4 to decimal form.

Punch	Display
5	5
+	5
4	4
[1/x]	0.25
=	5.25

[x ↔ y] This key allows you to correct the order in which you set up subtractions or divisions.

Example 4-6: What number is 5 less than 12?

Punch	Display
5	5
-	5
12	12
"oops, that's backwards"	
[x ↔ y]	5
=	7

The only way to become familiar with a calculator and to gain confidence in using it is to *use* it. In addition, be sure to read the owner's manual or instruction booklet that comes with your calculator, if there is one. If you do not have such a booklet, your instructor or your school's library should have a book which you could borrow. But, above all, *practice*.

Exercise Set 4-1

1. Use a calculator to evaluate these expressions. Watch out for order of operations!
   (a) 8 × (-2)   (b) (-35) ÷ 7
   (c) (15 1/2) ÷ (7/8)   (d) 12 + 18 ÷ 6
   (e) 1/(4.6 × 5 - 3)   (f) 16 - 2 × 4 + 10 ÷ 2
2. Evaluate these expressions, watching order of operations.
   (a) (-7) × (-5)   (b) 64 ÷ (-4)
   (c) (5 1/2) × (4 3/4)   (d) 14 - 4 × 3
   (e) 1/(6 + 12 ÷ 3)   (f) (4 + 7 × 3) ÷ 5 + 1

Answers

2. (a) 35   (b) -16   (c) 26.125   (d) 2   (e) 0.1   (f) 6

4-2 HALF OF MIXED NUMBERS

A common problem in carpentry, machine shop, and other technical areas is the need to find half of a mixed number. There is a special technique which skilled technicians have learned to use that allows this calculation to be done in your head.

If the whole number part is even, just take half of it and then double the denominator of the fractional part.

Example 4-7: Half of 2 5/8 is 1 5/16.

Example 4-8: Half of 28 3/16 is 14 3/32.

If the whole number part is odd, the technique is very interesting. Take half of the whole number, ignoring the remainder; then, the numerator of the new fraction will be the *sum* of the old numerator and the old denominator; finally, the denominator is again just doubled.

Example 4-9: What is 1/2 of 15 1/2?

Half of 15 is 7. (ignoring remainder)

1 + 2 = 3

Twice 2 is 4.

Answer: 7 3/4

Example 4-10: What is half of 27 3/8? 13 and (3 + 8)/16 = 13 11/16.

Example 4-11: Find half of 15 1/4.

Procedure	Calculation	Answer
Find half of the whole number.	Half of 15 is 7.	Whole number is 7.
Exact? (Yes / No)	Not exact.	
Yes: Keep old numerator. No: Add old numerator and denominator.	1 + 4 = 5	Numerator is 5.
Double denominator.	2 X 4 = 8	Denominator is 8. Half of 15 1/4 is 7 5/8.

The only way to master this special technique is to practice!

Exercise Set 4-2

Part A

1. Find half of

(a) 6 7/8 (b) 5 1/4 (c) 10 1/2
(d) 17 3/8 (e) 64 1/4 (f) 21 15/16

2. Find 1/2 of

   (a) 12 1/2 (b) 15 1/8 (c) 8 7/16
   (d) 3 1/2 (e) 32 3/4 (f) 45 3/4

3. A window is 31 3/4 in. wide. How far in from either side is the center?
4. Where along a 11 1/2-ft wall is the center?

Part B

5. What is the radius of a steel rod which is 3 7/8 in. in diameter?
6. Where is the center of a tube which is 58 3/4 in. long?
7. If you have 4 1/2 hr to do a job, when should you be half finished?
8. If you work from 7 A.M. to 3:30 P.M., when is your workday half over?
9. If you need four equal lengths cut from a 10 1/2-ft pipe, how long should each of them be? (Neglect saw kerf.)
10. If a pipe has an outside diameter of 3 1/2 in. and an inside diameter of 2 3/8 in., what is the wall thickness? (*Hint:* Draw a sketch.)

Answers

2. (a) 6 1/4 (b) 7 9/16 (c) 4 7/32 (d) 1 3/4 (e) 16 3/8 (f) 22 7/8
4. 5 3/4 ft
6. 29 3/8 in.
8. 11:15 A.M.
10. 9/16 in.

4-3 AUSTRIAN SUBTRACTION

There is a different way to do subtraction, called *Austrian subtraction*, which is similar to making change. This method is based on the fact that subtraction is the opposite process of addition. For example, $8 - 5 = 3$ because $3 + 5 = 8$. To reason out $8 - 5$ by the Austrian method, you ask "What can I *add* to 5 to get an answer of 8?"

A common use of this addition method for subtraction is found at many check-out counters. Many experienced clerks and cashiers make change by adding:

Clerk: "That will be $1.68."
Customer: "Here is $2.00."
Clerk: "OK. $1.68" (purchase price)
"$1.70" (clerk hands back 2 pennies; 1.68 + .02 = 1.70)
"$1.75" (clerk hands back 1 nickel; 1.70 + .05 = 1.75)
"$2.00" (clerk hands back 1 quarter; 1.75 + .25 = 2.00)

The customer has \$0.32 change, which is correct, and neither person did any subtraction the way it is normally done, with borrowing and all the other usual cumbersome details.

To show the logic of Austrian subtraction in another situation, begin with the addition problem

$$\begin{array}{r} 465 \\ +324 \\ \hline 789 \end{array}$$

Now, the corresponding subtraction problem begins with

$$\begin{array}{r} 789 \\ -465 \\ \hline \end{array}$$

Begin in the units column and ask what can be added to 5 to get a total of 9. The answer is 4.

$$\begin{array}{r} 789 \\ -465 \\ \hline 4 \end{array}$$

Now, in the tens column, determine what can be added to 6 to get a total of 8. The answer is 2.

$$\begin{array}{r} 789 \\ -465 \\ \hline 24 \end{array}$$

Finally, in the hundreds column, 4 plus what equals 7? The answer must be 3.

$$\begin{array}{r} 789 \\ -465 \\ \hline 324 \end{array}$$

If you are paying close attention, you have probably noticed that this example is just as easy by regular subtraction because there is no need for borrowing. In examples of subtraction that normally require borrowing, Austrian subtraction really saves effort.

Example 4-12: Try Austrian subtraction on 328 - 149.

What can be added to 9 to get a total of 8? That can't be done. *But*, if we add 9 to 9, we at least get an 8 in that column! 9 + 9 = 18

$$\begin{array}{r} 328 \\ -149 \\ \hline \end{array}$$

Carry the 1 into the next column and add it to the 4, producing a sum of 5.

$$\begin{array}{r} 32\,8 \\ -14_{1}9 \\ \hline 9 \end{array}$$

Now, what can be added to 5 to get a total that ends in 2? Well, we can only get that 2 by adding 7 and carrying. 5 + 7 = 12

$$\begin{array}{r} 328 \\ -1\overset{5}{\not{4}}9 \\ \hline 9 \end{array}$$

The 1 that is carried plus the 1 already there produces a sum of 2.

$$\begin{array}{r} 328 \\ - 1_{1}\overset{5}{\not 4}9 \\ \hline 79 \end{array}$$

Finally, we can add 1 to 2 to get 3.

$$\begin{array}{r} 328 \\ - \overset{2}{\not 1}\overset{5}{\not 4}9 \\ \hline 179 \end{array}$$

Final appearance of problem	*Corresponding addition*
$\begin{array}{r} 328 \\ - 1_{1}4_{1}9 \\ \hline 179 \end{array}$	$\begin{array}{r} 1_{1}4_{1}9 \\ + 179 \\ \hline 328 \end{array}$

Exercise Set 4-3

1. Use Austrian subtraction to find each answer.

   (a) $\begin{array}{r} 359 \\ -214 \\ \hline \end{array}$ (b) $\begin{array}{r} 625 \\ -584 \\ \hline \end{array}$ (c) $\begin{array}{r} 1007 \\ -385 \\ \hline \end{array}$ (d) $\begin{array}{r} 6201 \\ -5918 \\ \hline \end{array}$

2. Use Austrian subtraction to find each answer.

   (a) $\begin{array}{r} 59 \\ -35 \\ \hline \end{array}$ (b) $\begin{array}{r} 216 \\ -124 \\ \hline \end{array}$ (c) $\begin{array}{r} 688 \\ -499 \\ \hline \end{array}$ (d) $\begin{array}{r} 7001 \\ -5287 \\ \hline \end{array}$

Answers

2. (a) 24 (b) 92 (c) 189 (d) 1714

4-4 UNIT ANALYSIS

In applied arithmetic problems it is very important to watch and use the physical units (units of measure).

Example 4-13: How much is 1 dime plus 2 nickels?

Since dimes and nickels are different units, we must first convert to a common unit. One method would be to use "cents":

$$1 \text{ dime} + 2 \text{ nickels} = 10¢ + 2(5¢) = 10¢ + 10¢ = 20¢$$

Another method would be to use dimes:

$$1 \text{ dime} + 2 \text{ nickels} = 1 \text{ dime} + 1 \text{ dime} = 2 \text{ dimes}$$

Finally, we could use decimal dollar notation:

$$1 \text{ dime} + 2 \text{ nickels} = \$0.10 + 2(\$0.05) = \$0.10 + \$0.10 = \$0.20$$

Using the units of measure in a problem to help determine the correct procedure is called *unit analysis*.

Example 4-14: In a car that averages 24 miles per gallon, how many gallons should it take to travel 400 miles?

Most students just grab the numbers in a problem such as this and then guess what operation they need to perform with those numbers. To show how helpful the units can be, just take the units of the information given in this problem and look at the possibilities. We have miles per gallon, which can be written miles/gallons, and we also have miles. The answer should be in gallons. One possibility would be

$$\frac{\text{miles}}{\text{gallons}} \times \text{miles} = \frac{\text{square miles}}{\text{gallons}}$$

which is nonsense! Another possibility would be

$$\frac{\text{miles}}{\text{gallons}} \div \text{miles} = \frac{\text{miles}}{\text{gallons}} \times \frac{1}{\text{miles}} = \frac{1}{\text{gallons}}$$

which doesn't make much sense either. A third possibility is

$$\text{miles} \div \frac{\text{miles}}{\text{gallons}} = \frac{\text{miles}}{1} \times \frac{\text{gallons}}{\text{miles}} = \text{gallons}$$

which is exactly what we want. So,

$$400 \text{ miles} \div \frac{24 \text{ miles}}{\text{gallons}} = \frac{400 \text{ miles}}{1} \times \frac{1 \text{ gallon}}{24 \text{ miles}} = \frac{400 \text{ miles}}{24 \text{ miles}} (\text{gallons})$$

$$\doteq 16.7 \text{ gallons}$$

($\doteq$ means approximately equal to.)

Frequently, you will need to convert units and also do unit analysis.

Example 4-15: A custom paint color needs 1 ounce of red pigment per quart. How many ounces of pigment would you need for 6 gallons?

One input is ounces per quart, which we will write ounces/quarts, while the other is gallons. The first thing to be done is convert the gallons to quarts, 6 gallons =

6(4 quarts) = 24 quarts. Now, for unit analysis:

$$\frac{\text{ounces}}{\text{quarts}} \div \text{quarts} = \frac{\text{ounces}}{\text{quarts}} \times \frac{1}{\text{quarts}} = \frac{\text{ounces}}{\text{square quarts}}$$

That isn't it, so we might try

$$\frac{\text{ounces}}{\text{quarts}} \times \text{quarts} = \frac{\text{ounces} \times \text{quarts}}{\text{quarts}} = \text{ounces}$$

That will do the job, so we now bring in the numbers and get

$$\frac{1 \text{ ounce}}{\text{quarts}} \times 24 \text{ quarts} = 24 \text{ ounces of pigment needed}$$

Exercise Set 4-4

Part A

1. What is the combined value of 6 quarters, 7 dimes, and 4 nickels? (Do this problem in at least two different ways.)
2. How many ounces are there in a combined liquid with 3 cups of one ingredient, 6 teaspoons of another, and 1 pint of a third ingredient? (Use measurement facts from Appendix E if needed.)

Part B

3. What operation would produce the required units on the answer in each of these cases?

	Input	Input	Answer
(a)	Square inches	Inches	Cubic inches
(b)	Cubic feet	Square feet	Feet
(c)	Feet per second	Seconds	Feet

4. What operation would produce the required units on the answer in each of these cases?

	Input	Input	Answer
(a)	Miles	Gallons	Miles per gallon
(b)	Miles per hour	Miles	Hours
(c)	Foot-pounds	Pounds	Feet

Answers

2. 41 ounces or 5 cups and 1 ounce
4. (a) Division (miles ÷ gallons)
   (b) Division (miles ÷ miles per hour)
   (c) Division (foot-pound ÷ pounds)

4-5 DECIMALS AND COMMON FRACTIONS COMBINED

It is not unusual in many types of problems to need to do arithmetic on decimal and common fractions. The usual procedure is to convert one type of number to the other type. For instance, to take 1/2 of $3.68, you could do three different things:

1. $\frac{1}{2}$ of \$3.68 = 0.5 × \$3.68 = \$1.84

2. $\frac{1}{2}$ of \$3.68 = $\frac{1}{2} \times \frac{368}{100} = \frac{368}{200} = \frac{184}{100}$ = \$1.84

3. $\frac{1}{2}$ of \$3.68 = $\frac{1}{2} \times \frac{\$3.68}{1} = \frac{\$3.68}{2}$ = \$1.84

In the first step of each of these three approaches to this problem, it is necessary to convert decimals to common fractions or common fractions to decimals.

Example 4-16: What is 1/3 of 2.91 inches?

$$\frac{1}{3} \times \frac{2.91}{1} \text{ in.} = \frac{1}{3} \times \frac{291}{100} \text{ in.} = \frac{291}{300} \text{ in.} = \frac{97}{100} \text{ in.} = 0.97 \text{ in.}$$

Or, you could do the same problem by doing

$$\frac{1}{3} \times 2.91 \text{ in.} = \frac{1}{3} \times \frac{2.91}{1} \text{ in.} = \frac{2.91}{3} \text{ in.} = 0.97 \text{ in.}$$

Or, you could even convert 1/3 ≐ 0.333 and get

$$\frac{1}{3} \times 2.91 \text{ in.} \doteq 0.333 \times 2.91 \text{ in.} =$$

```
    2.91
  ×0.333
     873
    873
   873
 0.96903 in.
```

If you use this last approach, you must remember that 0.333 is only an approximation to the fraction 1/3, so that your answer will be slightly off. In this case, the answer should be rounded off to 0.97 in.

Example 4-17: A taxi charges \$1.25 plus \$0.30 for each 1/5 mi. How much would a ride of 4.2 mi cost you?

Cost per mile times miles will produce the mileage cost. Likewise, cost per 1/5 mi times the number of 1/5 mi will produce the mileage cost. So,

$$(\$0.30)\left(\frac{4.2}{1/5}\right) = (\$0.30)\left(\frac{4.2}{0.2}\right) = (\$0.30)(21) = \$6.30$$

Finally, add the basic charge of \$1.25 and you will find the total cost is \$6.30 + \$1.25 = \$7.55.

Exercise Set 4-5

Part A

1. If a \$16.95 shirt is on sale for 1/3 off, how much will you have to pay?
2. If you make a 1/10 down payment on a \$65,000 house, how much will you have to finance?
3. Many Americans end up taking home about two-thirds of their salary. If you have interviewed for a job paying \$12,600 per year, about how much of that will be take-home pay?
4. You need the radius of a small cylinder, and your friend in the machine shop measures its diameter for you and says "eight hundred sixteen and a half thousandths." What is the radius?

Part B

5. To mark five points equally spaced around a circle of circumference 3.82 in., you need to calculate 1/5 of 3.82 in. How much would that be?
6. To mark five points equally spaced along a metal rod that is 17.46 in. long, you need to calculate 1/6 of 17.46 in. How many inches would that be?

Part C

7. If a taxi ride costs \$1.75 plus \$0.35 for each 1/5 mi, how much would a ride of 3.6 mi cost?
8. If 1/2 of the employees in a plant each make \$11,000 per year, 1/4 of the employees each make \$16,500, 1/8 of them each make \$18,500, and 1/8 of them each make \$22,900, what is the average annual salary in the plant?

Answers

2. \$58,500
4. 0.40825 in.
6. 2.91 in.
8. \$14,800

Chapter 5

Systems of Measure

Human beings have used various systems of measure since the beginning of recorded history, many of them inaccurate and used by only one small tribe. In recent history, worldwide trade has brought about extensively used systems for more accurate measure, and we presently find almost everyone using one of two such systems. The *English system* is used in the United States and in a few other countries, and those of us living in the United States are familiar with it. The other system, called the *metric system*, is used in most countries of the world and is also the tool of all scientists.

5-1 ENGLISH SYSTEM MEASURES

Within this section, we shall discuss linear measures, liquid measures, and weight measures in the English system. Areas and volumes shall be discussed in Chapter 12.

Linear Measure: You probably remember the following:

TABLE 5-1 English Linear Measure Relationships

12 inches (in.) =	1 foot (ft)
3 feet (ft) =	1 yard (yd)
5,280 feet (ft) =	1 mile (mi)

These three facts will usually be enough to remember for English lengths. You should commit these to memory if you have not already. (See Appendix D for other English measures.)

We know that 1 in. is a smaller unit of measure than 1 ft. In the statement 12 in. = 1 ft, we see that the larger number (*12*) is with the smaller unit (*inch*). Likewise, the smaller number (*1*) is with the larger unit (*foot*). The same is true for 3 ft = 1 yd and 5,280 ft = 1 mi. Some other instances are stated below.

> To have the same amount of money, you must have more dimes (smaller unit) than quarters (larger unit).
>
> To travel the same distance, a car with a smaller amount of fuel must get better fuel mileage than one with a larger amount of fuel.
>
> To roll the same distance, a smaller wheel will turn more revolutions than a larger wheel.
>
> To balance two weights on a balance bar, the larger (heavier) weight must be closer to the balancing point than the smaller (lighter) weight. (You probably remember balancing a larger and smaller person on a teeter-totter.)

The following rules may be used as a guide to solve many of the problems in this chapter.

Rule 5-1.* $\left.\begin{array}{l}\text{smaller number}\\ \text{of Larger Unit}\end{array}\right\} = \left\{\begin{array}{l}\text{Larger Number}\\ \text{of smaller unit}\end{array}\right.$

When you change from a larger unit to a smaller unit, you must change from a smaller number to a larger number. So, you must *multiply* by the proper conversion factor.

Rule 5-2.* $\left.\begin{array}{l}\text{Larger Number}\\ \text{of smaller unit}\end{array}\right\} = \left\{\begin{array}{l}\text{smaller number}\\ \text{of Larger Unit}\end{array}\right.$

When you change from a smaller unit to a larger unit, you must change from a larger number to a smaller number. So, you must *divide* by the proper conversion factor.

Example 5-1: How many inches are in 13 ft?

We know that there are 13 ft and 1 ft is larger than 1 in. So we are changing from a larger unit to a smaller unit. We must *multiply* by 12 because 12 in. = 1 ft. (That is, 12 is the proper conversion factor for this example. See Rule 5-1.)

*When a conversion factor is less than 1, the use of *multiply* and *divide* will be reversed in these rules. This situation arises in converting kilometers to miles. (See Example 5-34.)

$$13 \text{ ft} = 13\ (1 \text{ ft}) = 13\ (12 \text{ in.}) = 156 \text{ in.}$$

smaller number of Larger Unit = Larger Number of smaller unit

There are 156 in. in 13 ft.

Example 5-2: Fifty-four feet equal how many yards?

We know that there are 54 ft and 1 ft is smaller than 1 yd. So we are changing from a smaller unit to a larger unit. We must *divide* by 3 because 3 ft = 1 yd. (See Rule 5-2.)

Notice that dividing by 3 is the same as multiplying by 1/3. Since 3 ft = 1 yd, 1 ft = 1/3 yd.

$$54 \text{ ft} = 54\ (1 \text{ ft}) = 54\left(\frac{1}{3} \text{ yd}\right) = \frac{54}{3} \text{ yd} = 18 \text{ yd}$$

Larger Number of smaller unit = smaller number of Larger Unit

We see that 54 ft equal 18 yd.

Example 5-3: Find the number of yards in 1 mi.

We must change from larger unit (mile) to smaller unit (yard). Table 5-1 does not show the conversion factor between mile and yard. We do know that 1 mi = 5,280 ft and 3 ft = 1 yd. So, we may ask

$$1 \text{ mi} = 5{,}280 \text{ ft} = \underline{\hspace{3cm}} \text{ yd}$$

Now we are changing from a smaller unit (feet) to a larger unit (yard). We must *divide* by 3. (Remember that 1 ft = 1/3 yd.)

$$1 \text{ mi} = 5{,}280 \text{ ft} = 5{,}280\ (1 \text{ ft}) = 5{,}280\left(\frac{1}{3} \text{ yd}\right)$$

$$= \frac{5{,}280}{3} \text{ yd} = 1760 \text{ yd}$$

smaller number of Larger Unit = Larger Number of smaller unit

You may find it helpful to remember that 1 mi = 1760 yd.

Example 5-4: Using a carpenter's tape measure, Bob found his board to be 11 ft 1 1/2 in. long. Mary used a yardstick to measure a length of 2 yd 14 in. If Mary cuts her length off Bob's board, how long is the remaining piece of board? The problem is:

Bob's measure - Mary's measure = ________

$$11 \text{ ft } 1\frac{1}{2} \text{ in.} - 2 \text{ yd } 14 \text{ in.} = \_\_\_\_\_\_$$

Since 11 ft = 3 yd 2 ft and 14 in. = 1 ft 2 in.,

$$\begin{array}{r} 11 \text{ ft } 1\frac{1}{2} \text{ in.} \\ \underline{-2 \text{ yd } 14 \text{ in.}} \end{array} \quad \text{becomes} \quad \begin{array}{r} 3 \text{ yd } \overset{1}{\not{2}} \text{ ft } \overset{13\frac{1}{2}}{\not{1}}\frac{\not{1}}{2} \text{ in.} \\ \underline{-2 \text{ yd } 1 \text{ ft } \; 2 \;\; \text{ in.}} \\ 1 \text{ yd } 0 \text{ ft } 11\frac{1}{2} \text{ in.} \end{array} \quad \left(\begin{array}{l} 1 \text{ ft} = 12 \text{ in. and} \\ 12 \text{ in.} + 1\frac{1}{2} \text{ in.} = 13\frac{1}{2} \text{ in.} \end{array} \right)$$

So the length of the remaining piece of board is 1 yd $11\frac{1}{2}$ in. or 3 ft $11\frac{1}{2}$ in. or $47\frac{1}{2}$ in.

Liquid Measure: When we speak of liquid measure, we are talking about amount of volume. For liquid volume, we have specific names to indicate certain volumes. The basic units of liquid measure (liquid volume) in the English system are shown below.

TABLE 5-2 English Liquid Measure Relationships

1 pint (pt) = 16 fluid ounces (fl oz)
1 quart (qt) = 2 pints (pt)
1 gallon (gal) = 4 quarts (qt)

You should commit these to memory if you have not already. (See Appendix D for other English liquid measures.)

Example 5-5: Five quarts of oil contain how many fluid ounces?

5 qt = ________ fl oz

Since a quart is larger than a fluid ounce, we are to change from a larger unit to a smaller unit. We shall change quarts to pints and then change pints to ounces. Each change is from a larger unit to a smaller unit. So we *multiply* each time. (See Rule 5-1.)

$$5 \text{ qt} = 5\ (1 \text{ qt}) = 5\ (2 \text{ pt}) = 10\ (1 \text{ pt}) = 10\ (16 \text{ fl oz}) = 160 \text{ fl oz}$$

smaller number of Larger Unit = Larger Number of smaller unit

We have found that 5 qt of oil contain 160 fl oz.

Example 5-6: In mixing a photographic developing solution, Jean needs 20 fl oz of developer. Using a measuring cup, how many cups of developer should she use?

The problem is to complete 20 fl oz = ________ cups. We are changing from a smaller unit to a larger unit, so we must *divide*. Recall that 1 cup = 8 fl oz (Appendix D).

$$20 \text{ fl oz} = 20\ (1 \text{ fl oz}) = 20\left(\frac{1}{8} \text{ cup}\right) = \frac{20}{8} \text{ cups} = 2\frac{1}{2} \text{ cups}$$

Larger Number of smaller unit = smaller number of Larger Unit

So, Jean should use 2 1/2 cups of the developer to mix the solution.

Weight Measures: In measuring weight, we measure how heavy an object is. The basic English units of weight are shown below.

TABLE 5-3 English Weight Measure Relationships

1 pound (lb) = 16 ounces (oz)
1 ton (t) = 2,000 pounds (lb)

It should be pointed out that the ounce in weight is not the same as the ounce in liquid measure. So a volume of one ounce of a liquid may or may not weigh one ounce.

Example 5-7: 2 7/8 lb = ________ oz

This requires a change from a larger unit to a smaller unit. So we must *multiply* by 16 because 1 lb = 16 oz.

$$2\frac{7}{8} \text{ lb} = \left(\frac{23}{8}\right)(1 \text{ lb}) = \left(\frac{23}{8}\right)(16 \text{ oz}) = 23(2) \text{ oz} = 46 \text{ oz}$$

smaller number of Larger Unit = Larger Number of smaller unit

Some scales may not indicate fractional parts of a pound. So you may have to think of 2 7/8 lb as 2 lb 14 oz [7/8 lb = (7/8)(16 oz) = 14 oz].

Example 5-8: Bill bought 6 oz of candy. If it cost him $1.45 per pound, how much did he have to pay (neglect tax)?

First, change 6 oz to pounds. Converting from a smaller unit to a larger unit means we *divide* by 16 because 16 oz = 1 lb.

$$6 \text{ oz} = 6\ (1 \text{ oz}) = 6\left(\frac{1}{16} \text{ lb}\right) = \frac{6}{16} \text{ lb} = \frac{3}{8} \text{ lb}$$

Second, multiply the number of pounds by the price per pound to get the total price.

$$\left(\frac{3}{8} \text{ lb}\right) \times (\$1.45/\text{lb}) = \left(\frac{3}{8}\right) \times (\$1.45) = \frac{\$4.35}{8} = \$0.54$$

So, Bill paid $0.54, or 54 cents (nearest cent), without tax.

Exercise Set 5-1

Part A

1. Convert
   (a) 2 mi to ft (b) 3 tons to lb
   (c) 6 qt to pt (d) 5 lb to oz
2. Convert
   (a) 36 ft to yd (b) 32 oz to lb
   (c) 6,000 lb to tons (d) 140 qt to gal
3. Convert each given length to inches. Use fractions or decimal numbers if necessary.
   (a) 1 1/2 ft (b) 4.2 ft (c) 2.5 yd
   (d) 10 1/3 ft (e) 2/3 yd (f) 1 3/8 ft
4. Convert each given length to feet. Use fractions or decimal numbers if necessary.
   (a) 3 1/2 yd (b) 58 in. (c) 16 in.
   (d) 2/3 yd (e) 134 in. (f) 2 3/4 yd
5. Convert each given length to yards. Use fractions or decimal numbers if necessary.
   (a) 14 ft (b) 96 in. (c) 10 ft 6 in.
   (d) 256 in. (e) 8.4 ft (f) 16 ft 9 in.

6. Convert
   (a) 1/2 mi to ft
   (b) 3.5 tons to lb
   (c) 34 ft to yd
   (d) 14 oz to lb
   (e) 3/4 lb to oz
   (f) 3,280 lb to ton

Part B

7. Complete the table below by converting the given liquid measures to the other liquid measures.

	Fluid ounce	Cup	Pint	Quart	Gallon
(a)	256				
(b)		12			
(c)				6	
(d)					3.5
(e)			18		
(f)		36			

8. Complete the table below for weight measures.

	Ounce	Pound	Ton
(a)			1/2
(b)		500	
(c)	3,760		
(d)			0.125
(e)	14,400		
(f)		125	

9. Complete the following addition problems.

(a)
```
  4 gal 2 qt 1 pt
  6 gal 3 qt 1 pt
 +1 gal 1 qt 1 pt
```

(b)
```
  3 lb 12 oz
  1 lb  5 oz
 +6 lb 11 oz
```

(c)
```
  3 yd 2 ft  4 in.
  1 yd 1 ft 10 in.
 +4 yd 1 ft  3 in.
```

(d)
```
  6 gal 3 qt 1 pt
  3 gal 2 qt
+10 gal      1 pt
```

(e)
```
 15 lb  6 oz
       15 oz
 +7 lb  3 oz
```

(f)
```
  5 yd      7 in.
  6 yd 2 ft
+      1 ft 9 in.
```

10. Complete the following subtraction problems.

(a)	6 tons 458 lb −2 tons 841 lb	(b)	13 gal 2 qt 1 pt −10 gal 3 qt	(c)	4 yd 2 ft 5 in. −1 yd 2 ft 9 in.
(d)	7 lb 3 oz −2 lb 10 oz	(e)	8 gal 1 qt −5 gal 2 qt 1 pt	(f)	5 yd 1 ft 8 in. −1 yd 2 ft 7 in.

11. How much would 4 oz of candy cost at $2.36 per pound?
12. How much will 55 gal of oil cost at $1.55 per quart?
13. How many 5 1/2-in. pieces of wire can be cut from 132 ft of wire?
14. How many pieces of wood 2 ft 1 1/2 in. long can be cut from a 10 ft 8 in. board if 1/8 in. waste is allowed for each cut?
15. Joan was able to print 800 pages in 15 min. At that rate, how long (hours and minutes) would it take Joan to print 12,000 pages? If Joan earns minimum wage, how much would she earn for the job?

Answers

2. (a) 12 yd (b) 2 lb (c) 3 tons (d) 35 gal
4. (a) 10 1/2 ft (b) 4 5/6 ft (c) 1 1/3 ft
   (d) 2 ft (e) 11 1/6 ft (f) 8.25 ft
6. (a) 2640 ft (b) 7000 lb (c) 11 1/3 yd
   (d) 7/8 lb (e) 12 oz (f) 1.64 tons
8. (a) 16,000 oz; 1000 lb (b) 8000 oz; 1/4 ton (c) 235 lb; 0.1175 ton
   (d) 4000 oz; 250 lb (e) 900 lb; 0.45 ton (f) 2000 oz; 1/16 ton
10. (a) 3 tons 1617 lb (b) 2 gal 3 qt 1 pt (c) 2 yd 2 ft 8 in.
    (d) 5 lb 9 oz (e) 2 gal 2 qt 1 pt (f) 3 yd 2 ft 1 in.
12. $341
14. 5 pieces; 4 cuts

5-2 POWERS OF 10 IN THE METRIC SYSTEM

It seems clear that the United States will eventually convert to the metric system, and this is one of the main reasons for our study of it. The basic advantage of the metric system is that it is based on powers of 10, just like our number system. As an example, while we must always remember 5,280 feet = 1 mile and 36 inches = 3 feet = 1 yard, the metric system has equivalents such as 100 centimeters = 1 meter and 1 000 meters = 1 kilometer. (NOTE: The space in 1 000 is used instead of the comma in 1,000. This is common practice when writing metric measures.) Computations in the metric system are really quite simple because of this use of powers of

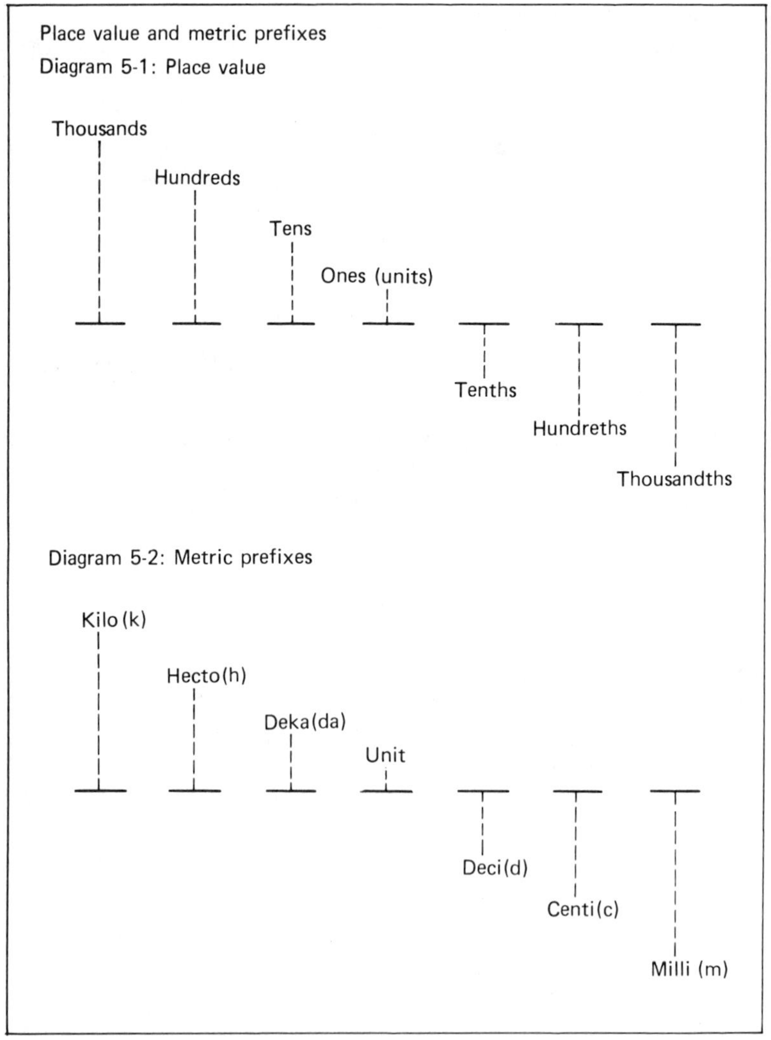

10. There are standard prefixes, each representing a different power of 10, and we need to learn the six prefixes listed in Diagram 5-2.

The basic unit of measure in our monetary system is the dollar.

\$1 000	= 1 kilodollar, 1 kilo means 1 000
\$100	= 1 hectodollar, 1 hecto means 100
\$10	= 1 dekadollar, 1 deka means 10
\$1	= 1 dollar, 1 dollar is the basic unit of money
\$0.1	= 1 decidollar, 1 deci means 1/10

$0.01 = 1 centidollar, 1 centi means 1/100

$0.001 = 1 millidollar, 1 milli means 1/1000

A tax of 1 millidollar ($0.001) has the same meaning as $1 tax per $1000.

The basic unit of length in the metric system is the meter. A meter is a little longer than a yard (1 m ≐ 1.09 yd) (≐ means approximately equal to).

1 kilometer (km) = 1 000 meters

1 hectometer (hm) = 100 meters

1 dekameter (dam) = 10 meters

1 meter (m) = 1 meter

1 decimeter (dm) = 0.1 meter

1 centimeter (cm) = 0.01 meter

1 millimeter (mm) = 0.001 meter

The metric measurement is based on the decimal system as was indicated in Diagram 5-2. Therefore, the units are related by powers of 10.

1 dekameter = 10 meters	(1 dam = 10 m)
1 dekameter = 100 decimeters	(1 dam = 100 dm)
1 meter = 10 decimeters	(1 m = 10 dm)
1 meter = 100 centimeters	(1 m = 100 cm)
1 meter = 1 000 millimeters	(1 m = 1 000 mm)
1 kilometer = 1 000 meters	(1 km = 1 000 m)
1 centimeter = 0.01 meter	(1 cm = 0.01 m)
1 centimeter = 0.1 decimeter	(1 cm = 0.1 dm)
1 millimeter = 0.001 meter	(1 mm = 0.001 m)

Diagram 5-3 Multipliers for converting units

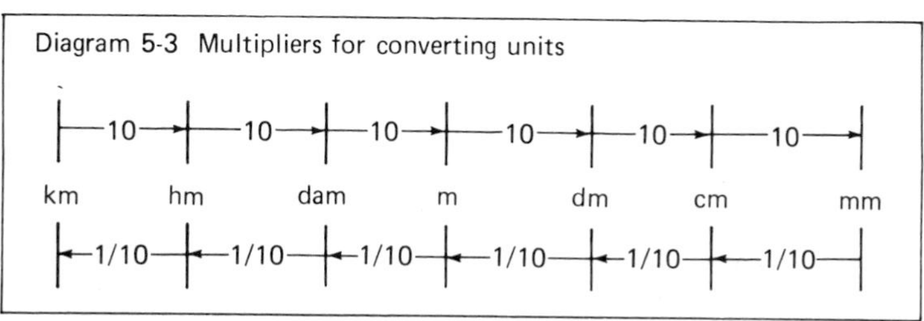

TABLE 5-4 Summary of Metric Linear Relationships

1 km = 1 000 m	(1 kilometer = 1 000 meters)
1 hm = 100 m	(1 hectometer = 100 meters)
1 dam = 10 m	(1 dekameter = 10 meters)
m	(1 meter)
1 dm = 0.1 m	(1 decimeter = 0.1 meter)
1 cm = 0.01 m	(1 centimeter = 0.01 meter)
1 mm = 0.001 m	(1 millimeter = 0.001 meter)

Case A: Any unit is 10 times the unit just to its right.

Illustration 5-1

1 km = 10 hm   1 hm = 10 dam

1 dam = 10 m   1 m = 10 dm

1 dm = 10 cm   1 cm = 10 mm

Case B: Any unit is 1/10 of the unit just to its left.

Illustration 5-2

1 mm = 0.1 cm   1 cm = 0.1 dm

1 dm = 0.1 m   1 m = 0.1 dam

1 dam = 0.1 hm   1 hm = 0.1 km

Case C: For units that are not next to each other, repeat Case A or Case B as many times as needed.

1 km = ___?___ m

|—10→|—10→|—10→|

1 km   hm   dam   m

1 km = 10 hm = 10 (10 dm) = 10(10)(10 m) = 1 000 m

Therefore, we may count the number of places (powers of 10) and then just write the result. (1 km = 1 000 m because km is 3 powers of 10 away from m.)

Example 5-9: Convert 5 km to meters.

Method A: 5 km = ___?___ m

5 km = 5 thousand meters = 5 000 m

Method B: 5 km = 5 000. m (See Diagram 5-3.)

= 5000 m (because km and m are separated by three powers of 10)

Example 5-10: Convert 22 mm to decimeters.

Method A: 22 mm = ___?___ dm

$$22 \text{ mm} = 22\left(\frac{1}{100} \text{ dm}\right) \qquad \left(\text{because } 1 \text{ mm} = \frac{1}{100} \text{ dm}\right)$$

$$= \frac{22}{100} \text{ dm}$$

$$= 0.22 \text{ dm}$$

Method B: 22 mm = ___?___ dm

22 mm = .22 dm = 0.22 dm (because mm and dm are two powers of 10 apart)

> If you change larger units of measure to smaller units of measure, then the number of smaller units must be increased by powers of 10.

Example 5-11: 5 m = ___?___ mm (See Diagram 5-3.)

Method A: 5 m = 5(10^3) mm = 5(1 000) mm = 5 000 mm

Method B: 5 mm = 5 000. mm = 5 000 mm

Example 5-12: Change \$5 to pennies.

\$5 = ___?___ ¢ = 5 00. ¢ = 500¢

> If you change smaller units of measure to larger units of measure, then the number of larger units must be reduced by powers of 10.

Example 5-13: 6 m = ___?___ km (See Diagram 5-3.)

$$6 \text{ m} = 6\left(\frac{1}{1000}\right) \text{ km} = 0.006 \text{ km}$$

Example 5-14: Change \$7000 to thousand-dollar bills.

$$\$7000 = 7.000 \text{ thousand dollars}$$
$$= 7 \text{ thousand dollars}$$
$$= 7 \text{ one-thousand-dollar bills}$$

Example 5-15: Change 100 m to kilometers.

$$100 \text{ m} = \underline{\ ?\ } \text{ km} = 0.100 \text{ km} = 0.1 \text{ km}$$

Example 5-16: $7\,000 \text{ cm} = \underline{\ ?\ } \text{ m} = 70.00 \text{ m} = 70 \text{ m}$

Example 5-17: $6.5 \text{ m} = \underline{\ ?\ } \text{ cm} = 6\,50. \text{ cm} = 650 \text{ cm}$

Example 5-18: $5 \text{ m} + 20 \text{ cm} = \underline{\quad} \text{ m}$

$= 5 \text{ m} + 0.2 \text{ m}$ (change to basic unit)

$= 5.2 \text{ m}$

Example 5-19: $5 \text{ km} + 200 \text{ m} = 5\,000 \text{ m} + 200 \text{ m} = 5\,200 \text{ m}$

Example 5-20: $2.5 \text{ m} - 21 \text{ cm} = 2.5 \text{ m} - 0.21 \text{ m} = 2.29 \text{ m}$

Exercise Set 5-2

1. Fill in the blanks with a whole number.
   (a) 1 cm = ________ mm (b) 1 m = ________ cm
   (c) 1 km = ________ cm (d) 6 m = ________ cm
   (e) 75 km = ________ m (f) 14 m = ________ mm
2. Fill in the blanks with whole or decimal numbers.
   (a) 24 cm = ________ m (b) 2 000 m = ________ km
   (c) 2 400 mm = ________ cm (d) 1 m = ________ km
   (e) 50 000 m = ________ km (f) 65 dm = ________ m
3. Using 1 dollar as a unit, express in everyday terms what we mean if we say the following.
   (a) 1 centidollar (b) 1 decidollar (c) 1 dekadollar
   (d) 1 hectodollar (e) 1 kilodollar (f) 1 millidollar
4. Simplify.
   (a) 1 milliunit + 5 centiunits (b) 3.4 units + 2 deciunits
   (c) 2.5 hectounits + 1.2 kilounits (d) 2.1 units − 21 milliunits

5. Simplify.

(a) 13 dm - 1 m
(b) 24 cm + 5 mm
(c) 2.24 km - 162 dam
(d) 0.005 m + 9 cm

Answers

2. (a) 0.24 m (b) 2 km (c) 240 cm
(d) 0.001 km (e) 50 km (f) 6.5 m
4. (a) 51 mu or 5.1 cu (b) 36 du or 3.6 u
(c) 14.5 hu or 1.45 ku (d) 2 079 mu or 2.079 u

5-3 METRIC WEIGHT

In the English system, the pound (lb) is the basic unit of weight, with ounces and tons as derived units. In the metric system, the basic unit for weight is the gram (g), with the derived units following the same prefix pattern as other metric measurements.

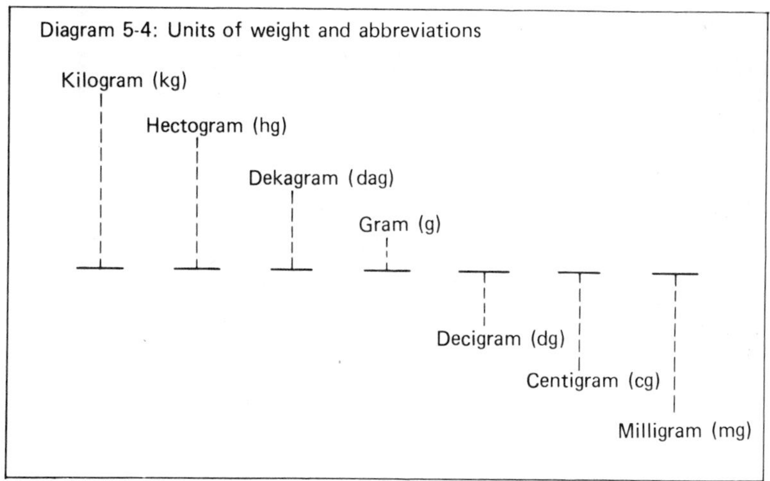

Example 5-21: 5 500 g = __?__ kg

5 500 g = 5.500 kg = 5.5 kg

Example 5-22: 0.04 g = __?__ mg

0.04 g = 0 040. mg = 40 mg

In the metric system of weights, the most often used units are the kilogram, gram, and milligram. In this section, we will deal with problems that use these units of measure. The standard unit of measure in international trade is the kilogram (1 kg = 1 000 g). The metric ton is 1 000 kilograms, which is just a little larger than the English ton (1 metric ton ≐ 2204 lb).

Example 5-23: How many kilograms are in 75 000 g?

$$75\ 000 \text{ g} = 75.000 \text{ kg} = 75 \text{ kg}$$

Example 5-24: How many 2.5-g screws are there in a 5-kg box of screws?

$$5 \text{ kg} = 5\ 000 \text{ g} \quad \text{and} \quad \frac{5\ 000 \text{ g}}{2.5 \text{ g}} = 2{,}000$$

So there are 2,000 screws in the box.

Exercise Set 5-3

1. Complete the following.
   (a) 2.5 kg = __________ g
   (b) 0.6 kg = __________ g
   (c) 6 000 mg = __________ g
   (d) 250 g = __________ kg
   (e) 5.5 kg = __________ g
   (f) 1 450 mg = __________ kg
2. How many 5-mg weights are in a gram?
3. How many 5-g lock washers are there in a 0.75-kg box of washers?
4. If an aspirin is 0.6 g, how many aspirins are in a bottle that nets 0.06 kg?
5. How many 50-kg boxes of material are there in a metric ton?
6. What is the weight of a man in kilograms if his weight is 70 000 g?
7. How many 45-kg boxes should we carry on a 1.35 metric ton truck?

Answers

2. 200 weights
4. 100 aspirins
6. 70 kg

5-4 METRIC LIQUID VOLUME

In the metric system, volume is measured in cubic centimeters, which is abbreviated cc. Volume is also measured in liters (ℓ). There is a direct relationship between the cubic centimeter and the milliliter (1 mℓ = 1 cc). A liter is just a little larger than a quart (1 ℓ = 1.06 qt). Throughout the metric system, the same prefixes are used as indicated in Diagram 5-5.

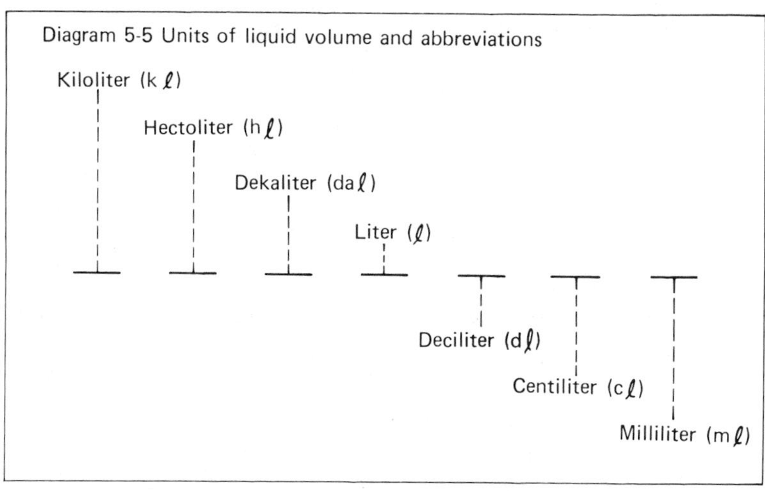

The basic unit of measure is the liter, as indicated in Diagram 5-5. The units that are most often used are the kiloliter, liter, centiliter, and milliliter. In this section, we shall deal with the units that are most often used.

Example 5-25: Convert 550 mℓ to liters.

$$550 \text{ m}\ell = \underline{\ ?\ }\ \ell = .\underline{550}\ \ell = 0.55\ \ell$$

Example 5-26: Convert 2.5 ℓ to milliliters.

$$2.5\ \ell = \underline{\ ?\ }\ \text{m}\ell = 2\,\underline{500}.\ \text{m}\ell = 2\,500\ \text{m}\ell$$

Example 5-27: How do you change 5 mℓ to liters?

Divide 5 by three powers of 10.

$$5\ \text{m}\ell = 0.\underline{005}\ \ell \qquad \text{(divide by 1000)}$$
$$= 0.005\ \ell$$

Example 5-28: How do you change 6 kℓ to liters?

Multiply by three powers of 10.

$$6\ \text{k}\ell = 6\,\underline{000}.\ \ell \qquad \text{(multiply by 1000)}$$
$$= 6\,000\ \ell$$

Example 5-29: How do you change 50 cc to liters?

First change 50 cc to 50 mℓ. Then

$$50 \text{ cc} = 50 \text{ m}\ell = 0.050\ \ell = 0.05\ \ell$$

Exercise Set 5-4

1. Complete the following.
   (a) 800 mℓ = ________ ℓ
   (b) 0.0006 ℓ = ________ mℓ
   (c) 50 kℓ = ________ ℓ
   (d) 50 kℓ = ________ mℓ
   (e) 6 700 mℓ = ________ kℓ
   (f) 4.67 mℓ = ________ cℓ
2. How many liters are in 5 500 mℓ?
3. How many kiloliters are in 5 500 mℓ?
4. How many milliliters are in 50 cℓ?
5. How many liters are in 50 cℓ?
6. If 1 cc = 1 mℓ, how many cubic centimeters are in 300 mℓ?
7. What part of a liter is 2 200 cc?
8. How many cubic centimeters are there in 0.6 ℓ?
9. A 2.6-ℓ engine has how many cubic centimeters of displacement?

Answers

2. 5.5 ℓ
4. 500 mℓ
6. 300 cc
8. 600 cc

5-5 CONVERTING METRIC AND ENGLISH MEASURES

Appendix E contains conversion facts for converting metric measures to English measures and for converting English measures to metric measures. This section illustrates the use of conversion facts but does not cover all of them, so you may want to look at Appendix E to see what types of conversions are included.

1 inch = 2.54 centimeters (exactly)

An inch is larger than a centimeter; a centimeter is smaller than an inch. It takes a little more than 2½ cm to make 1 in.

Example 5-30: Six inches are how many centimeters?

We need to change from a larger unit (inch) to a smaller unit (centimeter), so we must *multiply*.

$$6 \text{ in.} = 6\,(2.54 \text{ cm}) = 15.24 \text{ cm}$$

Example 5-31: One centimeter is how many inches?

We need to change from a smaller unit (centimeter) to a larger unit (inch), so we must *divide*.

$$1 \text{ cm} = \frac{1}{2.54} \text{ in.} \doteq 0.3937 \text{ in.}$$

This answer, where 1 cm $\doteq$ 0.3937 in., is also listed in Appendix E, but you can see that if you know the basic fact, 1 in. = 2.54 cm, this conversion fact is easy to derive.

Example 5-32: Which is larger, 100 mm or 4 in.?

To compare millimeters and inches, you could look in Appendix E for a suitable conversion fact. However, this comparison can be done quickly using 1 in. = 2.54 cm. First, convert 100 mm to centimeters (smaller unit to larger unit, *divide*).

$$100 \text{ mm} = \frac{100}{10} \text{ cm} = 10 \text{ cm}$$

Second, convert 4 in. to centimeters.

$$4 \text{ in.} = 4\,(2.54 \text{ cm}) = 10.16 \text{ cm}$$

So, finally, 4 in. is slightly bigger than 100 mm.

Example 5-33: One mile is how many meters?

Again, Appendix E might give a quick comparison, but we can do it by converting to centimeters.

$$1 \text{ mi} = 5280 \text{ ft} = 5280\,(12 \text{ in.}) = 63360 \text{ in.}$$

$$= 63\,360\,(2.54 \text{ cm}) = 160\,934.4 \text{ cm} = 1\,609.344 \text{ m}$$

Now, Example 5-33 was rather lengthy, so let's look at a quick comparison between miles and kilometers.

1 kilometer $\doteq$ 5/8 mile

This conversion is *not* exact, but it is accurate enough for everyday comparisons. Notice that kilometers are smaller; miles are larger.

Example 5-34: Ninety kilometers are how many miles?

We will be converting from a smaller unit (kilometer) to a larger unit (mile), so there will be less miles than there are kilometers. Multiplying by 5/8 will produce a smaller number, so multiplication is the correct procedure.

$$90 \text{ km} = 90\left(\frac{5}{8}\right) \text{ mi} = \frac{450}{8} \text{ mi} = 56 \text{ mi}$$

Example 5-35: What is 55 mph in kilometers per hour?

Both speeds are "per hour," so we are really asking how many kilometers are in 55 mi. Kilometers are smaller than miles, so there will be more kilometers. Multiplying by 5/8 would produce *less*, so we need to divide by 5/8.

$$55 \text{ mi} = \frac{55}{5/8} \text{ km} = \left(\frac{55}{1}\right)\left(\frac{8}{5}\right) \text{ km} = 88 \text{ km}$$

So, 55 mph = 88 km per hour.

Example 5-36: How close to a mile run is the 1 500-m run?

We know that 1 500 m is 1.5 km, and the conversion fact we are using says that each kilometer is about 5/8 mi. So, we have

$$1\,500 \text{ m} = 1.5 \text{ km} \doteq (1.5)\left(\frac{5}{8}\right) \text{ mi} \doteq 0.94 \text{ mi}$$

Now, since the 1 500-m run is just slightly shorter than 1 mi, sports fans and announcers have come to call this distance the "metric mile."

1 kilogram ≐ 2.2 pounds

This conversion is not exact, but it is close enough for most types of weight problems. Notice that 1 kg is more than 2 lb, so kilograms are heavier than pounds. This means that the weight of any object will involve less kilograms than pounds.

Example 5-37: Is a 1 300-kg car lighter than a 3000-lb car?

Each kilogram is equivalent to about 2.2 lb, so the 1 300-kg car would weigh about

$$1300(2.2)\text{ lb} = 2860\text{ lb}$$

Notice that the weight in pounds is always greater than the weight in kilograms. So, to answer the question, the 1 300-kg car is lighter than the 3000-lb car.

Example 5-38: What is the metric equivalent of 1 ton (2000 lb)?

There will be less kilograms than pounds, so we must divide by 2.2.

$$1\text{ ton} = 2000\text{ lb} \doteq \frac{2000}{2.2}\text{ kg} \doteq 909\text{ kg}$$

1 liter $\doteq$ 1.06 quarts

Example 5-39: Twenty liters are how many quarts?

The liter is larger than the quart, so there will be more quarts than liters.

$$20\ \ell \doteq 20(1.06)\text{ qt} = 21.2\text{ qt}$$

Example 5-40: One gallon is how many liters?

First, we must change 1 gal to 4 qt (see Appendix D). Now, the liter is larger than the quart, so there will be *less* liters. Division is the appropriate procedure.

$$1\text{ gal} = 4\text{ qt} \doteq \frac{4}{1.06}\ \ell \doteq 3.77\ \ell$$

Example 5-41: If gasoline sells for 42.9¢ per liter, what is the price per gallon?

The preceding example shows that 1 gal is about 3.77 ℓ. So, if each liter costs 42.9¢, 1 gal would cost

$$(3.77\ \ell)(42.9¢/\ell) = 161.733¢$$

The cost per gallon is about 162¢ or $1.62.

Exercise Set 5-5

Part A

1. Convert the following English lengths to the indicated metric unit.
   (a) 6.2 in. to millimeters (b) 2.4 ft to meters

(c) 0.001 yd to millimeters
(d) 220 yd to meters
(e) 16 mi to kilometers
(f) 20 ft to meters
(g) 200 mi to kilometers

2. Convert the following metric lengths to the indicated English unit.
   (a) 1 m to inches
   (b) 1 m to yards
   (c) 100 m to yards
   (d) 1 km to miles
   (e) 1.2 cm to inches
   (f) 9 mm to inches
   (g) 16.2 cm to inches

3. Convert the following English weights to the indicated metric unit.
   (a) 2 lb to kilograms
   (b) 15 oz to grams
   (c) 3.2 oz to grams
   (d) 184 1/2 lb to kilograms
   (e) 2 lb 3 oz to grams
   (f) 8 lb 8 1/2 oz to grams

4. Convert the following metric weights to the indicated English unit.
   (a) 16 kg to pounds
   (b) 380 g to ounces
   (c) 45 g to ounces
   (d) 3.021 kg to pounds
   (e) 4 mg to ounces
   (f) 100 mg to ounces

5. Express your weight in pounds and in kilograms (sometimes called *kilos*).

Part B

6. If 6 picas equal 1 in., express 1 pica in millimeters.

7. Is the 100-m dash longer or shorter than the 100-yd dash? By how many meters? By how many yards?

8. What would be the reading on a European speedometer if you were traveling at 60 mph? (Remember that time measurements are the same in metric and English.)

9. Convert each dimension in Fig. 5-1 to inches, rounding off to the nearest 1/1000 in.

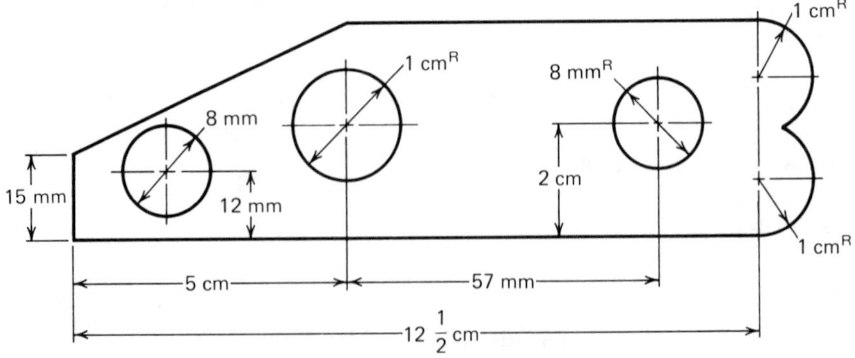

Figure 5-1

10. Convert the following.
    (a) 15 ℓ to gallons
    (b) 7.2 ℓ to quarts
    (c) 180 mℓ to pints
    (d) 3.2 qt to liters
    (e) 2 pt to liters
    (f) 1/2 pt to milliliters
    (g) 4 1/2 gal to liters
11. Is meat cheaper at $4.19 per pound or at $9.19 per kilogram?
12. If U.S. airlines allow 44 lb of baggage on overseas flights and European airlines allow 16 kg, which policy is more restrictive?
13. If you want to mail a 60-g package first class to a friend in New York, how much will the postage cost?
14. A European publisher ordered 600 kg of newsprint and neglected to specify the weight in pounds. How many pounds should you send her?
15. You have contracted to move First National Bank's German-made vault into their new building. The vault has "55 000 kg" printed on the inside of the door. What size crane can lift the vault?

Part C

16. If your wrench set is based on sixteenths of an inch, what is the best wrench to use on a 19-mm nut?
17. The owner's manual of a German car lists the piston stroke length as 72 mm. What is that length to the nearest sixteenth of an inch?
18. While you were in Europe for a convention, you and your wife fell in love with a country cottage. You were able to obtain a blueprint that was prepared in the metric system. The basic floor plan is shown in Fig. 5-2. Convert

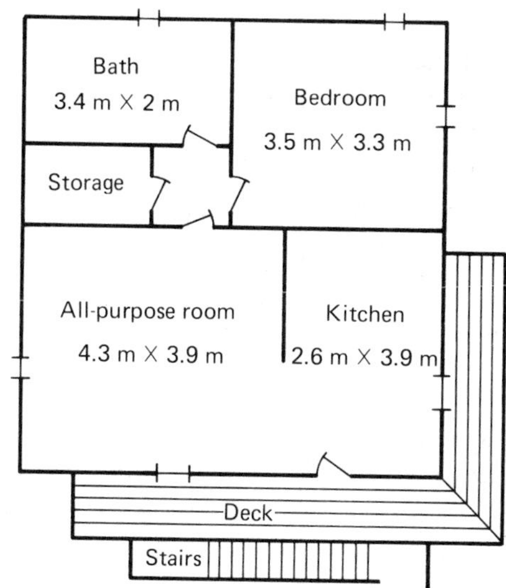

Figure 5-2

these dimensions to the English system so that you can begin estimating construction costs.

19. A body-shop manager needs 10 gal of paint. If he can buy the paint at $35.60 per gallon or at $9.39 per liter, which should he buy? Remember that he cannot buy part of a liter.

Answers

2. (a) 39.37 in. (b) 1.09 yd (c) 109.36 yd (d) 0.62 mi
   (e) 0.47 in. (f) 0.35 in. (g) 6.38 in.
4. (a) 35.2 lb (b) 13.38 oz (c) 1.58 oz
   (d) 6.65 lb (e) 0.00014 oz (f) 0.0035 oz
6. 4.23 mm
8. 96 km per hour
10. (a) 3.975 gal (b) 7.632 qt (c) 0.382 pt (d) 3.019
    (e) 0.943 (f) 235.75 m (g) 16.974
12. European, because 16 kg = 35.2 lb
14. 1320 lb
16. 12/16 in. or 3/4 in.
18. 3.4 m = 11 ft 2 in. 2 m = 6 ft 7 in. 3.5 m = 11 ft 6 in.
    3.3 m = 10 ft 10 in. 4.3 m = 14 ft 1 in. 3.9 m = 12 ft 10 in.
    2.6 m = 8 ft 6 in. all to nearest inch

Review Exercise

Part A

1. Convert
   (a) 2 in. to centimeters (b) 2 cm to inches
   (c) 64 ft to meters (d) 1.7 in. to millimeters
   (e) 14 m to yards (f) 6.2 in. to centimeters
   (g) 16 lb to grams (h) 3.2 kg to pounds
2. In metric units, what are the dimensions of a "two by four" piece of lumber?
3. On a metal strip 15 cm long, I want to drill five equally spaced holes, with 16 mm from the center of the end holes to the ends of the strip. How far apart should the centers of the holes be, in centimeters?

Part B

4. If one 3-cm bolt weighs 54 g, how many bolts would it take to weigh 1 kg? How many bolts would it take to weigh 100 lb?
5. What is the area of a 24 cm by 18 cm page with a 6 mm by 8 mm rectangle cut off each corner?

6. You need to fence in your business parking lot, which measures 200 ft × 150 ft. You can buy Polish-made wire fencing in 20-m rolls at $59.40 per roll or you can buy American wire fencing in 50-ft rolls at $47.70 per roll. Which wire would be cheaper for you to use?
7. "The suspect is 5 ft 10 in., 230 lb, with a 2 1/2-in. scar on the right side of the jaw." Translate this description into metric units for the benefit of French authorities watching for him at the Paris airport.

Part C

8. A foreign-made car averages 10 km per liter and a similar American-made car averages 25 mi per gallon. If they have the same rental charge, not including gas, which car would you rent?
9. Is gas cheaper at $1.60 per gallon or 40¢ per liter?
10. Would your gas costs be less using gas costing 40¢ per liter in a car averaging 10 km per liter or using gas costing $1.60 per gallon in a car averaging 25 mi per gallon?
11. Your auto repair shop can buy transmission fluid locally for 68¢ per quart. If you order it from a European Common Market exporter, it costs 65¢ per liter plus $6.80 for shipping and handling for each 100-ℓ container. Which is cheaper?

Answers

2. 5.08 cm by 10.16 cm
4. 19 bolts; 841 bolts
6. Polish
8. American
10. $1.60/gal at 25 mi/gal

Chapter 6

Measurement and Measuring Devices

Measurement is a language of science that has been used for thousands of years. We are affected by many different units of measure and various measuring devices. For instance, we buy lumber for a house in terms of feet (board feet). We buy cloth in terms of yards. We buy screws, nuts, and bolts in terms of length, diameter, and threads per inch. In our car we install parts whose dimensions are accurately measured to thousandths or ten thousandths of an inch or hundredths of a millimeter.

In order to become somewhat more self-sufficient and to be able to do more and better work, each of us needs to learn to use the various measuring devices. In general, the more accurate the measurement must be, the more sophisticated the measuring device should be.

In this chapter we shall discuss and demonstrate the use of the more commonly used measuring devices. We shall be concerned primarily with linear measure, although angular measure will be considered.

6-1 RULERS

The ruler is probably the most commonly used measuring device. There are various types of rulers. Although rulers are made of many different types of materials, the units of measure are the same within the limitations of accuracy.

English and Metric Rulers: For the English (or British) ruler, the smallest whole unit of linear measure is the inch. This unit may be broken down into subdivisions by dividing it into two equal halves and then repeating this "halving" process several times until we have 64 equal subdivisions. This would produce halves, fourths, eighths, sixteenths, thirty-seconds, and sixty-fourths of an inch. See Fig. 6-1.

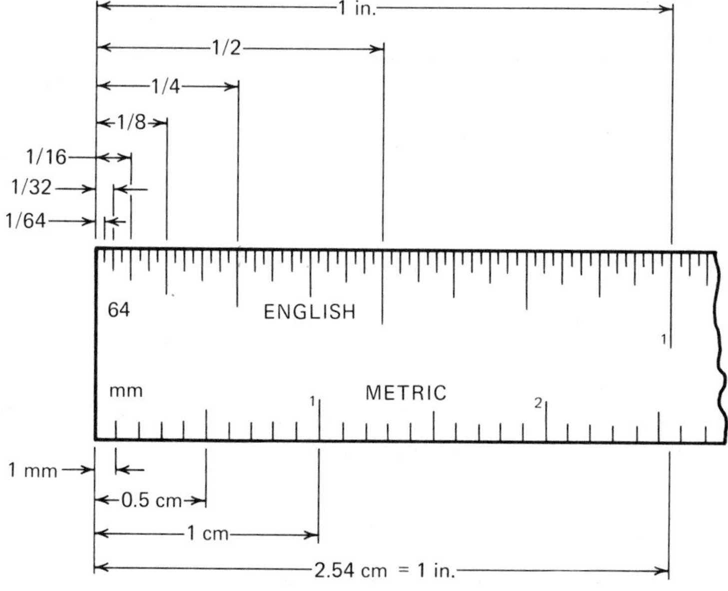

Figure 6-1. English-metric ruler.

The metric ruler has the millimeter as its smallest whole unit of linear measure. In comparison to the English ruler, **2.54 cm = 1 in.** or 25.4 mm = 1 in. This means that 1/16 of an inch is approximately equal to 1.6 mm as shown in Fig. 6-1.

Decimal Ruler: The decimal ruler is subdivided into tenths of an inch instead of fourths, eighths, and so forth. The more common decimal rulers are graduated like the one in Fig. 6-2.

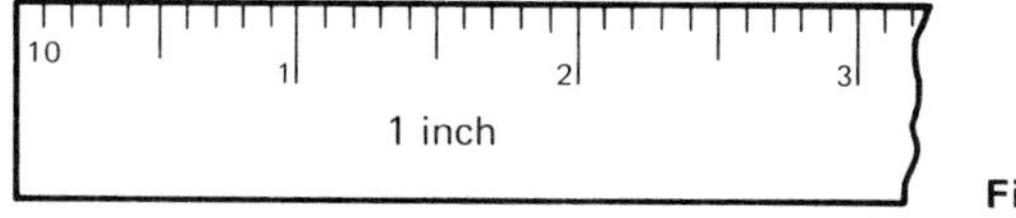

Figure 6-2. Decimal ruler.

Read the measurements in Fig. 6-3 to the nearest fractional or decimal part of an inch or centimeter.

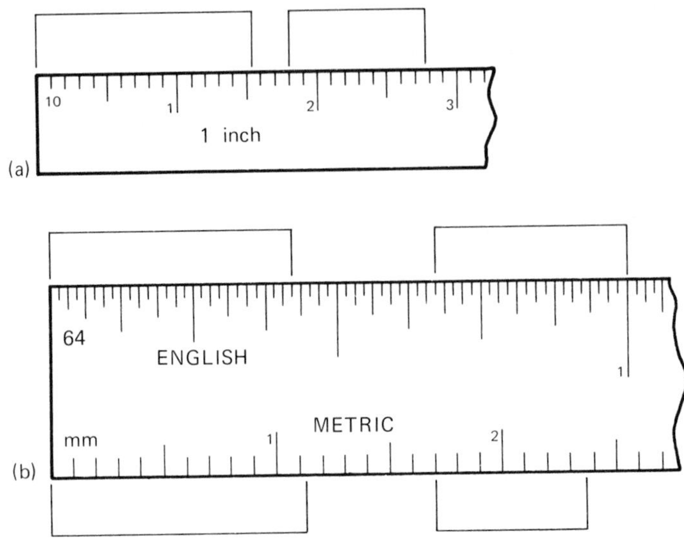

Figure 6-3

Printer's Ruler: The printer's ruler is graduated in inches on one edge and in *points* and *picas* on the other edge as shown in Fig. 6-4.

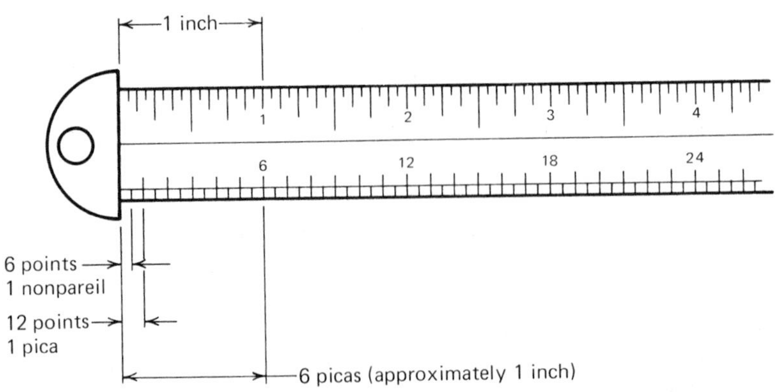

Figure 6-4. Printer's ruler.

By studying Fig. 6-4, we can see that the following relations are true.

6 picas $\doteq$ 1 inch	12 points = 1 pica
72 points $\doteq$ 1 inch	6 points = 1 nonpareil
1 point $\doteq \frac{1}{72}$ inch	6 points = 1 half pica

Using the printer's ruler shown in Fig. 6-5, indicate the readings to the nearest fractional part of an inch or in terms of picas and points.

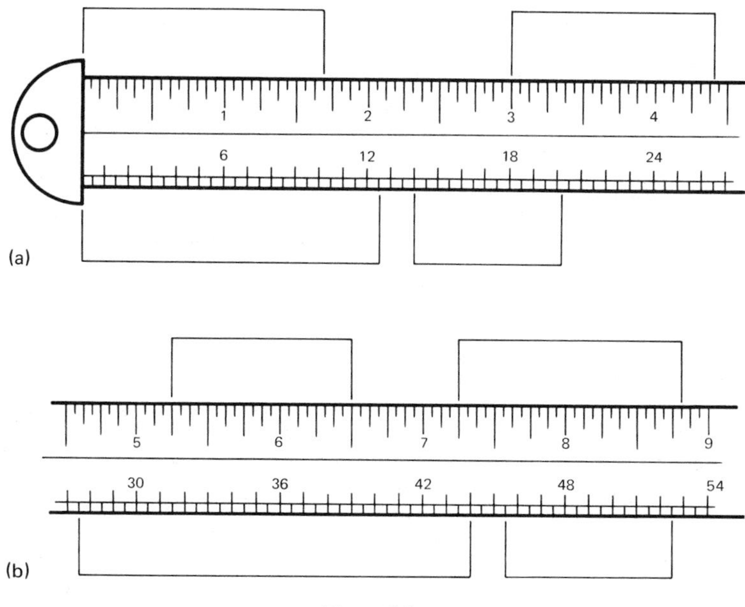

Figure 6-5

Exercise Set 6-1

Part A

1. Use a ruler or steel scale to measure each of the lines in Fig. 6-6 to the accuracy indicated.

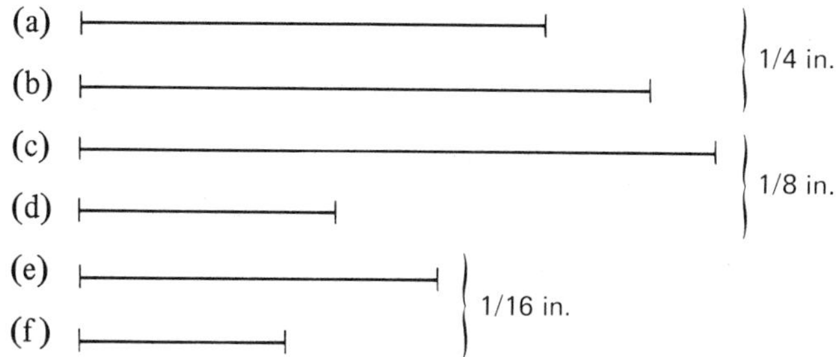

Figure 6-6

(g) Construct a line of length (a) + (d) and indicate its length.
(h) Construct a line equal in length to (e) − (f). How long is it?

2. Use a decimal ruler to measure each of the lines in Fig. 6-7 with the indicated accuracy.

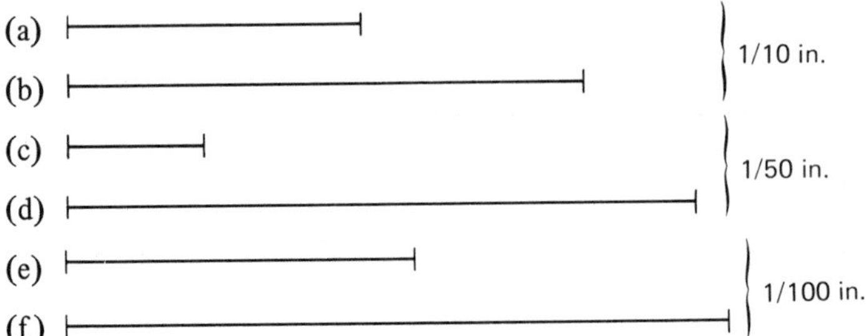

Figure 6-7

(g) Draw a line of length (c) + (f). How long is it?
(h) Line (c) is how much shorter than line (e)?

3. Use a metric ruler to measure the lines in Exercise 2 to the nearest millimeter.

4. Use a printer's ruler to measure each line in Fig. 6-8. Express your answer in terms of whole picas and points and to the nearest 1/8 in.

	Picas and Points	Inches
(a)	______	______
(b)	______	______
(c)	______	______
(d)	______	______
(e)	______	______
(f)	______	______

Figure 6-8

(g) How many picas and points long would the lines (a) + (c) be?

Part B

5. Compare the measurements from Exercise 2 with those of Exercise 3 by converting the millimeters to inches. Does each of your answers come within 1/10 in.?

6. On the straight lines in Fig. 6-9, lay out the following lengths one after another: 1/2 in., 2 3/4 in., 1 7/8 in., and 9/16 in.

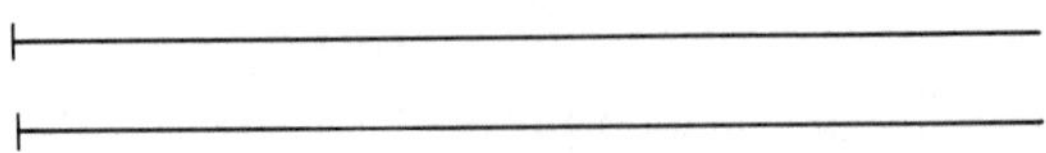

Figure 6-9

Add these lengths together to find the total length of the line laid out. Measure the line laid out. Does it come within 1/16 in. of this length?

Part C

7. Measure lengths A, B, and C and lengths D and E and check the sums against the overall dimension F in Fig. 6-10.

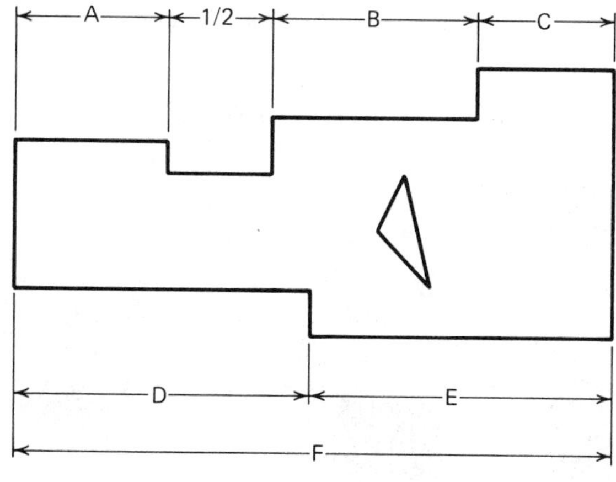

Figure 6-10

8. Fig. 6-11 is a sketch of a cutout for a piece of paper. Indicate the length of each edge (i.e., DE = 3/8 in.). How many inches is it around the cut edge (total perimeter) of the cutout?

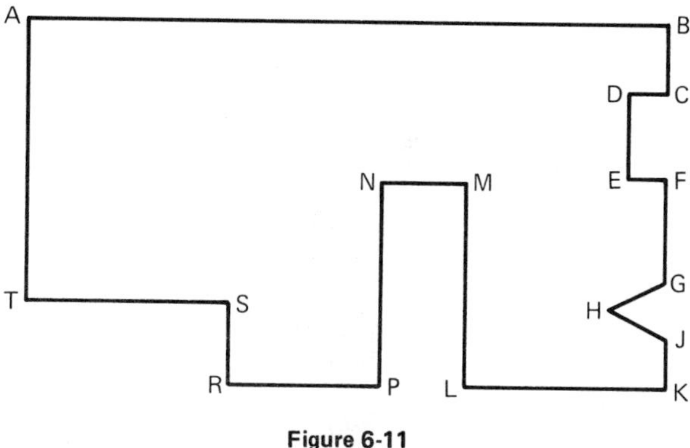

Figure 6-11

6-2 VERNIERS

Even though there are many different types of verniers, we wish to discuss only the vernier caliper (see Fig. 6-12).

Vernier Caliper: The vernier caliper may be used to measure either outside or inside dimensions. All vernier calipers are designed to be accurate to 0.001 in. The one shown in Fig. 6-12 is also accurate to 1/128 in.

The vernier caliper, and any other measuring device with a vernier scale, uses two scales. The main scale has an inch divided in 40 equal parts. Each part represents 1/40 in. or 0.025 in. The other scale is the vernier scale. On the vernier scale, there are 25 divisions covering the same length as 24 divisions on the main scale (see Fig. 6-12). Thus the 25 divisions cover 0.600 in. because 24 X 1/40 = 24 X 0.025 =

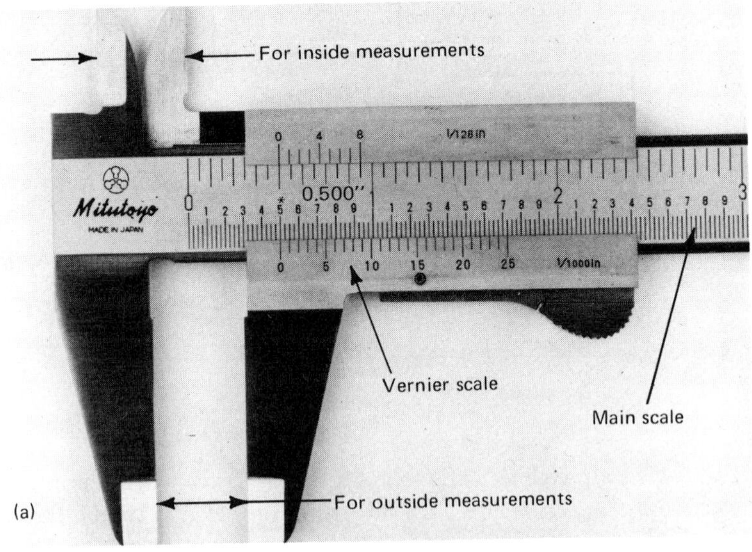

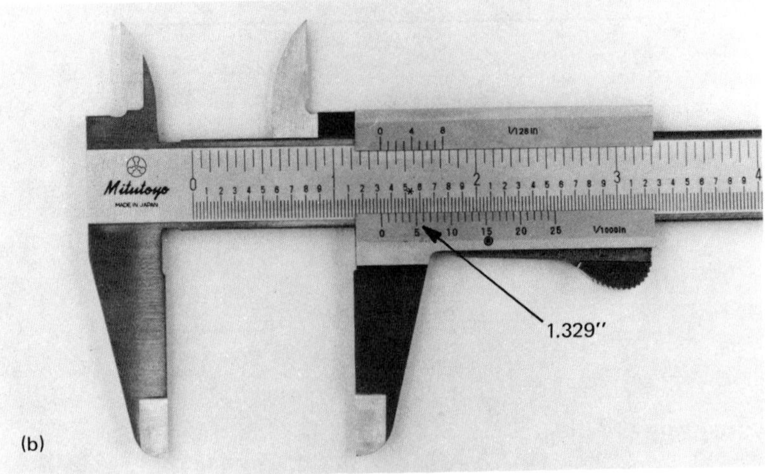

Figure 6-12. (a) Vernier caliper. (b) Reading a vernier scale. Courtesy Black & Co. Hardware.

0.600 in. So we see that one division on the vernier scale equals 0.024 in. because 1/25 × 0.600 in. = 0.024 in. Therefore, the difference between a division on the main scale and the vernier scale is 0.025 in. - 0.024 in. = 0.001 in.

In order to read the vernier scale [see Fig. 6-12(b)] we may follow the steps listed below:

Steps for Reading the Vernier Scale	
Step 1. Count the number of inches to the left of the zero on the vernier.	1.000 in.
Step 2. Count the tenths of an inch (or 0.100 in.) to the left of the zero on the vernier.	0.300 in.
Step 3. Count the smaller subdivisions (0.025 in.) to the left of the zero on the vernier.	0.025 in.
Step 4. Count the marks (0.001 in.) on the vernier from zero to the line that coincides with a line on the main scale.	0.004 in.
Step 5. Add the figures obtained in Steps 1 through 4 for the correct reading.	1.329 in.

Sample Readings

(a) Reading 0.516″

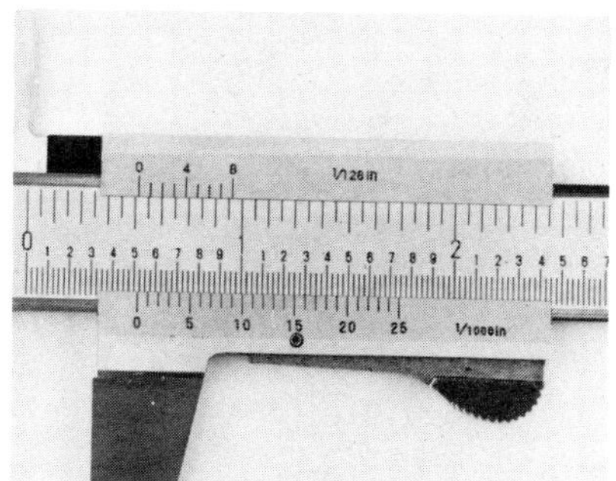

Figure 6-13. Courtesy Black & Co. Hardware.

(b) Reading 2.094″

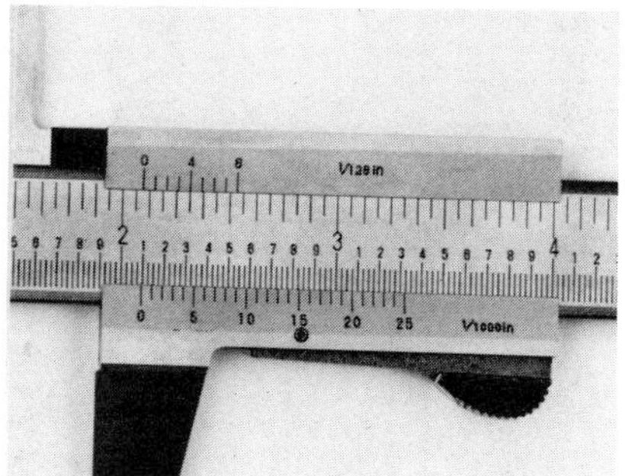

Figure 6-14. Courtesy Black & Co. Hardware.

Exercise Set 6-2

Part A

1. If you have access to a vernier caliper, set it at each of the settings in Fig. 6-15 and indicate each of these readings.

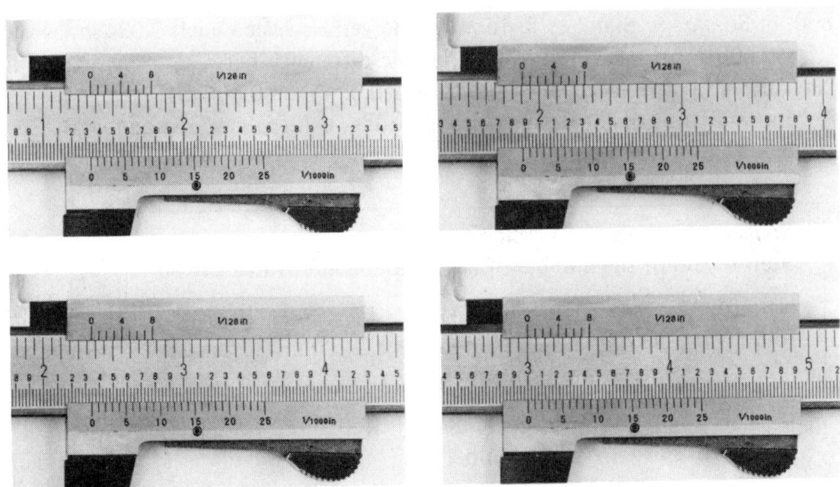

Figure 6-15. Courtesy Black & Co. Hardware.

2. Using a vernier caliper, measure the outside diameter of each of the bolts pictured in Fig. 6-16.

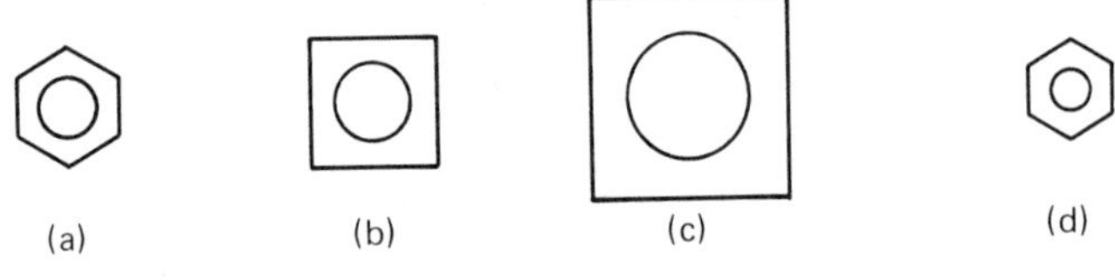

Figure 6-16

3. Measure the inside diameter of the holes shown on the template in Fig. 6-17.

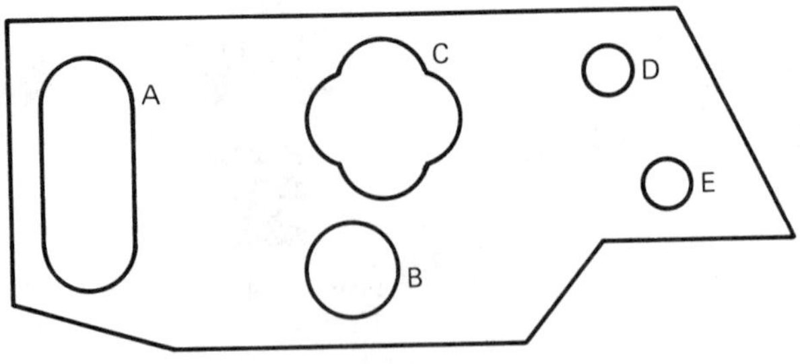

Figure 6-17

Part B

4. Discuss differences between a decimal ruler and a vernier caliper. List some advantages and disadvantages of each.

5. Using a vernier caliper, lay off segments of the indicated lengths.
   (a) 1.234 in.
   (b) 4.010 in.
   (c) 0.575 in.
   (d) 2.196 in.

Part C

6. Make a sketch of a vernier caliper for each measurement indicated.
   (a) 3.628 in.
   (b) 1.999 in.
   (c) 0.489 in.
   (d) 4.701 in.
7. With a vernier caliper, measure the thickness of 10 pages in this chapter. Using this measurement, what is the thickness of 1 page? Now measure 20 pages with the vernier caliper and use this measurement to derive the thickness of 1 page. Finally, measure 50 pages with the vernier caliper and use this measurement to derive the thickness of 1 page. Which method do you think is most accurate for determining the thickness of a page?
8. Give at least five (5) examples of where and how a vernier caliper could be used in your area of study.

6-3 MICROMETER

The micrometer is a measuring device that can be used when something must be measured accurately to a thousandth or ten thousandth of an inch or a tenth or hundredth of a millimeter. There are various kinds of micrometers. We wish to discuss each of the following: the outside micrometer, the inside micrometer, and the vernier micrometer.

The basic structure of the micrometer is as shown in Fig. 6-18 (p. 120). The standard micrometer is made up of five parts: (1) the frame, (2) the anvil, (3) the spindle, (4) the sleeve (barrel), and (5) the thimble. As the thimble is rotated, the spindle moves toward or away from the anvil, which is stationary.

English Micrometer: The English micrometer has 40 vertical graduations on the sleeve, or barrel (see Fig. 6-19). Each of these 40 graduations represents 1/40 or 0.025 in. Since one complete rotation of the thimble moves the spindle and thimble 0.025 in., the thimble is divided into 25 horizontal graduations. Each time a new line on the thimble crosses the horizontal line on the barrel, the spindle has moved toward or away from the anvil by 1/25 of a revolution or 0.001 in. Each fifth division of the thimble is numbered. The 5 represents 0.005 in., 10 represents 0.010 in., and so forth.

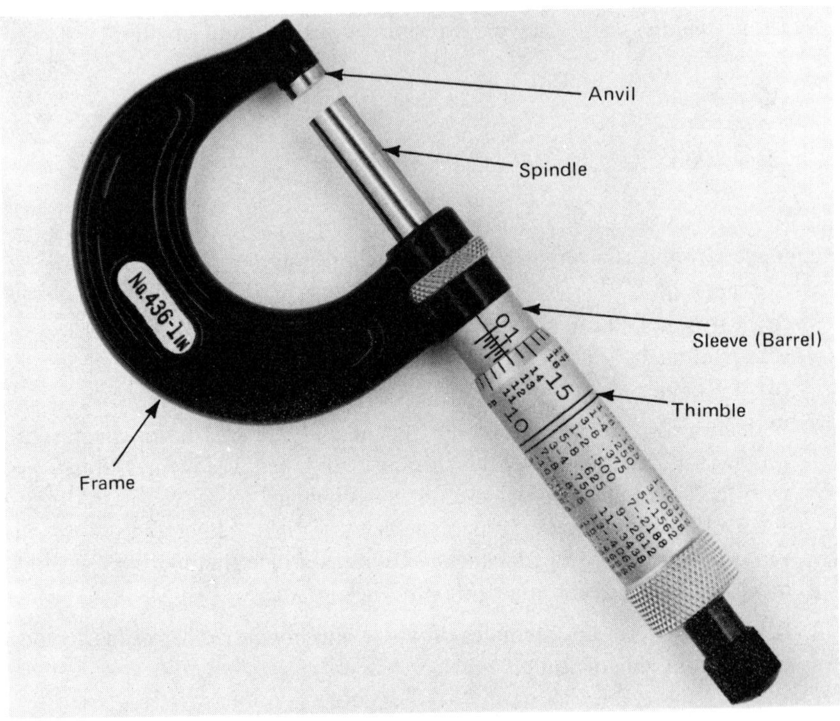

Figure 6-18. Parts of a standard micrometer. Courtesy Champaign Ace Hardware.

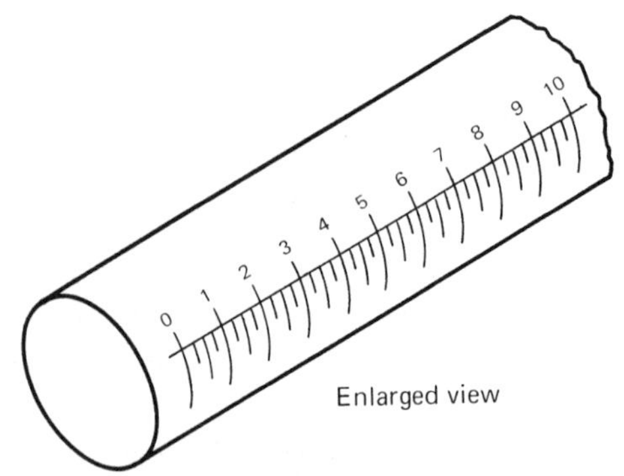

Figure 6-19. Graduated barrel of an English micrometer.

Metric Micrometer: The metric micrometer operates on the same principle as the English micrometer. The only difference is the unit of measure and the graduations. The barrel is divided into 25 vertical graduations, each of which represents 1 mm. Beneath the main graduations, you will find small graduations indicating 1/2 mm (see Fig. 6-20). The thimble is divided into 50 horizontal graduations, each of which represents 0.01 mm. Thus two complete revolutions of the thimble cause the spindle to move a distance of 1 mm.

Outside Micrometer: The outside micrometer is used to measure the outside diameter, thickness, width, and length of objects (see Fig. 6-18 for a picture of the outside micrometer). In order to make a measurement with an outside micrometer, place the anvil against one side or surface of the object to be measured and gently turn the thimble until the spindle rests against the opposite side or surface of the object.

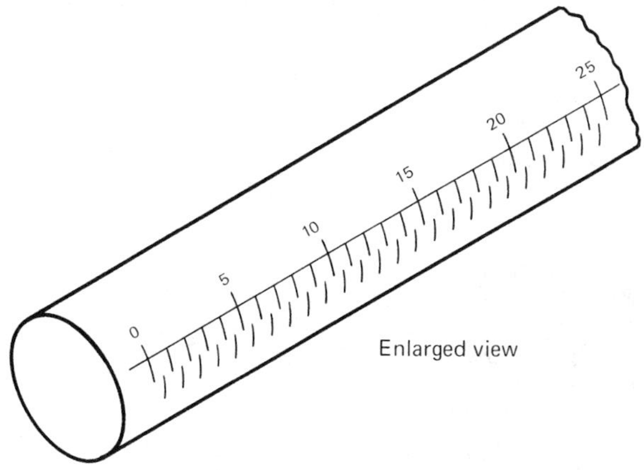

Figure 6-20. Graduated barrel of a metric micrometer.

Reading the English Micrometer

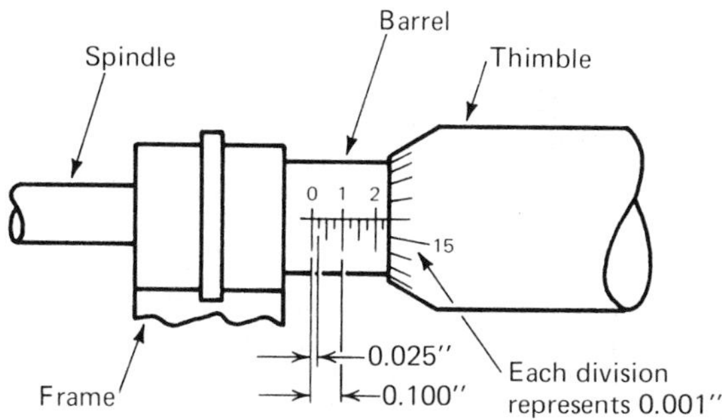

Figure 6-21

We may determine the correct reading in Fig. 6-21 by following the four steps shown on p. 122.

Steps for Reading the English Micrometer	
Step 1. Count the numbered lines on the barrel and multiply by 0.100 in.	2 X 0.100 in. = 0.200 in.
Step 2. Count all visible shorter (nonnumbered) lines between the last numbered line and the thimble and multiply by 0.025 in.	1 X 0.025 in. = 0.025 in.
Step 3. Count on the thimble from zero to the mark nearest the horizontal line on the barrel and multiply by 0.001 in.	16 X 0.001 in. = 0.016 in.
Step 4. Add all answers from Steps 1 through 3 for the correct reading.	0.200 in. 0.025 in. +0.016 in.
Correct reading is	0.241 in.

Sample Readings

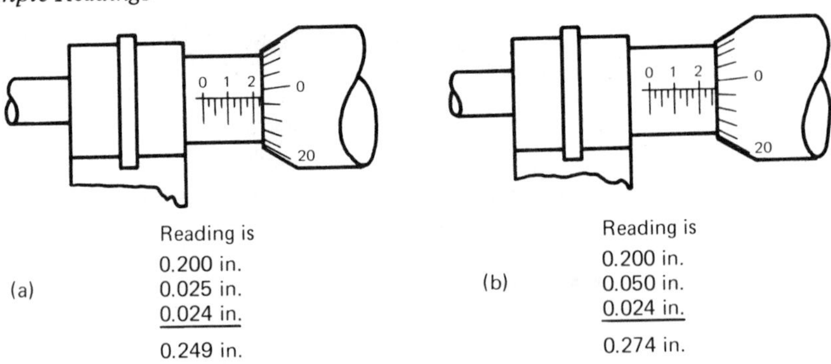

Figure 6-22

Reading the Metric Micrometer

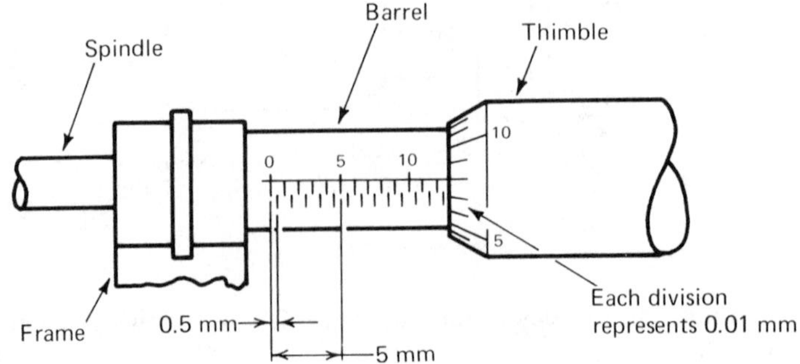

Figure 6-23

Steps for Reading the Metric Micrometer	
Step 1. Count millimeter lines visible on the barrel.	12 lines = 12.00 mm
Step 2. If the next half-millimeter line is visible, write 0.50 mm.	1 line = 0.50 mm
Step 3. Count to the mark on the thimble nearest the long line on the barrel and multiply by 0.01 mm.	8 lines = 0.08 mm
Step 4. Add the answers from Steps 1 through 3 for the correct reading.	12.58 mm

Sample Readings

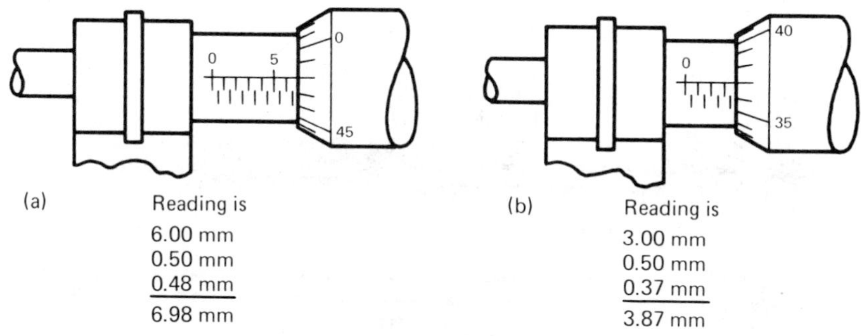

Figure 6-24

Examples of micrometer readings

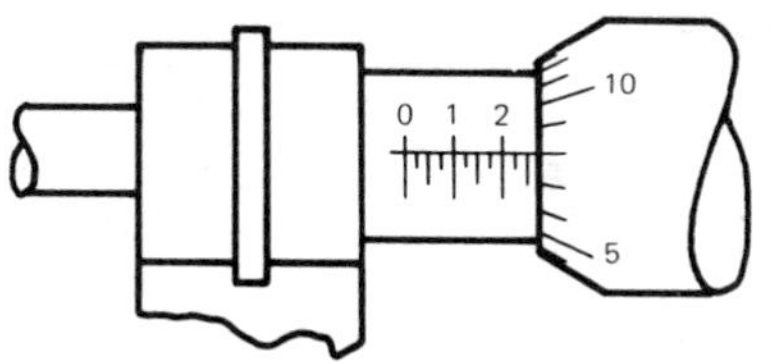

Figure 6-25(a)

Since the barrel is numbered 0, 1, 2, . . . , this is an English micrometer.

The reading is	0.200 in.
	0.050 in.
	+0.008 in.
Actual reading:	0.258 in.

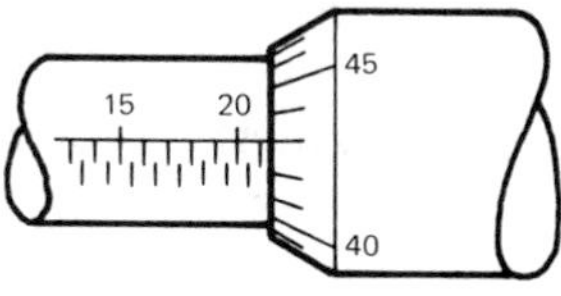

Figure 6-25(b)

Since the barrel is numbered 0, 5, 10, 15, 20, 25, this is a metric micrometer.

The reading is:	21 upper lines =	21.00 mm
	0 lower lines =	0.00 mm
	43 on thimble =	0.43 mm
	Actual reading:	21.43 mm

Inside Micrometer: The inside micrometer is used to measure inside dimensions such as the bore of a cylinder or the inside diameter of a hollow shaft. The inside micrometer is read the same as the outside micrometer discussed above. The more common inside micrometers have movements up to half an inch. A typical inside micrometer set is pictured in Fig. 6-26.

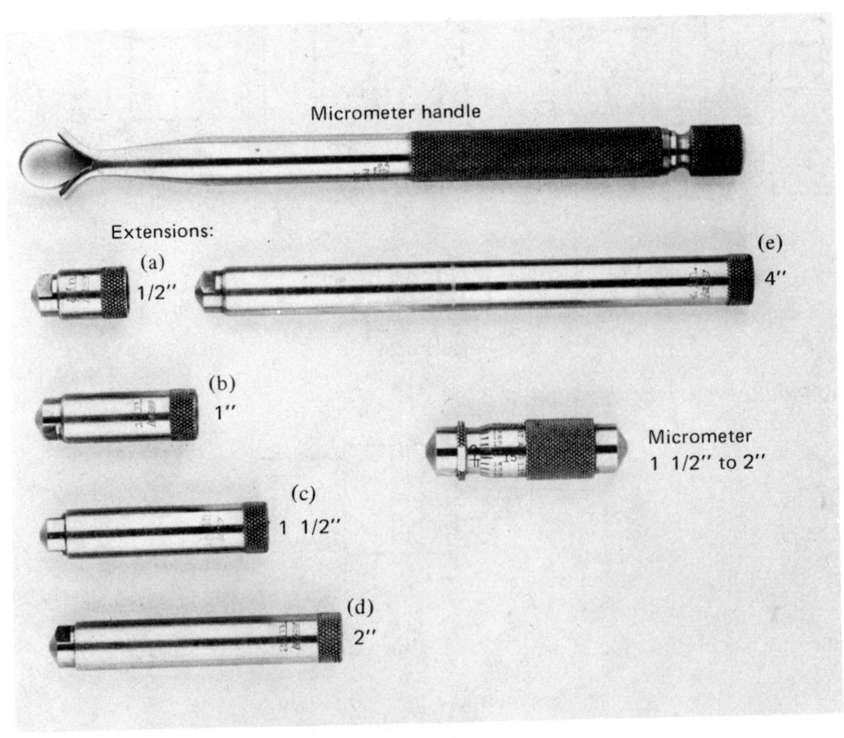

Figure 6-26. Inside micrometer set.

Using this inside micrometer set, shown in Fig. 6-26, we can make any measurement between 1 1/2 and 8 in. There are similar sets which are smaller than this one and others which are larger. These sets would enable us to measure inside dimensions less than 1 1/2 in. and those greater than 8 in.

Vernier Micrometer: The vernier micrometer operates on the same basis as the standard micrometer. However, the vernier micrometer will measure accurately to a ten thousandth of an inch (0.0001 in.) or two thousandths of a millimeter (0.002 mm). In addition to the main scale that all micrometers have, the vernier micrometer has a vernier scale on the barrel (Fig. 6-27).

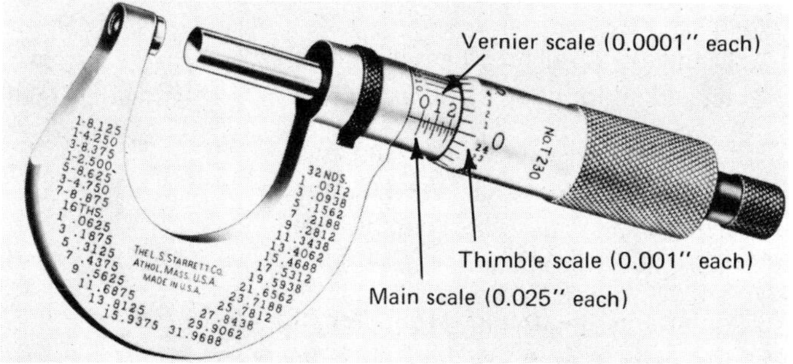

Figure 6-27. Vernier micrometer. Courtesy of the L. S. Starrett Company.

On the English vernier micrometer, the vernier is divided into 10 equal parts, which are equal in total length to nine divisions on the thimble. Since each division on the thimble equals 0.001 in., nine divisions would be 0.009 in. Thus, one division on the vernier is equal to 1/10 of 0.009 in. or 0.0009 in. So the difference between a division on the thimble and a division on the vernier is 0.0010 in. - 0.0009 in. = 0.0001 in. See Fig. 6-28.

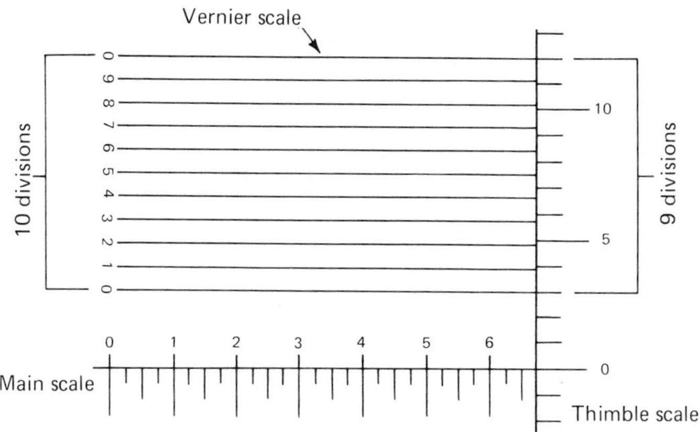

Figure 6-28

Reading the English Vernier Micrometer: Figure 6-27 shows an English vernier micrometer. The reading shown in Figure 6-29 is used to illustrate the following steps.

Steps for Reading the English Vernier Micrometer	
Step 1. Count the numbered lines on the barrel and multiply by 0.100 in.	4(0.100 in.) = 0.400 in.
Step 2. Count all visible shorter (nonnumbered) lines between the last numbered line and the thimble and multiply by 0.025 in.	2(0.025 in.) = 0.050 in.
Step 3. Count on the thimble from zero to the mark nearest the horizontal line on the barrel and multiply by 0.001 in.	19(0.001 in.) = 0.019 in.
Step 4. Add to the above the number of ten thousandths indicated where the line on the vernier coincides with the line on the thimble.	0.0007 in.
	0.4697 in.

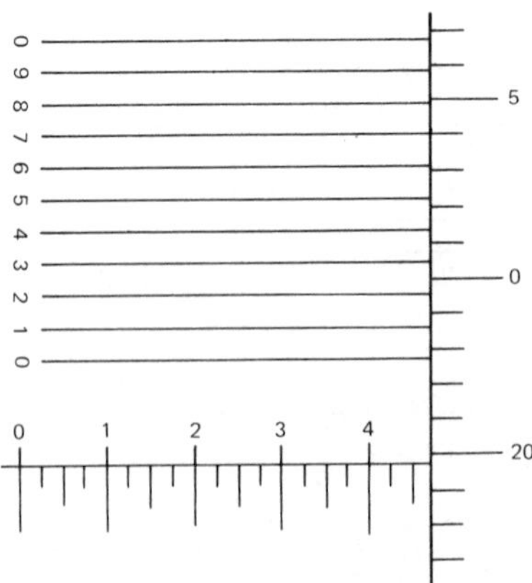

Figure 6-29

Sample Readings

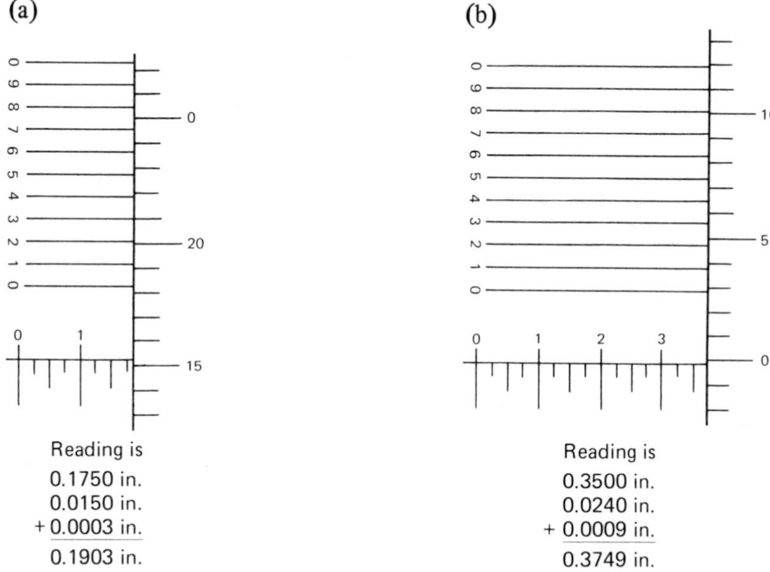

Figure 6-30

On the metric vernier micrometer, the vernier is divided into five equal parts that are equal in total length to four divisions on the thimble. Since each division on the thimble equals 0.01 mm, four divisions would equal 0.04 mm.

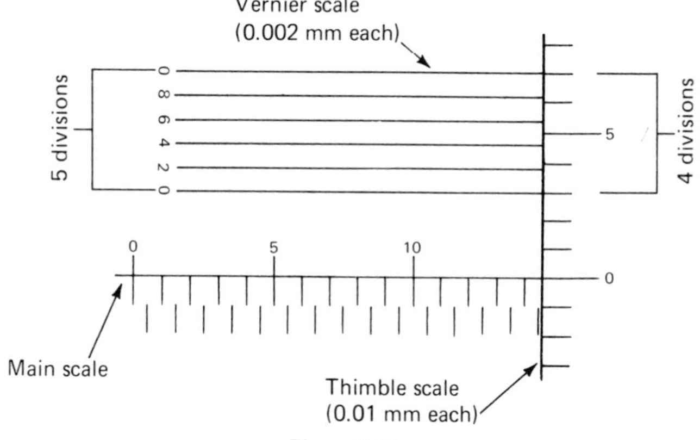

Figure 6-31

Thus, one division on the vernier is equal to 1/5 of 0.04 mm, or 0.008 mm. So the difference between a division on the thimble and one on the vernier is 0.010 mm - 0.008 mm = 0.002 mm. (See Fig. 6-31.)

Reading the Metric Vernier Micrometer: The reading shown in Figure 6-32 is used to illustrate the following steps.

Steps for Reading the Metric Vernier Micrometer	
Step 1. Count millimeter lines visible on the barrel.	11.**000** mm
Step 2. If the next half-millimeter line is visible, write 0.5 mm.	0.5**00** mm
Step 3. Count to the mark on the thimble at or below the horizontal line on the barrel and multiply by 0.01 mm.	0.28**0** mm
Step 4. Add to these the multiple of 0.002 mm indicated where the line on the vernier scale coincides with the line on the thimble.	0.006 mm
	11.786 mm

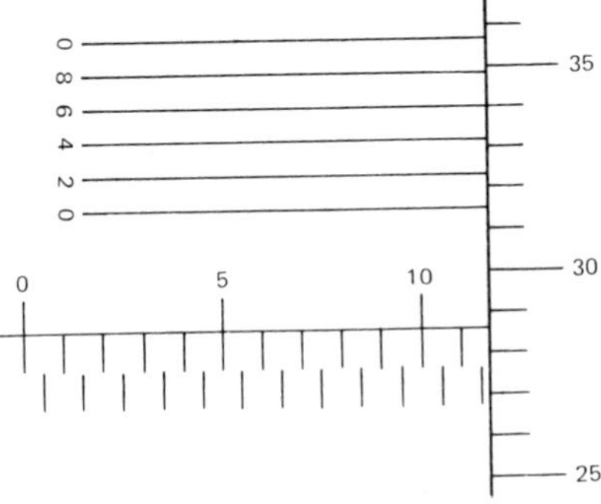

Figure 6-32

Sample Readings

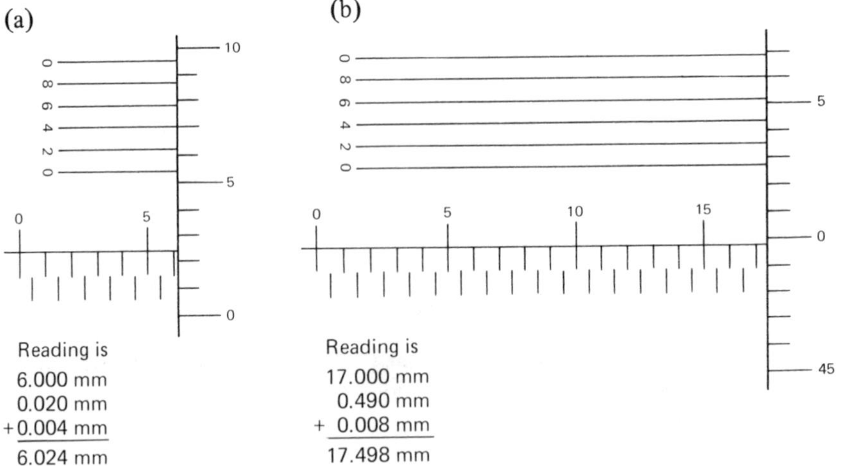

(a) Reading is

6.000 mm
0.020 mm
+0.004 mm
6.024 mm

(b) Reading is

17.000 mm
0.490 mm
+ 0.008 mm
17.498 mm

Figure 6-33

Exercise Set 6-3

Part A

1. Obtain an English micrometer, if possible; set it to match each setting in Fig. 6-34; indicate each reading.

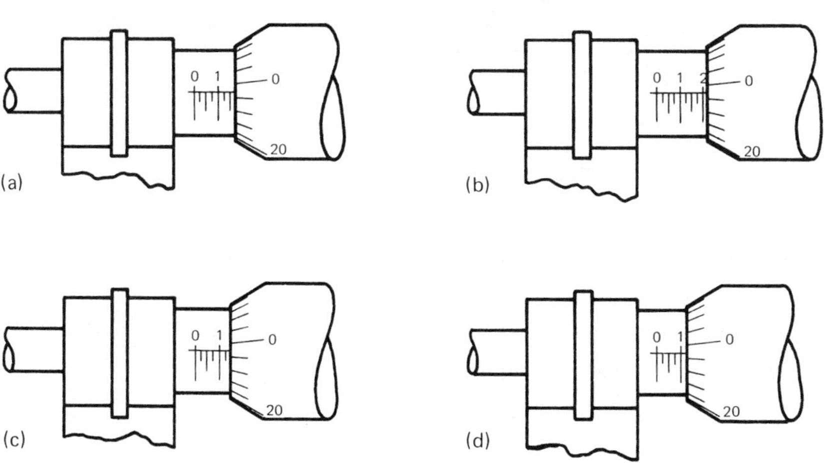

Figure 6-34

2. Using a metric micrometer, if available; match the settings in Fig. 6-35; indicate those readings.

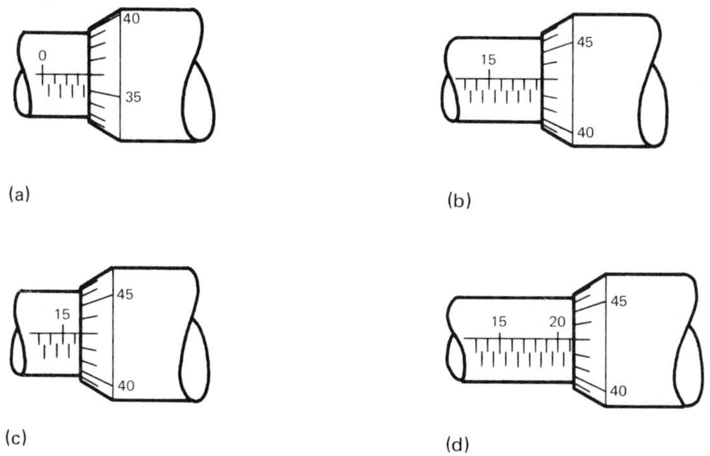

Figure 6-35

3. Classify the micrometer in Fig. 6-18 as English or metric, outside or inside, and standard or vernier.

Part B

4. Refer to Fig. 6-26 and fill in the blanks.

	Measured Length	Micrometer Setting	Extensions Used
(a)	3.785 in.	0.285 in.	d (or a and c)
(b)	________	0.500 in.	a and e
(c)	5.575 in.	0.075 in.	________
(d)	________	0.193 in.	c
(e)	7.932 in.	0.432 in.	________

5. If you have an English vernier micrometer available, set it to match the settings in Fig. 6-36 and indicate each reading.

6. If possible, use a metric vernier micrometer to match the setting indicated in Fig. 6-37 and write the readings shown.

Part C

7. Use a vernier micrometer to measure the thickness of this page. How does this measurement compare with that calculated in Exercise 7 on p. 119? If your answers are not the same, discuss possible reasons why they are different.

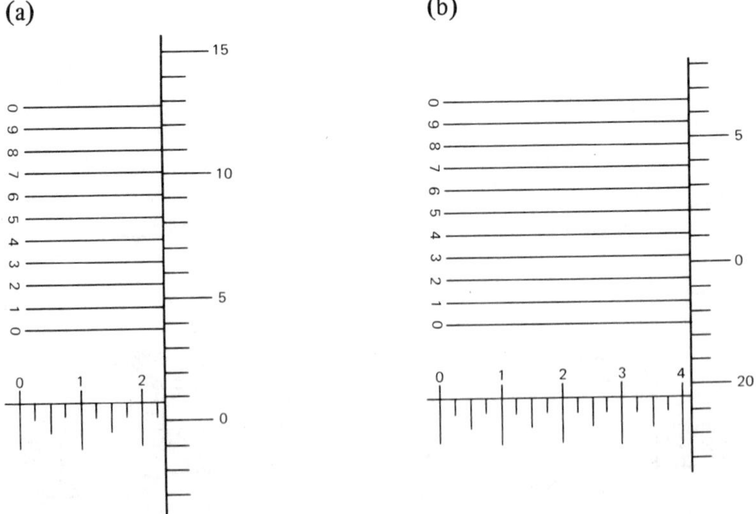

Figure 6-36

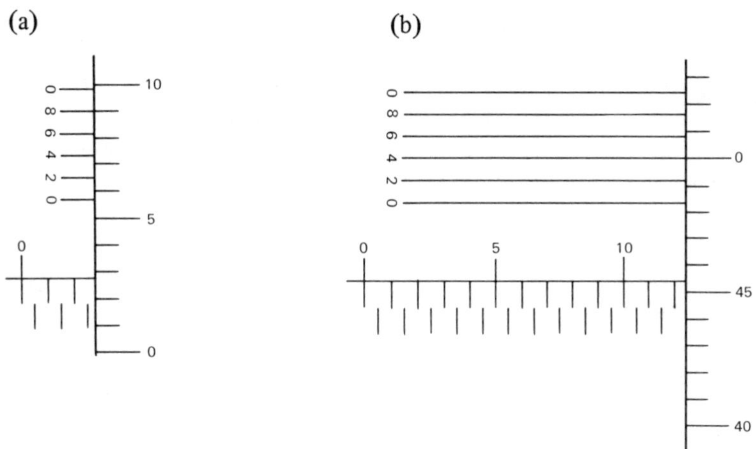

Figure 6-37

8. Without referring to your text, make a sketch of each of the following.
   (a) Standard outside English micrometer
   (b) An inside metric micrometer
   (c) The three scales of an English vernier micrometer
   (d) Standard outside metric micrometer
9. Set a vernier micrometer to match each reading below and make a sketch of each setting.
   (a) 0.4734 in. (b) 0.7902 in. (c) 0.7831 in.
   (d) 0.634 mm (e) 5.102 mm (f) 23.928 mm
10. Fill in each blank with the correct reading.

	Main Scale	Thimble Scale	Vernier Scale	Reading
(a)	11.5 mm	0.47 mm	0.004 mm	11.974 mm
(b)	17.0 mm	0.32 mm	0.008 mm	________
(c)	0.325 in.	0.022 in.	0.0007 in.	________
(d)	0.775 in.	0.013 in.	0.0005 in.	________
(e)	5.5 mm	0.18 mm	0.002 mm	________

11. Give at least five (5) examples of where and how some type of micrometer is used within your area of interest.

Answers

2. 3.86; 19.43; 16.43; 21.43
4. 6.500 in.; e, 3.193; d and e
6. 2.528 mm; 12.454 mm
10. (b) 17.328 mm (c) 0.3477 in. (d) 0.7885 in. (e) 5.682 mm

6-4 OTHER MEASURING DEVICES

Calipers: Measurement by transfer is one of the oldest types of measurement. Human beings probably used this method many years before a standard for length ever came into existence. A basic instrument used to transfer a measurement is the caliper. There are four types of simple calipers (Fig. 6-38):

1. Divider
2. Outside caliper
3. Inside caliper
4. Hermaphrodite caliper

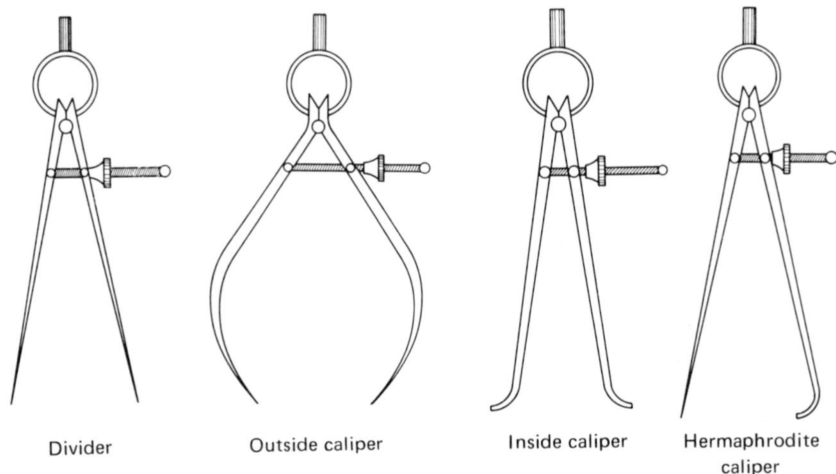

Figure 6-38. Simple caliper instruments.

The divider is a member of the caliper family although the word *caliper* is not contained in its name. The vernier caliper, which we have already discussed, is a special caliper and will not be discussed in this section.

Simple caliper instruments duplicate the separation between the reference point and the measured point (see Fig. 6-39). They are used to transfer this separation to a standard of measure such as a scale.

When using any of the simple calipers mentioned in Fig. 6-39, you should be extremely careful because there are two measurements involved. One measurement is made in setting the caliper on the object to be measured. The second one is made when the caliper is transferred to a scale or standard of measurement.

Compass: Another instrument that is a special type of divider is the compass (Fig. 6-40). The compass is used as a divider but is also capable of marking the transferred measurement. By using the pencil in one leg of the compass, it is possible to construct drawings, in particular, circles and arcs.

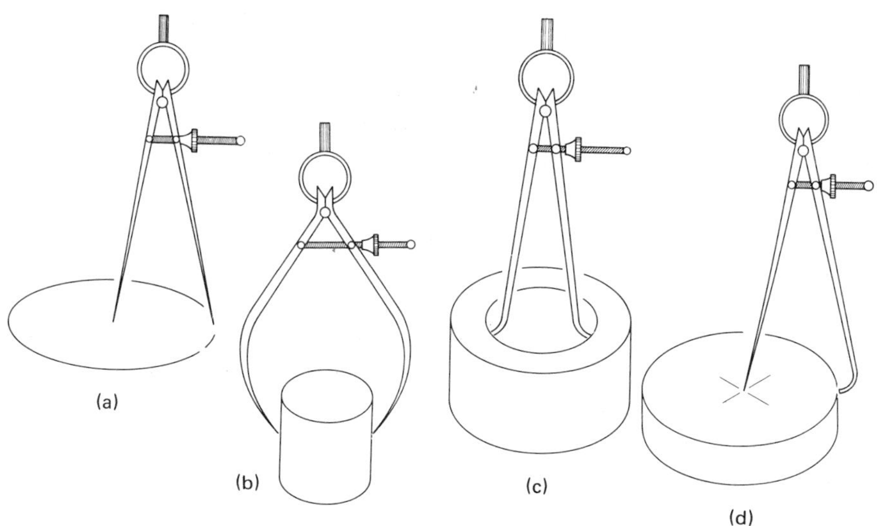

Figure 6-39. (a) Dividers are best used for line-to-line measures. (b) Outside calipers are best suited for measures across surfaces. (c) Inside calipers are best suited for measures between surfaces. (d) Hermaphrodite calipers are best suited for line-to-end measurements.

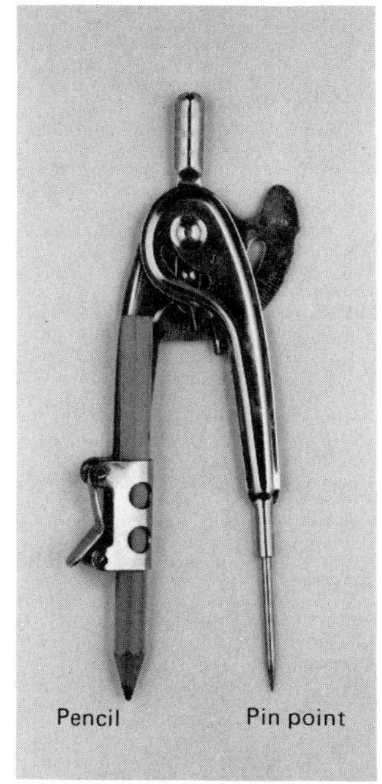

(a) Simple compass.

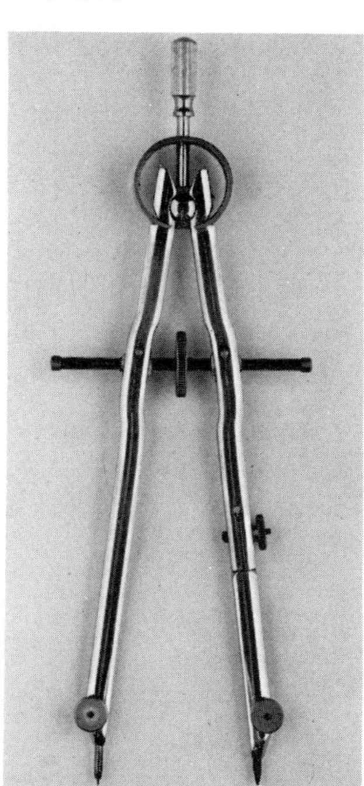
(b) Engineer's compass.

Figure 6-40. Two types of compasses.

6-5 ANGULAR MEASURE

In many instances you may need to cut some particular angle on an object. For example, an angle must be cut on a valve so that it will fit into the head of an engine. You must cut a certain angle when cutting threads on a rod or when cutting rafters for a housetop.

A *circle* is a closed curve such that all points on the curve are the same distance from a fixed point called the *center*. Two things of major importance when working with a circle are the diameter and the radius. The *diameter* of a circle is a straight-line segment that passes through the center and has its end points on the circle. The *radius* of a circle is a straight-line segment that has one end point at the center and the other on the circle. There are three other terms that are useful in discussing circles. The distance around the circle is called the *circumference length*; an *arc* is a segment of the circle; and a *chord* is any straight-line segment that has its end points on the circle. The sketches in Fig. 6-41 give a pictorial view of the discussion above.

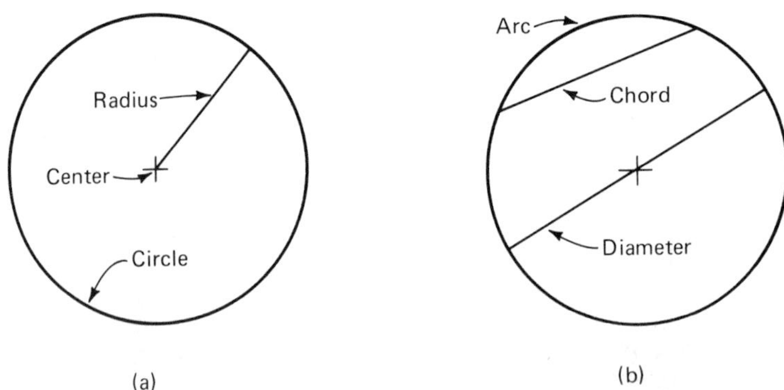

Figure 6-41. (a) On a circle all points are same distance from center. (b) Diameter goes through the center of the circle.

An *angle* is an open figure formed by the intersection of two straight lines that open as they extend from a common starting point called the *vertex*. An angle may be denoted by ∠ABC, ∠B, ∠2, or similar designation as in Fig. 6-42.

Figure 6-42

An angle is measured in terms of degrees, minutes, and seconds. When a complete circle is divided into 360 equal parts, each of these represents one *degree*, denoted 1°. Each degree is divided into 60 equal parts, each of which is called one *minute* (1′) of angular measure. Finally, each minute is divided into 60 equal parts, each of which is called a *second* (1″) of angular measure. In summary, we see that

1 revolution = 360°	1° = 60′	1′ = 60″

A circle may be divided into two equal parts, each of which contains a 180° angle; this is a *straight angle*. When the circle is divided into four equal parts, each part contains a 90° angle; this is a *right angle*.

As mentioned earlier, the compass may be used to construct circles. In order to measure angles, an instrument called a *protractor* is used (see Fig. 6-43). Most protractors measure angles from 0 to 180°, accurate to the nearest degree or half degree (30′). Some protractors have a vernier scale and can be used to measure an angle accurate to the nearest 5′.

On the *universal bevel protractor* (Fig. 6-43), the dial reads from 0 to 90° in each direction and back to 0°. The vernier scale reads from 0 to 60′ in each direction by 5′ intervals. An important rule to remember is always to read the vernier in the same direction as the dial and add the minutes to the degrees.

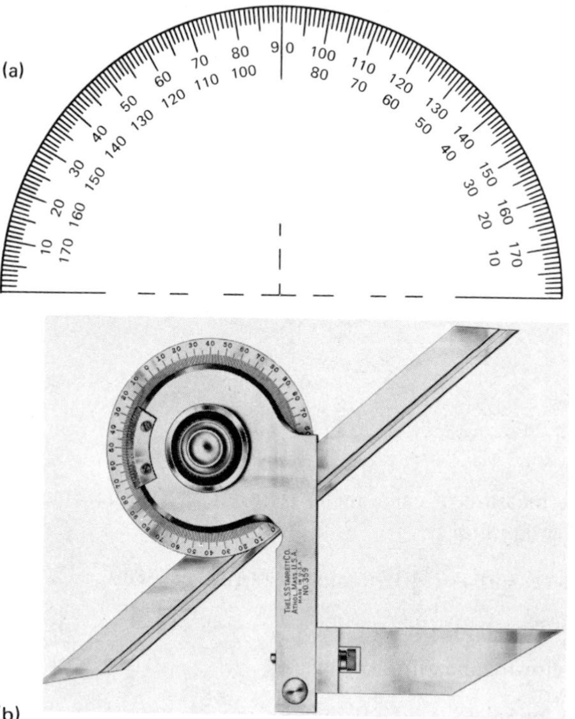

Figure 6-43. (a) Semicircular protractor and (b) universal bevel protractor. Courtesy of the L. S. Starrett Company.

Sample Readings

When set to a whole degree, the zero and two 60′ graduations line up with dial divisions. In Fig. 6-44, since the zero on the vernier scale lines up with 17 on the dial, the reading is 17°.

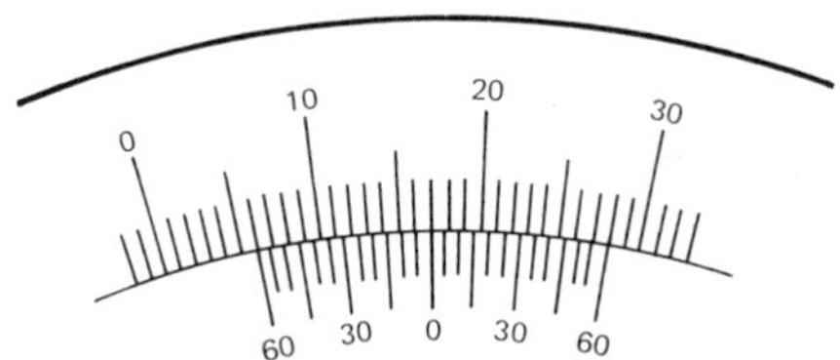

Figure 6-44

Care must be used to read the minutes from the correct pair of lined-up graduations. In Fig. 6-45, the reading is 12° plus 50′ = 12°50′.

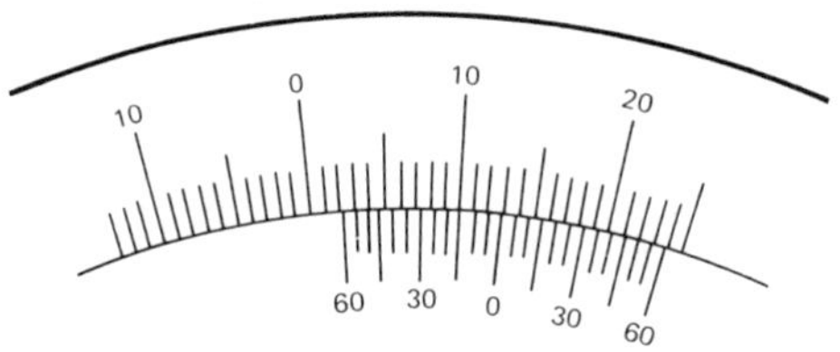

Figure 6-45

In Fig. 6-46, the reading at the star is 50°20′.

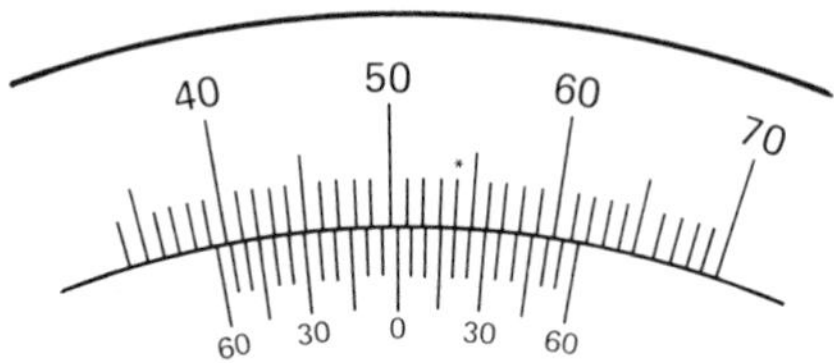

Figure 6-46

Exercise Set 6-5

Part A

1. Change the measure of each angle from minutes to the largest whole number of degrees and minutes.

   Example 6-1: 160′ = 2°40′ because 2° = 120′ and 160′ - 120′ = 40′.

   (a) 65′ = (b) 130′ = (c) 385′ = (d) 1,082′ =

2. Add the following measures of angles.

   Example 6-2: 58°25′ + 37°40′28″ = 95°65′28″ = 96°5′28″ because 65′ = 1°5′

   (a) 39°32′ + 14°2′ = (b) 121°16′31″ + 2°41′52″ =
   (c) 84°30″ + 16°3′30″ = (d) 45°18′ + 28′59″ =

3. Find the indicated differences below.

Example 6-3:

$$\begin{array}{r} 79^\circ 24' 15'' \\ -45^\circ 30' \ 5'' \\ \hline \end{array} \qquad \begin{array}{r} 78^\circ 84' 15'' \\ -45^\circ 30' \ 5'' \\ \hline 33^\circ 54' 10'' \end{array}$$

(a) $\begin{array}{r} 181^\circ 51' \\ -90^\circ 59' \\ \hline \end{array}$ (b) $\begin{array}{r} 75^\circ \ 2' 14'' \\ -52^\circ 10' 10'' \\ \hline \end{array}$ (c) $\begin{array}{r} 59' 48'' \\ -26' 59'' \\ \hline \end{array}$ (d) $\begin{array}{r} 98^\circ 43' 16'' \\ -41^\circ 50' 22'' \\ \hline \end{array}$

4. What are the readings A through E as indicated in Fig. 6-47?

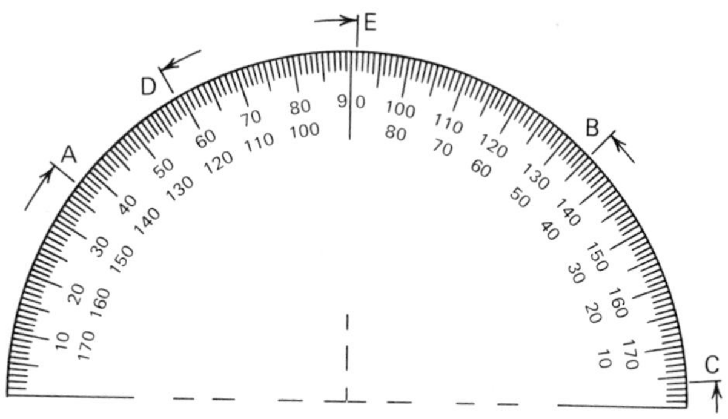

Figure 6-47

5. Measure each angle in Fig. 6-48.

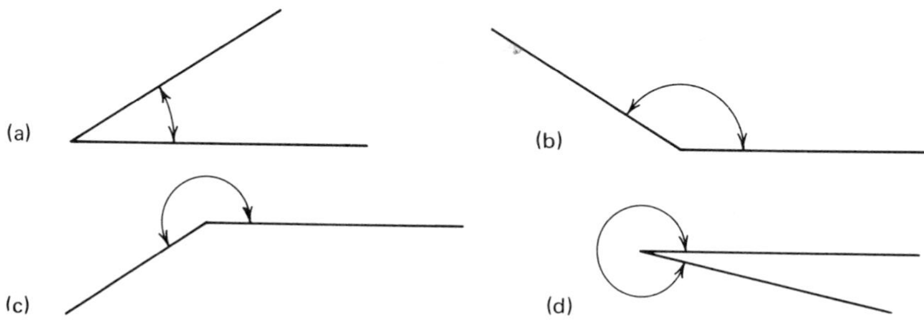

Figure 6-48

Part B

6. Draw an angle to represent each of the following measures.
(a) 59° (b) 102° (c) 41° (d) 168°

7. Using dividers (or compass if dividers are not available), transfer the length of each line in Fig. 6-49 to a ruler and indicate the proper length.

(a)

(b)

Figure 6-49

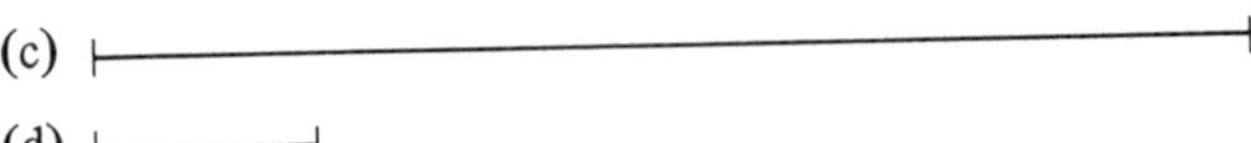

Figure 6-49. continued.

8. With an inside caliper (or divider), determine the inside diameter (ID) of each circle in Fig. 6-50 by transferring the measurement to a ruler.

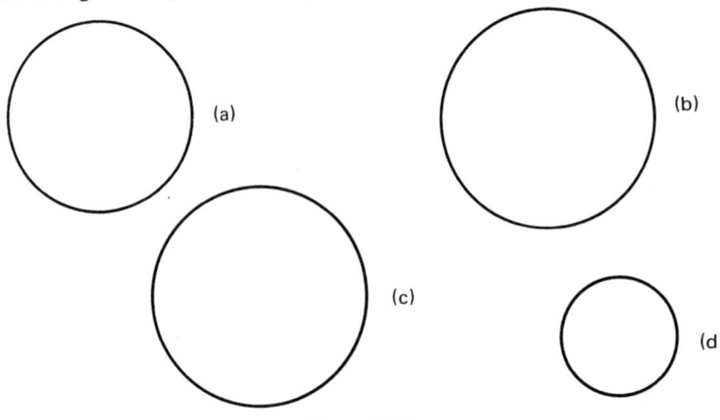

Figure 6-50

9. Use an outside caliper (or divider) to determine the outside diameter (OD) for each circle in Fig. 6-51.

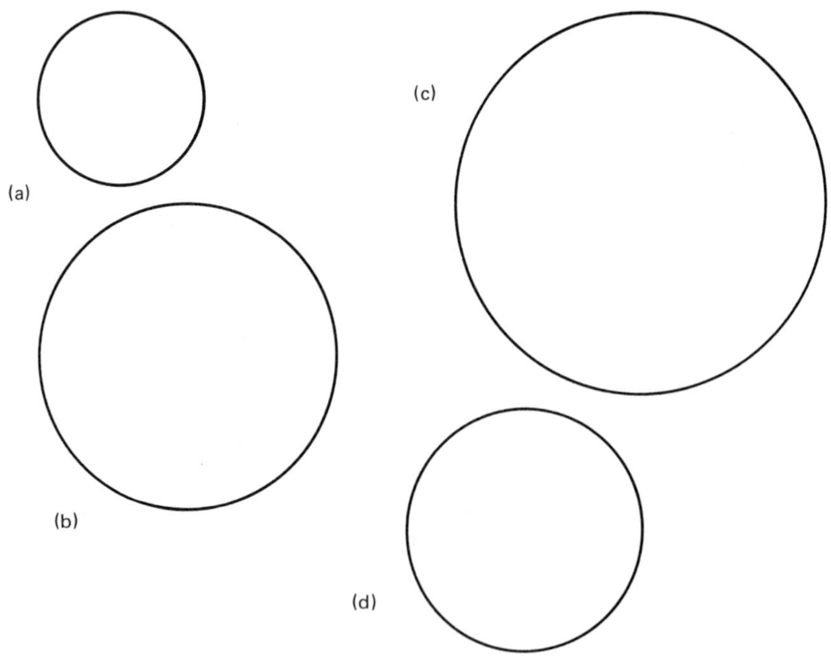

Figure 6-51

Part C

10. Read each part of Fig. 6-52 to the nearest 5′ and indicate the readings.

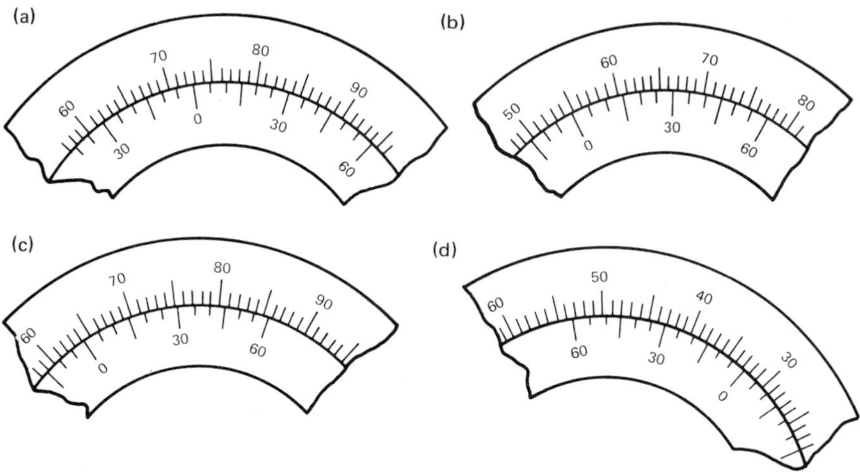

Figure 6-52

11. Make a sketch of a protractor dial and vernier scale for each reading below.
 (a) 81°45′ (b) 15°20′ (c) 55°15′ (d) 47°55′
12. Explain how the protractor with a vernier scale works. Why is it accurate to the nearest 5′ of a degree?

Answers

2. (a) 53°34′ (b) 123°58′23″ (c) 100°4′0″ (d) 45°46′59″
4. A = 37°; B = 45°; C = 4°; D = 121°; E = 91°
10. (a) 72°20′ (b) 54°45′ (c) 64°40′ (d) 32°45′

6-6 BASIC CONSTRUCTIONS

In this section you should receive some information that will be of help in making basic constructions and scale drawings. All the constructions here should be done with the aid of only a compass and a straightedge unless instructed otherwise. We shall assume that all points, lines, and angles will be coplanar (on the same flat surface). Before beginning our constructions, we need some basic notations and terminology.

A point will be named by a capital letter, such as P for point P. Line segment AB will be denoted by $\overline{AB}$ and the length of $\overline{AB}$ by AB. The notation for angle ABC (with vertex at B) will be ∠ABC, and the measure of that angle will be denoted by m∠ABC. A line will be denoted by a small letter, such as line k.

Two words, *perpendicular* and *parallel*, need defining. *Two lines are perpendicular if the angles formed at their point of intersection are 90°* (right angles). *Two straight lines are parallel if they are everywhere the same distance apart*. Another definition for parallel lines is *two straight lines are parallel if each of the two are perpendicular to a third line*. Note that the distance between two parallel lines is measured along the common perpendicular.

Construction 1: A line segment equal in length to a given line segment.

Given line segment AB in Fig. 6-53, construct line segment DE such that AB = DE; that is, the two segments are equal in length.

Figure 6-53

Using a compass, set the pinpoint on point A and adjust the pencil to point B. With this setting, transfer the pinpoint to point D and mark an arc for point E. (This could be done with dividers.) By using transferred measurement, we have the two segments the same length without knowing the exact length of either.

Construction 2: Adding or subtracting the length of line segments.

Given Fig. 6-54.

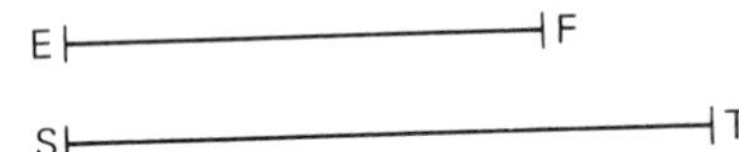

Figure 6-54

1. Construct (see Fig. 6-55) $\overline{AB}$ such that AB = EF + ST.

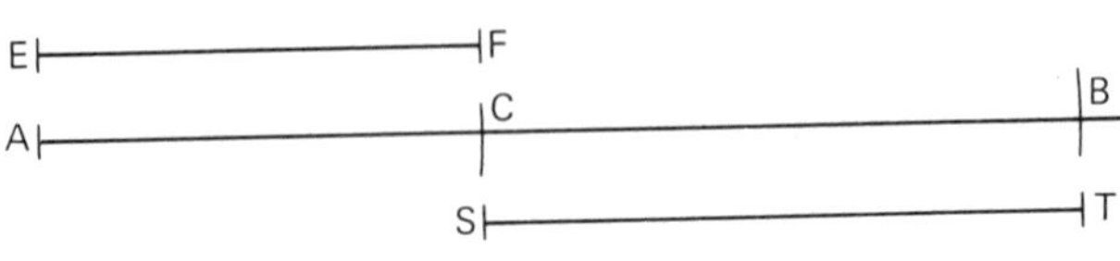

Figure 6-55

Start at point A and lay off a length equal to EF marking the end point C. From point C, lay off a length equal to ST and mark the end B. Segment AB is now equal in length to EF + ST.

2. Construct (see Fig. 6-56) MN such that MN = ST - EF.

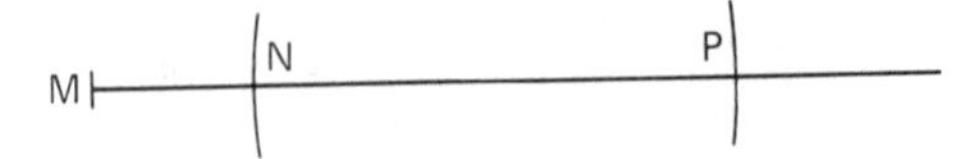

Figure 6-56

Starting at point M, lay off MP = ST. With the compass set for EF, set the pinpoint at point P and lay off PN = EF back toward M to point N. Hence, MN = ST - EF.

Construction 3: Angle equal to a given angle.

Given ∠ABC (see Fig. 6-57).

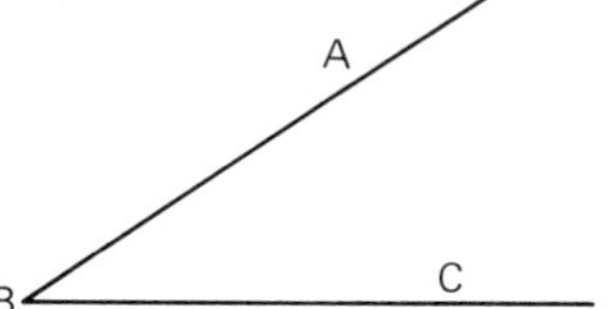

Figure 6-57

Construct (see Fig. 6-58) ∠DEF such that m∠DEF = m∠ABC.

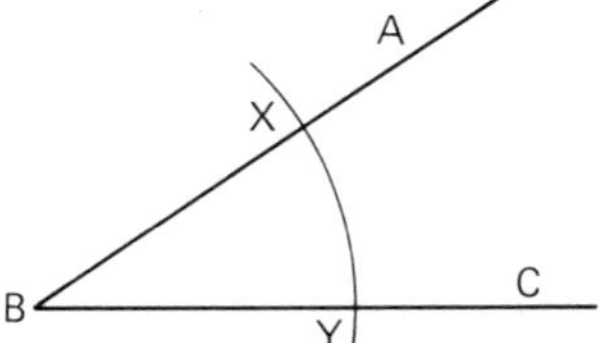

Figure 6-58

On ∠ABC, set the pinpoint on point B. Make an arc crossing side BA and side BC, marking the intersection points X and Y. With the same setting, place the pinpoint on point E and make an arc as indicated, labeling point F at the intersection with the line EF. [See Fig. 6-59(a).] Set the compass point on point X on ∠ABC and adjust the compass to point Y. With this setting, place the point at point F and cross the large arc with a small arc for point D. Now draw a line from E through D, forming ∠DEF as desired. [See Fig. 6-59(b).]

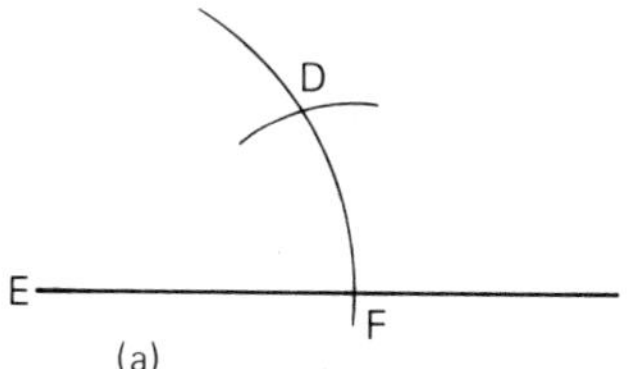

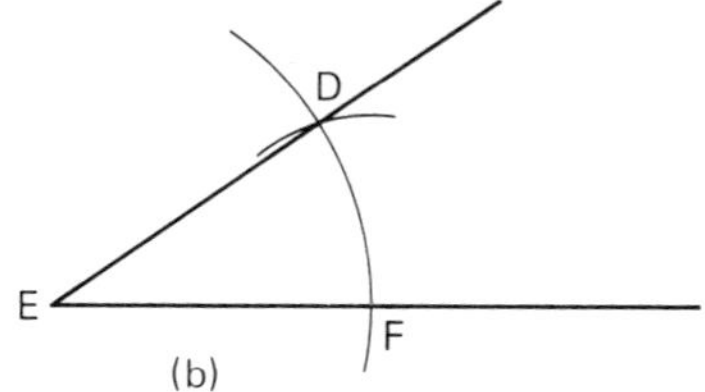

Figure 6-59

Construction 4: Adding angles.

Given ∠ABC and ∠DEF (see Fig. 6-60).

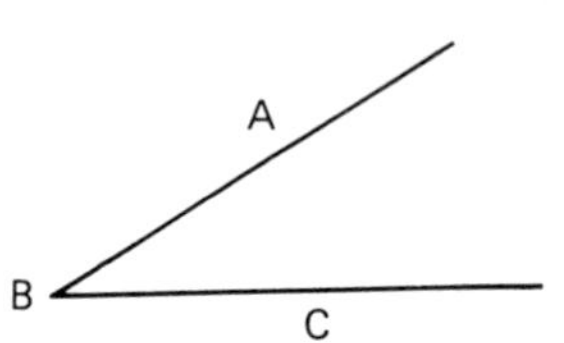

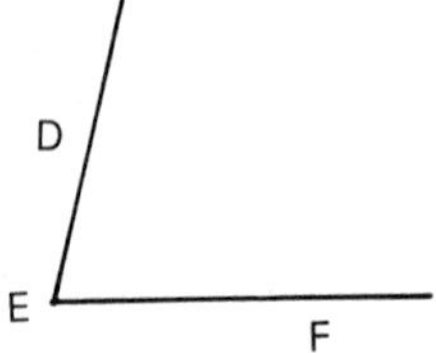

Figure 6-60

Construct ∠DBC such that m∠DBC = m∠ABC + m∠DEF.
Use Construction 3 to construct ∠ABC (see Fig. 6-61).

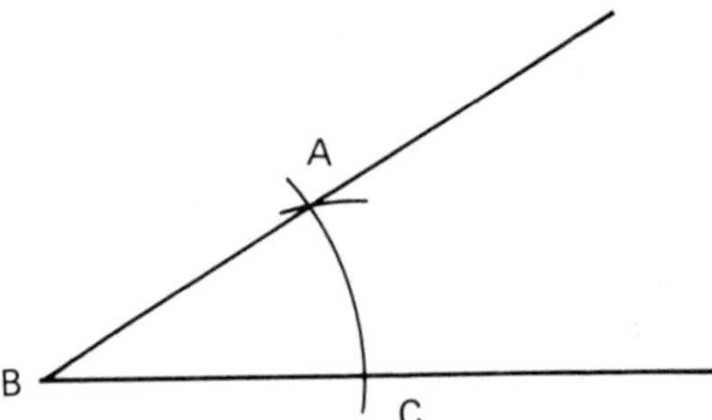

Figure 6-61

Using line segment $\overline{BA}$ as the initial side, construct ∠DEF (see Fig. 6-62).

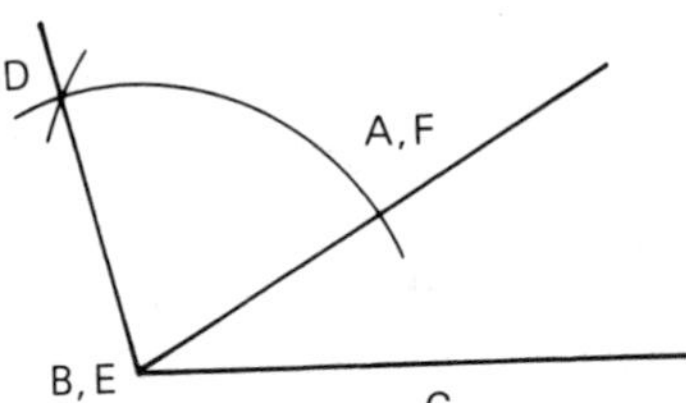

Figure 6-62

So, ∠DBC is the same as ∠DEC and m∠DBC = m∠ABC + m∠DEF.

Construction 5: Perpendicular lines (P not on *l*).

Given line *l* and point P not on *l* (see Fig. 6-63).

l

Figure 6-63

Construct line m perpendicular to *l* and through P.

Set the compass point at P and swing an arc crossing *l* at two points, called R and S. See Fig. 6-64. With the same setting, place the point at R and mark an arc below l. With the same setting, place the point at S and make another arc below l. The point of intersection of the two arcs is labeled Y.

Draw line m through points P and Y. Line m is perpendicular to l, as required. See Fig. 6-65. (The phrase *m is perpendicular to l* can be denoted by $m \perp l$.)

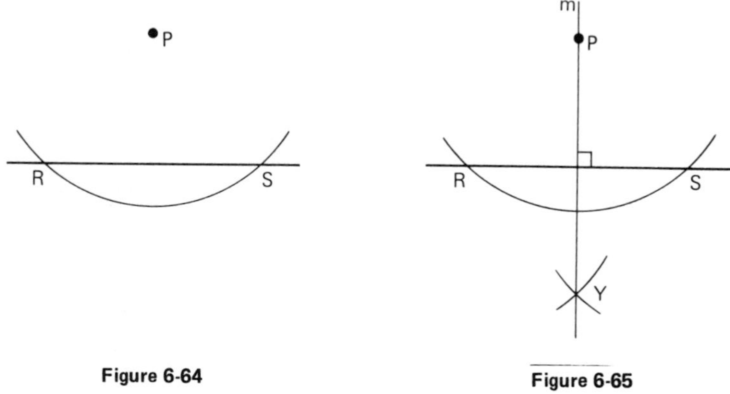

Figure 6-64

Figure 6-65

Construction 6: Perpendicular lines (P on l).

Given line l and point P on l (see Fig. 6-66).

Figure 6-66

Construct line m perpendicular to l through P.

Set the point of the compass on point P and swing an arc cutting l at two points, one on each side of P. Call these A and B.

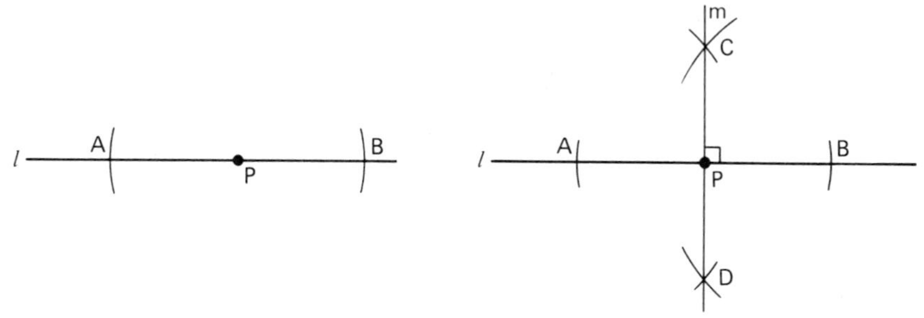

Figure 6-67

Increase the setting of the compass to more than AP. Set the point on A and swing an arc above P and below P. With the same setting, repeat this process with the point set at B. Label the points of intersection C and D. Draw line m through points C, P, and D. Line m is perpendicular to line l (see Fig. 6-67).

Construction 7: Parallel lines using perpendicular lines.

Given line l and point P not on l (see Fig. 6-68).

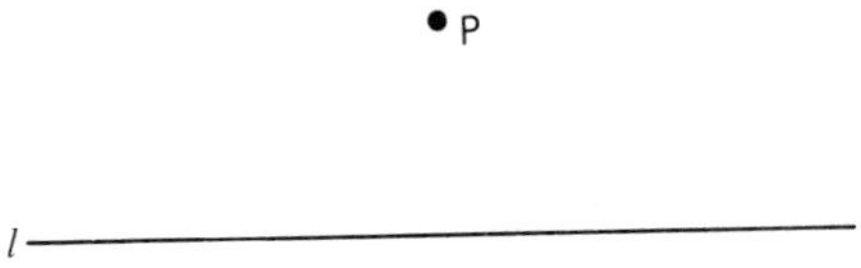

Figure 6-68

Construct line k parallel to l through P.
First, using Construction 5, construct $m \perp l$ through P (see Fig. 6-69).

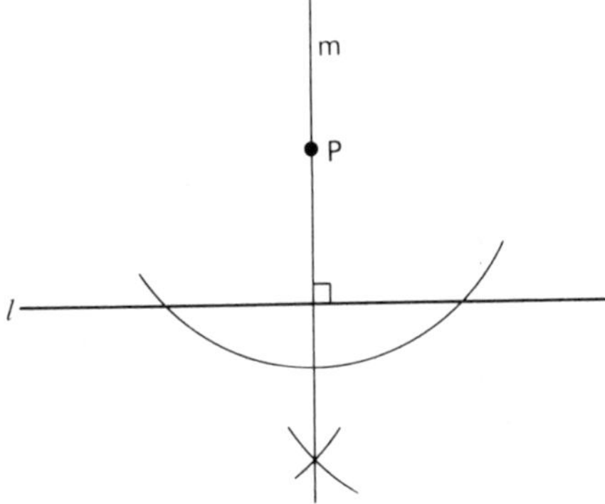

Figure 6-69

Second, use Construction 6 to construct $k \perp m$ through P (see Fig. 6-70).

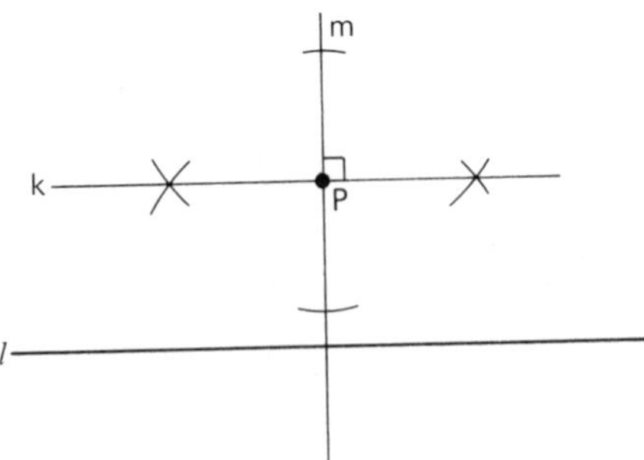

Figure 6-70

From our definition and discussion of perpendicular and parallel lines and Construction 7, it follows that line k is parallel to line l, denoted $k \parallel l$.

Construction 8: Bisect an angle.

Given $\angle ABC$ (see Fig. 6-71).

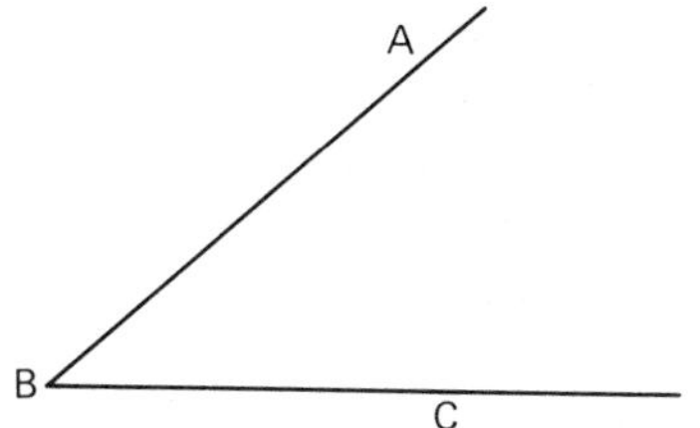

Figure 6-71

Construct $\angle ABD$ and $\angle DBC$ such that $m\angle ABD = m\angle DBC = (m\angle ABC)/2$. See Fig. 6-72. Set the compass point at point B and swing an arc that crosses $\overline{BA}$ and $\overline{BC}$. Label the points of intersection X and Y. With a radius equal to XY, set the point at X and swing an arc in $\angle ABC$ as indicated. With the same setting, repeat the process with the point at Y. Label the point of intersection of the two arcs D. Now, draw a line from B through D. This makes $m\angle ABD = m\angle DBC = (m\angle ABC)/2$. Therefore, line BD bisects $\angle ABC$.

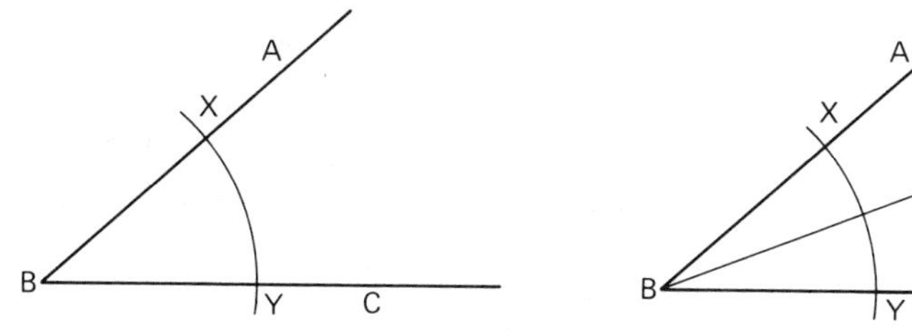

Figure 6-72

Exercise Set 6-6

All construction problems should be done using only a pencil, compass, and straightedge unless you are instructed otherwise.

Part A

Use Fig. 6-73 with Exercises 1, 2, and 3.

1. Construct a line segment equal in length to AB.
2. Construct a line segment equal in length to AB + CD.
3. Construct a line segment equal in length to EF + CD − AB.

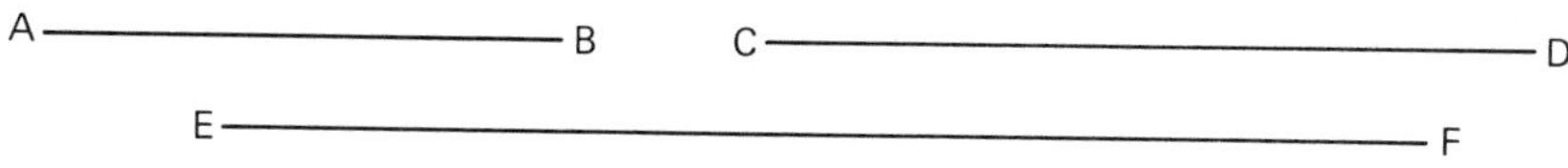

Figure 6-73

4. Construct angles equal to each of the angles given in Fig. 6-74.

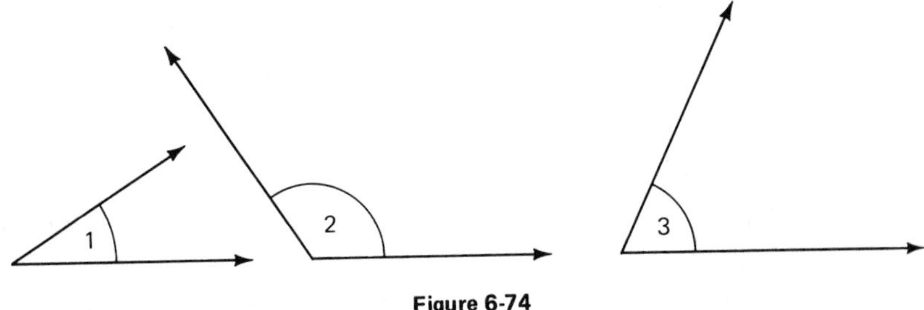

Figure 6-74

5. Bisect ∠ABC as given in Fig. 6-75.

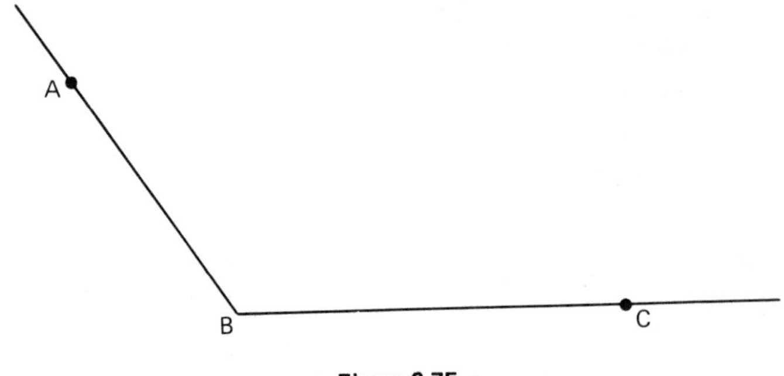

Figure 6-75

6. Construct a line through point P and parallel to line AB. See Fig. 6-76.

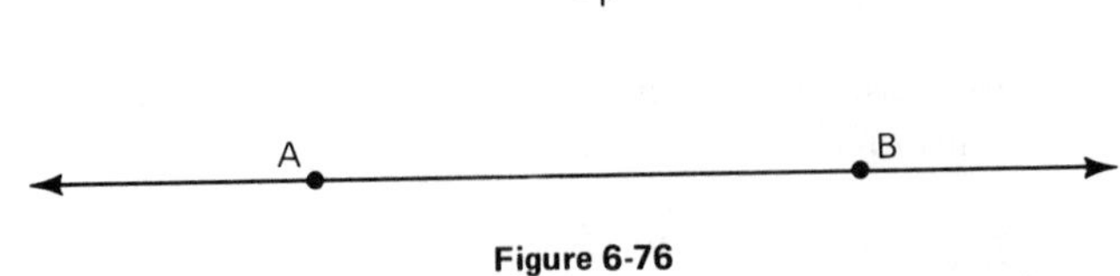

Figure 6-76

Part B

7. Construct a line parallel to l at a distance AB from it. See Fig. 6-77.

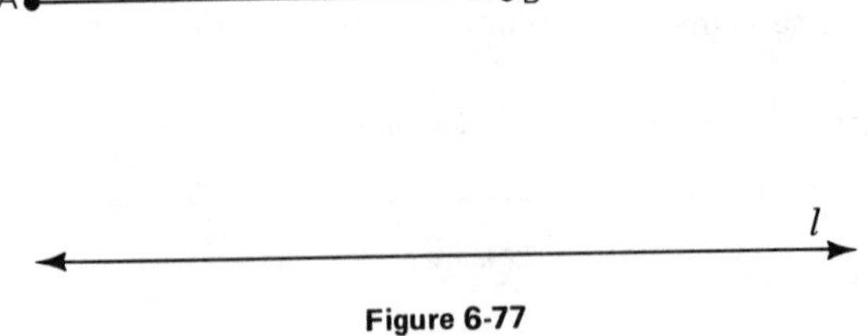

Figure 6-77

8. Construct an angle equal to m∠1 + m∠3 of Exercise 4.

9. Construct an angle equal to m∠2 - m∠1 of Exercise 4.

10. Construct a 45° angle. Check it with a protractor. What is the actual measure of the angle constructed?

11. Using the line segments in Fig. 6-73, construct a rectangle with one side equal to one-half of EF and the other side equal to AB.

Part C

12. Transfer the floor plan in Fig. 6-78 to your paper, by construction. (All angles are 90°.)

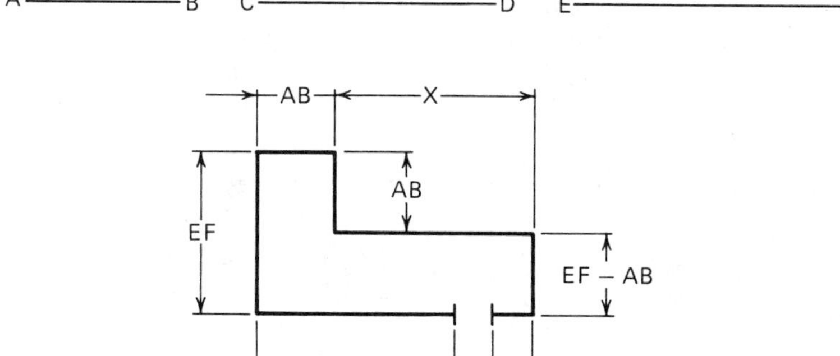

Figure 6-78

13. Using $\overline{PQ}$ in Fig. 6-79, construct an equilateral triangle with each side equal to (1/2) · PQ.

Figure 6-79

14. Using line segment PQ from Exercise 13, construct a square with each side equal to one and one-half times PQ.

15. Construct an angle whose measure is
(a) 30° (b) 45° (c) 75° (d) 135° (e) 150°

6-7 SCALE DRAWINGS

You are probably familiar with the use of a *scale* to indicate mileage on maps. This scale usually is a line divided into several segments and includes a notation comparing a length on the map to a distance on the highway. For instance, you might see "1 inch = 20 miles," which tells you that the distance from one town to another is 20 mi multiplied by the number of inches separating them on the map.

The idea of a scale is not limited to maps. Scales are used in blueprints, house plans, photography, wind-tunnel tests, and works of art. The use of scales that we are most interested in is *scale drawings*, of which blueprints and house plans are good examples. These scale drawings are representations of physical objects in a size larger or smaller than actual size. On a house plan we might see the scale indicated "1 inch = 4 feet," which says that every inch in the scale drawing represents 4 ft in the actual house. A blueprint of a valve lifter might indicate "1 inch = 1/2 inch," which says that every inch in the scale drawing represents only 1/2 in. in the actual valve lifter.

There are various scales, such as an engineering scale and an architect's scale. An ordinary ruler is commonly used as a scale. To do this, simply let each unit (or fraction of a unit) on the ruler represent a convenient number of full-size units.

Draw the template sketched in Fig. 6-80. Use a scale of 1/16 in. = 1/8 in. This scale means that 1/16 in. on your scale drawing for the object pictured in Fig. 6-80 represents 1/8 in. in terms of the actual size of the object whose figure is drawn. In order to draw the figure to scale, we must determine how long each segment should be when drawn to the scale being used.

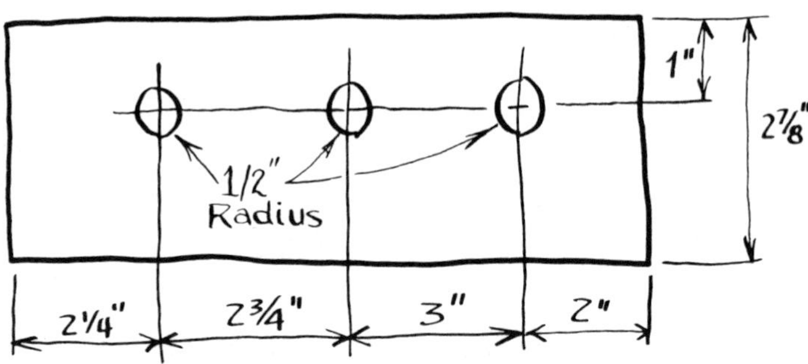

Figure 6-80. Rough drawing.

TABLE 6-1 Converting Actual Dimensions to Scale Dimensions

From Given Dimensions		For Scale Drawing	
Actual Size (in.)	Number of 1/8 in.	Number of 1/16 in.	Scaled Drawings (in.)
2 1/4	18	18	1 1/8
2 3/4	22	22	1 3/8
3	24	24	1 1/2
2	16	16	1
1/2	4	4	1/4
1	8	8	1/2
2 7/8	23	23	1 7/16

Table 6-1 shows the actual (given) dimensions converted to scale dimensions. From the table we see that a length 2 1/4 in. should be represented by a length of 1 1/8 in. on the scale drawing.

Figure 6-81 shows the completed scale drawing.

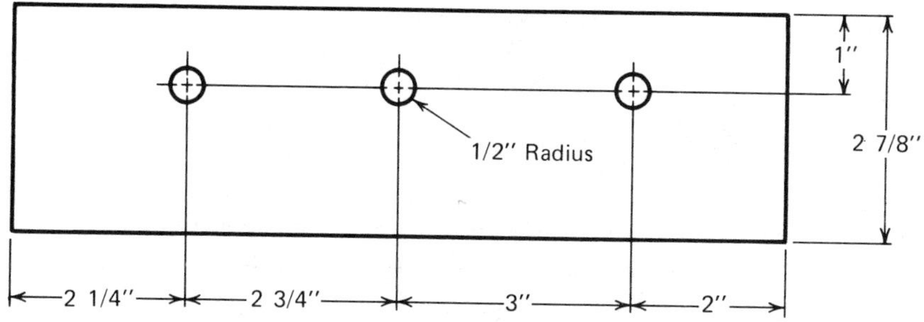

Figure 6-81. Completed scale drawing.

Exercise Set 6-7

Part A

Draw each of the figures below using the indicated scale unless instructed otherwise. (These figures are not scale drawings.)

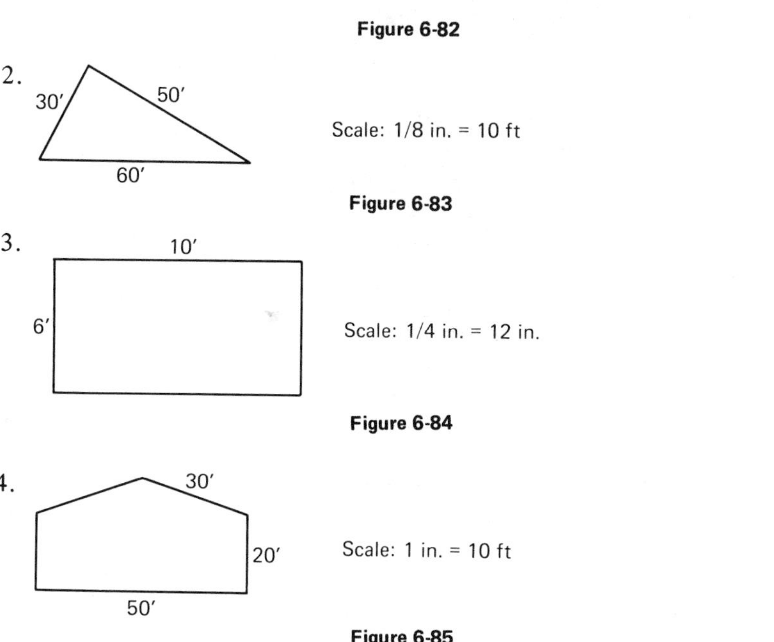

Figure 6-82

Figure 6-83

Figure 6-84

Figure 6-85

5.

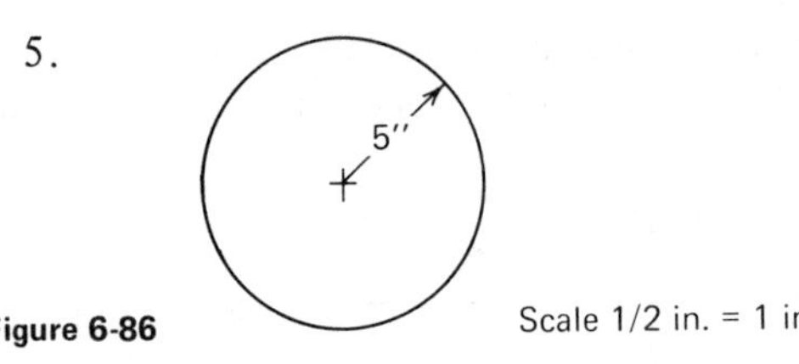

Figure 6-86

Part B

6. Use a scale of 1/2 in. = 1 in. to draw the template in Fig. 6-87.

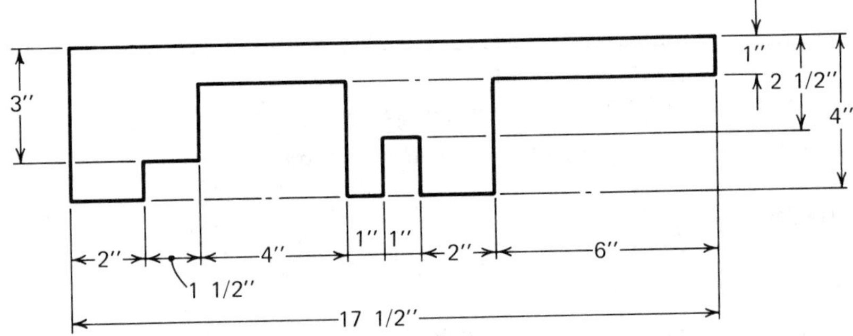

Figure 6-87

7. Fig. 6-88 is a sketch of a layout for a page that you may be asked to print. Draw this to scale using 1/2 in. = 1 in.

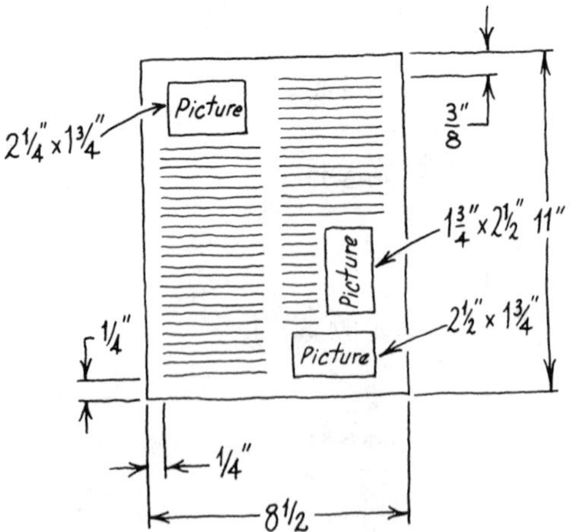

Figure 6-88

8. Make a scale drawing of the crankshaft sketched in Fig. 6-89. Use a scale of 1/2 in. = 1 in.

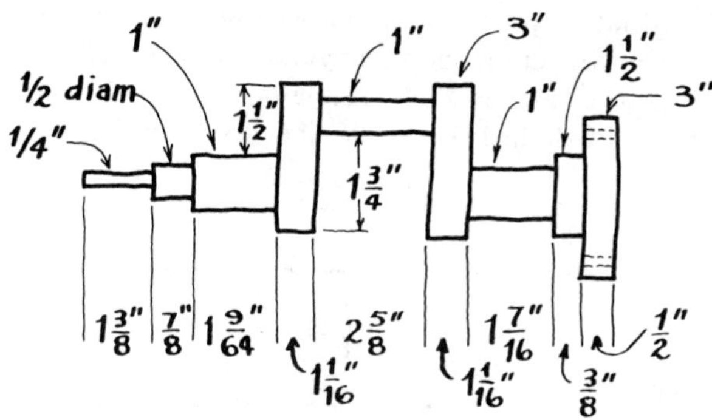

Figure 6-89

9. Sketched in Fig. 6-90 is a roof truss. Using a scale of 1/8 in. = 1 ft, make a scale drawing of the roof truss. The pitch is 1/4.

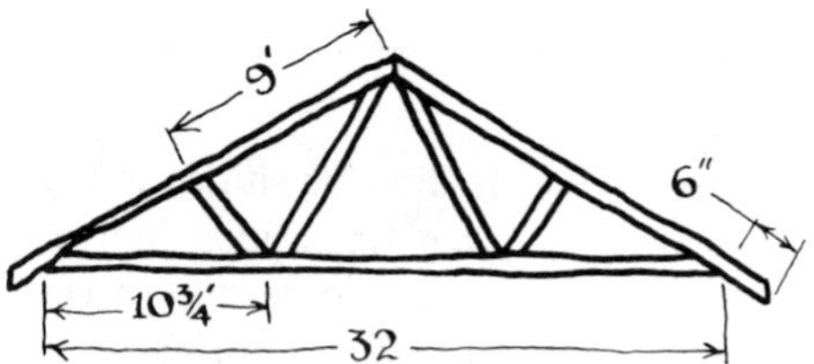

Figure 6-90

Part C

10. While plans were being made for a new graphic arts building, you were asked to make a layout for the darkroom area. Suppose your rough sketch resembled that shown in Fig. 6-91. Now, make a scale drawing with a scale of 1/8 in. = 1 ft. Neglect wall thickness.

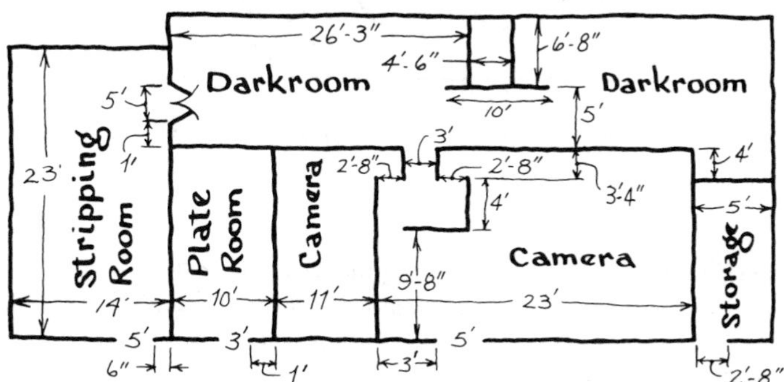

Figure 6-91

11. Suppose a friend of yours bought an older house. Since there was no blueprint for the house, your friend asked you to make him a scale drawing. Using the information in Fig. 6-92 and a scale of 1/8 in. = 1 ft, make the desired drawing. Neglect wall thickness. Make all doors 3 ft wide.

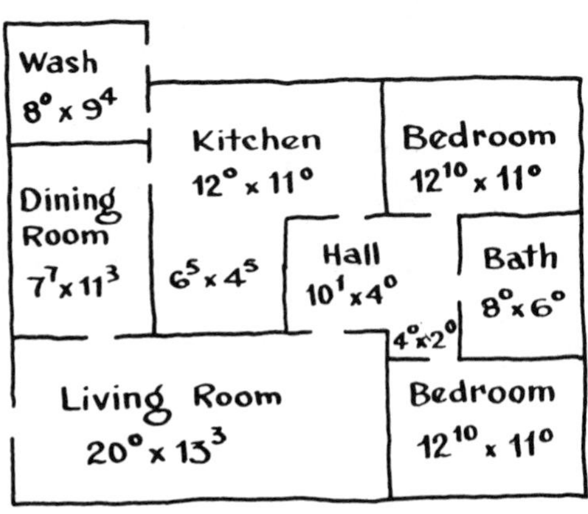

Figure 6-92

12. A customer gave you the drawing shown in Fig. 6-93 and asked you to make 25 templates using the given dimensions. Before setting up your machine, you decided to do a scale drawing to be sure of what the customer wants. Use a scale of 1 in. = 1 in. to make your drawing.

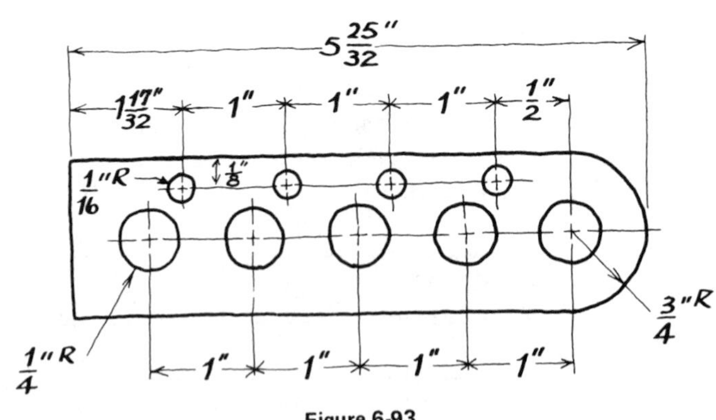

Figure 6-93

13. As part of your automotive shop management course you were asked to design and draw a layout of an automotive shop. In Fig. 6-94 is a rough sketch of the layout. Use a scale of 1/8 in. = 1 ft to make a scale drawing of the layout. Neglect wall thickness.

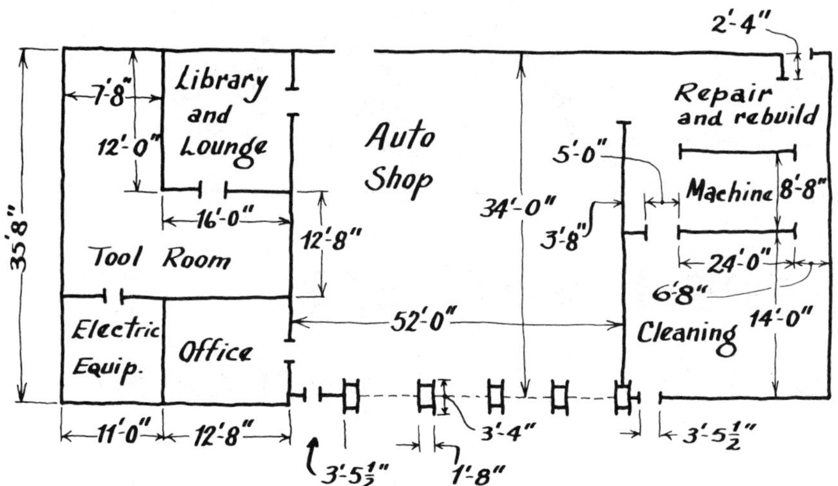
2'-4"
Library and Lounge
Auto Shop
Repair and rebuild
7'-8"
12'-0"
5'-0"
Machine
8'-8"
16'-0"
34'-0"
3'-8"
12'-8"
35'8"
Tool Room
24'-0"
6'8"
52'-0"
14'-0"
Electric Equip.
Office
Cleaning
3'-4"
11'-0"
12'-8"
3'-5½"
1'-8"
3'-5½"

Figure 6-94

Chapter 7

Signed Numbers and Simple Equations

In this chapter you will see how mathematical expressions can be simplified and then become tools that are applicable to many problems that require the solution of equations. Frequent applications of mathematical expressions can be found in layout, design, inventory, scheduling, and many other vocational-technical areas.

7-1 EXPRESSIONS AND TERMS

The basic idea behind mathematics is the need to express physical quantities using numerals or letters. These are the simplest forms of *mathematical expressions*. More complicated mathematical expressions are formed by combining letters and numerals and other symbols to express the operations of arithmetic.

The number of pennies equal in value to a nickel is usually denoted by the mathematical expression **5**.

The physical idea of *five women and three men* can be denoted by the mathematical expression **5w + 3m.**

The average grade for a class of six students whose individual grades are denoted by the mathematical expressions u, v, w, x, y, and z can be denoted by the generalized mathematical expression

$$\frac{u + v + w + x + y + z}{6}$$

> *Definition:* A *term* of a mathematical expression is any part of the expression that is being added or subtracted.

Example 7-1: $5ab/c$ is an expression with only one term. $3x + 4$ has two terms. $5 \cdot 2 + 3 \cdot 4 + 2$ has three terms. $y + 5a/b + 6b + 3$ has four terms.

7-2 COLLECTING TERMS

In this section we shall discuss collecting terms of an expression by using the distributive law.

Example 7-2: To show $2(3 + 5) = 2(3) + 2(5)$, examine the left side and then the right side of the equation. We said in Sec. 1-7 that operations inside enclosures are to be done first.

$$2(3 + 5) = 2(8) = 16$$

In Sec. 1-7, we also said that after the operations in enclosures are completed multiplication is to be performed before addition.

$$2(3) + 2(5) = 6 + 10 = 16$$

Hence we see that the two sides are equal. This one example does not prove the distributive law but only suggests its truth by illustration.

Distributive Law: $a(b + c) = a \cdot b + a \cdot c$

Example 7-3: To add 5b and 7b,

$$5b + 7b = (5 + 7)b = 12b$$

Exercise Set 7-2

Collect and simplify.

Part A

1. $3 \cdot 5 + 3 \cdot 7$
2. $2a + 3a$
3. $5n + 7n$
4. $d + 5d$
5. $7d - 5d$

Part B

6. $5a + 3a - 2a$
7. $1/2d + 1/3d$

8. $3/2d + 5/3d$
9. $5.8m - 2.7m$
10. $2\ 1/3y - 1\ 2/3y$

Part C

11. $8 + 3(4 + x)$
12. $2(a + b) + 3a$
13. $7(a + 2) + 9(2a + 3)$
14. $P = 3a + 5a + 4a$ (See Fig. 7-1.)

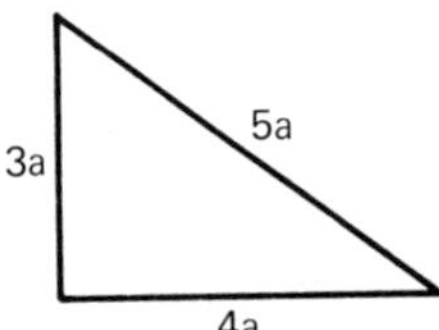

Figure 7-1

15. $P = 2(2a) + 2(2a + 6)$. (See Fig. 7-2.)

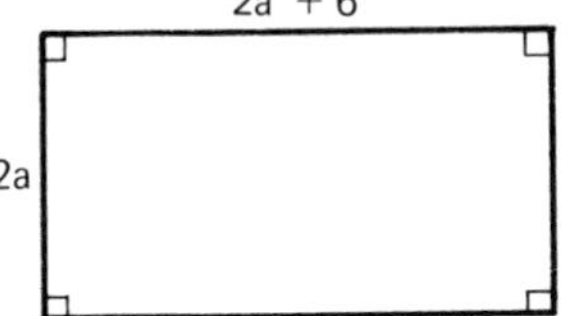

Figure 7-2

Answers

2. 5a
4. 6d
6. 6a
8. 19/6d
10. 2/3y
12. 5a + 2b
14. 12a

7-3 SIGNED NUMBERS

When adding, subtracting, multiplying, or dividing numbers, many problems crop up as a result of the misuse of signed numbers. It is very important that you have full command of the four basic operations listed above. There are several properties that we can use as guides when we work with signed numbers.

> Property 1. To add two numbers with like signs, find their sum and place before the result the sign that they share.

Example 7-4

1. $4 + 7 = 11$
2. 4 yd + 7 yd = 11 yd
3. $(-4) + (-7) = -11$
   [In part (3) suppose -4 and -7 denote consecutive withdrawals from your checking account, then -11 would represent the total withdrawal from your account.]
4. $(-4$ yd$) + (-7$ yd$) = -11$ yd
   [Denote a 4-yd loss as -4 yd and a 7-yd loss as -7 yd; then the 11-yd loss is denoted by -11 yd.]

> Property 2. To add two numbers with unlike signs, remove the signs and find the difference between these unsigned numbers. Place before the answer the sign that was removed from the larger unsigned number.

Example 7-5

1. (4 yd) + (-15 yd) = -11 yd
   [Denote a 4-yd gain as 4 yd and denote a 15-yd loss as -15 yd. Now denote the net loss of 11 yd by -11 yd. Notice that the difference between 15 and 4 is 11, and since 15 is the larger number, its sign ($-$) is attached to the difference.]
2. (-4 yd) + (15 yd) = 11 yd
   [Denote the 4-yd loss as -4 yd and denote the 15-yd gain as 15 yd. Now denote the net gain of 11 yd as 11 yd.]

> Property 3. To multiply two positive numbers, find the product of the two numbers. This product is a positive number.

Example 7-6

1. $3(4) = 12$
   [We know from Sec. 1-4 that multiplication is a short form of addition. Since 3 tells how many times 4 is used under addition, $3(4) = 4 + 4 + 4 = 12$.]

2. $5(2a) = 10a$
[5 tells us how many 2a's are to be added: $5(2a) = 2a + 2a + 2a + 2a + 2a = 10a$. Note again that multiplication is a short form of addition.]

> Property 4. To multiply a positive number and a negative number, find the product and place a negative sign in front of your result.

Example 7-7

$$3(-5) = -15$$

[3 tells us how many −5's are to be added: $3(-5) = (-5) + (-5) + (-5) = -15$.]

> Property 5. To multiply a negative number and a positive number, find the product and place a negative sign in front of your result.

Example 7-8

$$-5 \cdot 3 = -15$$

[Think of the sum of 5 reversals of 3 being added. The reversal of 3 is −3. The 5 reversals of 3 would be $(-3) + (-3) + (-3) + (-3) + (-3) = -15$.]

> Property 6. To multiply a negative number by a negative number, find the product and place a positive sign in front of your result.

Example 7-9

$$(-5) \cdot (-3) = +15$$

[The reversal of −3 is +3. The −5 tells us that we have 5 reversals of −3 to be added. Hence, $3 + 3 + 3 + 3 + 3 = +15$.]

> Property 7. To divide two numbers whose signs are alike, find the quotient and place a positive sign in front of your result. (*Note:* The quotient obtained by dividing any nonzero number by itself is 1.)

Illustration

$$\frac{5}{5} = 1 \quad \text{because} \quad 5 = 1 \cdot 5$$

$$\frac{-5}{-5} = 1 \quad \text{because} \quad -5 = 1(-5)$$

$$\frac{a}{a} = 1 \quad \text{because} \quad a = 1(a)$$

$$\frac{-a}{-a} = 1 \quad \text{because} \quad -a = 1(-a)$$

Example 7-10

1. $\dfrac{10}{5} = 2$

   $10 = 2 \cdot 5$

   Therefore,

   $$\frac{10}{5} = \frac{2 \cdot 5}{5} = 2\left(\frac{5}{5}\right) = 2(1) = 2$$

2. $\dfrac{-10}{-5} = 2$

   $-10 = 2 \cdot (-5)$

   Therefore,

   $$\frac{-10}{-5} = \frac{2 \cdot (-5)}{-5} = 2\left(\frac{-5}{-5}\right) = 2(1) = 2$$

Property 8. To divide two numbers whose signs are unlike, find the quotient and place a negative sign in front of your result.

Example 7-11

1. $\dfrac{-10}{5} = -2$

   $-10 = -2 \cdot 5$

   by Property 5. Therefore,

   $$\frac{-10}{5} = \frac{(-2) \cdot 5}{5} = -2\left(\frac{5}{5}\right) = -2(1) = -2$$

2. $\dfrac{10}{-5} = -2$

$10 = -2 \cdot -5$

by Property 6. Therefore,

$$\frac{10}{-5} = \frac{-2 \cdot (-5)}{-5} = -2\left(\frac{-5}{-5}\right) = -2(1) = -2$$

Exercise Set 7-3

Perform the computations.

1. $2 \cdot 6a$
2. $3 \cdot (-3)$
3. $(-5) \cdot (-7)$
4. $(-10) \cdot 6a$
5. $(-12) \cdot (5/6)$
6. $(-4) + (-10)$
7. $(-6) + 10$
8. $6 + (-10)$

Answers

2. -9 4. $-60a$ 6. -14 8. -4

7-4 ANOTHER APPROACH TO COLLECTING TERMS

Example 7-12: Simplify $25n - 3n$. If the letter n stands for a nut, then the problem could be stated as follows: 25 nuts - 3 nuts = 22 nuts. Now the physical idea suggested by these nuts leads to $25n - 3n = 22n$, with no further mention of the nuts.

Example 7-13: Simplify $3a + 4a$. If the letter a stands for an apple, then the problem could be written as follows: 3 apples + 4 apples = 7 apples. Using this physical idea, it should seem reasonable that $3a + 4a = 7a$.

Example 7-14: Simplify $16n - 3a - 6n + 4a$. If the letters represent the same physical objects as in Examples 7-12 and 7-13, we can collect $16n - 6n = 10n$, and we can collect $4a - 3a = 1a$. By doing both collections, we obtain

$$\begin{aligned} 16n - 3a - 6n + 4a &= (16n - 6n) + (4a - 3a) \\ &= 10n + a \end{aligned}$$

Notice that a term possessing **a** cannot be combined with a term possessing **n**, just as apples and nuts cannot be added together. Only like terms can be combined under addition or subtraction.

These three examples illustrate a physical approach to the collecting of terms. They achieve the same results as the distributive law, but you might find this approach useful.

Exercise Set 7-4

Part A

Simplify.

1. $5d + 7 + 6d$
2. $7a + 3(r + 2a)$
3. $-10a + 5(2a + 4)$
4. $3(a + 3) + 4a$
5. $3(3a + 7) - 5a$
6. $1/2a + (2/3)(5a + 6)$

Part B

7. $5d + 7 - 10d$
8. $7a + 3(2a - 3)$
9. $-10a + 5(2a - 5)$
10. $-3(3a + 2) + 4a$
11. $-5(2a + 3) + 3(a - 3)$
12. $(1/2)(3a - 5) - 10(2a + 3)$
13. $(1/2)a - (2/3)(5a + 6)$
14. Each leg of an isosceles triangle is five times the base. Find the perimeter, if b represents the base.
15. If the sides of a triangle are x, x + 1, and x + 2, find the perimeter of a second triangle whose sides are four times that of the first.

Part C

16. Express the sum of the sides of the triangle and collect terms in Fig. 7-3.

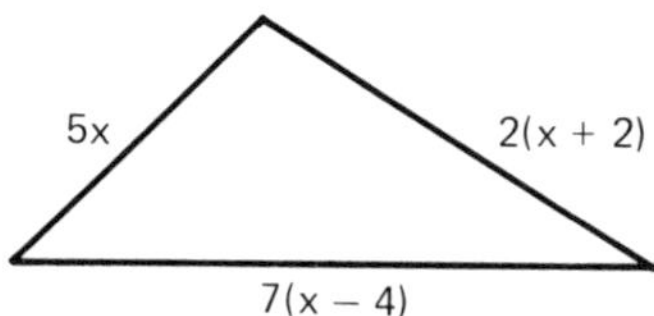

Figure 7-3

17. The sides of a lot are given in Fig. 7-4. Find the perimeter of the polygon. (Be sure to collect terms.)
18. The sides of a rectangle are given in Fig. 7-5. Find the perimeter of the rectangle. (Be sure to collect terms.)

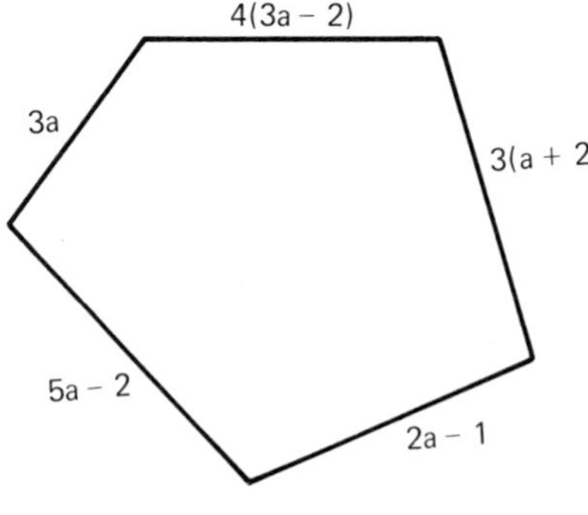

Figure 7-4

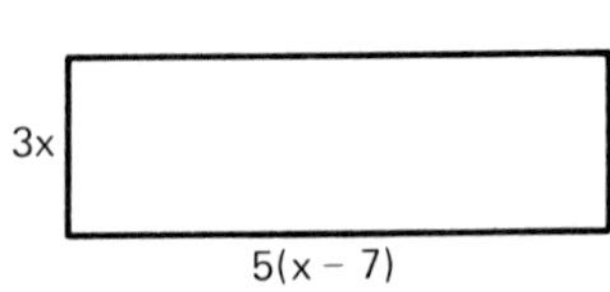

Figure 7-5

Answers

2. $9a + 3r$
4. $7a + 9$
6. $(23/6)a + 4$
8. $13a - 9$
10. $-5a - 6$
12. $-(18\ 1/2)a - 32\ 1/2$
14. $11b$
16. $14x - 24$
18. $16x - 70$

7-5 SIMPLE EQUATIONS

A basic problem in mathematics is finding a solution for an equation.

> *Definition:* An *equation* is a statement of equality between two mathematical expressions.

In previous chapters we found solutions by methods of general mathematics. Now we shall discuss a more efficient method of finding a solution. This method may use any of the following four procedures to simplify an equation without changing its solution.

Add Procedure. A mathematical expression may be added to both sides of an equation.

Subtract Procedure. A mathematical expression may be subtracted from both sides of an equation.

Multiply Procedure. Both sides of an equation may be multiplied by a nonzero mathematical expression.

Divide Procedure. Both sides of an equation may be divided by a nonzero mathematical expression.

Example 7-15: Find the solution of the equation $a + 2 = 20$.

Step 1. Copy the equation: $a + 2 = 20$

Step 2. Subtract 2 from both sides: $a + 2 - 2 = 20 - 2$

Step 3. Simplify: $a + 0 = 18$

Step 4. Simplify: $a = 18$

Pictorial illustration of Example 7-15: $a + 2 = 20$
Let a = weight of Al.

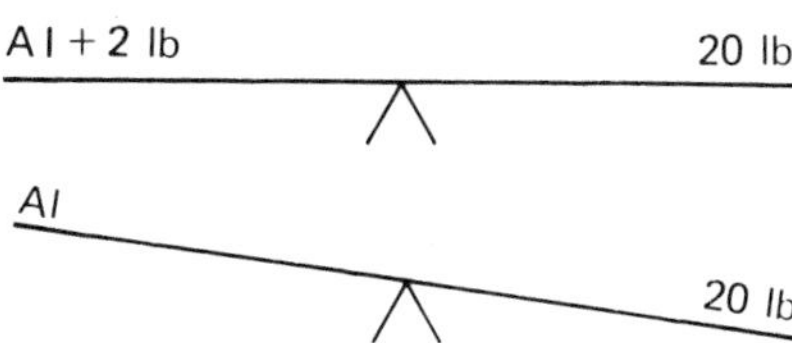

Result of taking 2 lb off one end, but not both ends.

Result of taking 2 lb off both ends. Therefore, $a = 18$.

Example 7-16: Find the solution of $3N + 2N + N = 90$.

Step 1. Copy the equation: $3N + 2N + N = 90$

Step 2. Collect terms: $6N = 90$

Step 3. Divide both sides by 6: $\frac{6N}{6} = \frac{90}{6}$

Step 4. Reduce: $N = 15$

Pictoral illustration of Example 7-16: $3N + 2N + N = 90$

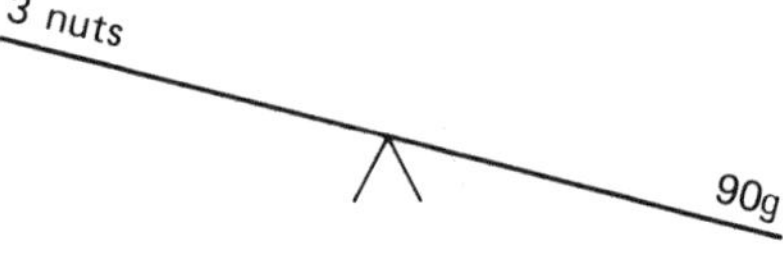

Let

$$3N = 3 \text{ nuts}$$

$$2N = 2 \text{ nuts}$$

$$N = 1 \text{ nut}$$

The result of placing 3 nuts on one end and 90 g on the other end is unbalanced.

The result of placing 5 nuts on one end and 90 g on the other end is unbalanced.

The result of placing 6 nuts on one end and 90 g on the other end is balanced.

The result of dividing both ends by 6 is balanced. Therefore, $1N = 15$ or $N = 15$.

To check, replace N with 15: $3(15) + 2(15) + (15) \stackrel{?}{=} 90$

Multiply: $45 + 30 + 15 \stackrel{?}{=} 90$

Simplify: $90 \stackrel{\checkmark}{=} 90$

Exercise Set 7-5

Solve each of the following.

1. $4x = 16$
2. $5x = 35$

3. $y + 8 = 19$
4. $x - 9 = 0$
5. $2p + 7 = 9$
6. $3y + 7 = 19$
7. $2x + 3(x - 4) = 10$
8. $6y - 5(3 - y) = 18$
9. $5.2x + 5(x - 2) = 14$
10. $3y + 5(y - 2) = 14$
11. $11x + 8 = x + 88$
12. $5m - 18 = -2m + 2$

Answers

2. 7
4. 9
6. 4
8. 3
10. 3
12. 2 6/7

7-6 LINEAR TERMS WITH FRACTIONAL COEFFICIENTS

In the first part of this chapter, we collected terms with whole number coefficients. Now we shall collect terms that may have fractional coefficients.

Example 7-17: Collect $5b - 38b$.

If b stands for a bolt, then the problem could be written as follows: Let $-38b$ represent 38 bolts that you lost that belonged to your friend, and let $5b$ represent 5 of your own bolts that you gave her in partial repayment. Now you see that your friend is still short 33 bolts, which may be represented by $-33b$. Therefore, the problem above could be simplified by subtracting $5b$ from $38b$. The difference would be negative $33b$ or $-33b$. Hence,

$$5b - 38b = -33b$$

Example 7-18: Collect $40h - 6\ 1/2h + 4(1\ 2/3h)$.

If h stands for an hour worked, the expression above may represent the following situation: A man normally works a 40-hr week (8:00 A.M. to 4:00 P.M. for 5 days). Monday it began raining at 9:30 A.M. so he quit for the rest of that day.

The remainder of that week he worked until 5:40 P.M. Did he work his 40 hrs? Let 40h represent the 40-hr week. Let −6 1/2h represent the 6 1/2 hr he did not work Monday. Let 1 2/3h represent the additional 1 2/3 hr he worked each day for the rest of the week. Since there were four such days, 4(1 2/3h) represents the total additional hours.

$$
\begin{aligned}
40h - 6\frac{1}{2}h + 4\left(1\frac{2}{3}h\right) &= 40h - 6\frac{1}{2}h + 6\frac{2}{3}h \\
&= 33\frac{1}{2}h + 6\frac{2}{3}h \\
&= 33\frac{3}{6}h + 6\frac{4}{6}h \\
&= 39\frac{7}{6}h \\
&= 40\frac{1}{6}h
\end{aligned}
$$

So he worked 40 hr 10 min.

Exercise Set 7-6

Simplify, by combining terms.

1. $9x - 15x$
2. $(5/2)y + (7/3)y - (11/3)y$
3. $3(x + y) - 7(x - y)$
4. $(2/5) \cdot 5y - (3/2) \cdot 2y$
5. $3(x + 2) - 2(x - 4)$
6. $(4/3)y - (7/3)y$

Answers

2. $(7/6)y$ 4. $-y$ 6. $-y$

7-7 SIMPLE EQUATIONS WITH FRACTIONAL COEFFICIENTS

Now we are presenting more simple equations, some having fractional coefficients and numbers.

Example 7-19: $-3 + (5/2)x = 17$

Step 1. Copy the problem: $-3 + \left(\frac{5}{2}\right)x = 17$

Step 2. Add +3 to both sides: $-3 + 3 + \left(\frac{5}{2}\right)x = 17 + 3$

Step 3. Simplify: $\left(\frac{5}{2}\right)x = 20$

Step 4. Multiply both sides by 2: $(2)\left(\frac{5}{2}\right)x = (2)(20)$

Step 5. Simplify: $5x = 40$

Step 6. Divide both sides by 5: $\frac{5x}{5} = \frac{40}{5}$

Step 7. Simplify: $x = 8$

Example 7-20: $x - 7/12 = 5/4$

Step 1. Copy the problem: $x - \frac{7}{12} = \frac{5}{4}$

Step 2. LCM(12, 4) = 12.

Step 3. Multiply both sides by 12: $x(12) - \left(\frac{7}{12}\right)(12) = \left(\frac{5}{4}\right)(12)$

Step 4. Simplify: $12x - 7 = 15$

Step 5. Add +7 to both sides: $12x - 7 + 7 = 15 + 7$

Step 6. Simplify: $12x = 22$

Step 7. Divide both sides by 12: $\frac{12x}{12} = \frac{22}{12}$

Step 8. Simplify: $x = \frac{11}{6}$ or $1\frac{5}{6}$

Example 7-21: $(3/4)x + 1/8 = 2/3$

Step 1. Copy the problem: $\left(\frac{3}{4}\right)x + \frac{1}{8} = \frac{2}{3}$

Step 2. LCM(4, 8, 3) = 24.

Step 3. Multiply both sides by 24: $(24)\left(\frac{3}{4}\right)x + (24)\left(\frac{1}{8}\right) = (24)\left(\frac{2}{3}\right)$

Step 4. Simplify: $18x + 3 = 16$

Step 5. Subtract +3 from both sides: $18x + 3 - 3 = 16 - 3$

Step 6. Simplify: $18x = 13$

Step 7. Divide both sides by 18: $\frac{18x}{18} = \frac{13}{18}$

Step 8. Simplify: $x = \frac{13}{18}$

Example 7-22: $x/7 = 30/42$

Step 1. Copy the problem: $\frac{x}{7} = \frac{30}{42}$

Step 2. $\text{LCM}(7, 42) = 42$.

Step 3. Multiply both sides by 42: $(42)\left(\frac{x}{7}\right) = (42)\left(\frac{30}{42}\right)$

Step 4. Simplify: $6x = 30$

Step 5. Divide both sides by 6: $\frac{6x}{6} = \frac{30}{6}$

Step 6. Simplify: $x = 5$

Example 7-23: $x/2 + x/3 = 5/6$

Step 1. Copy the problem: $\frac{x}{2} + \frac{x}{3} = \frac{5}{6}$

Step 2. $\text{LCM}(2, 3, 6) = 6$.

Step 3. Multiply both sides by 6: $\left(\frac{x}{2}\right)(6) + \left(\frac{x}{3}\right)(6) = \left(\frac{5}{6}\right)(6)$

Step 4. Simplify: $3x + 2x = 5$

Step 5. Combine like terms: $5x = 5$

Step 6. Divide both sides by 5: $\frac{5x}{5} = \frac{5}{5}$

Step 7. Simplify: $x = 1$

Example 7-24: $2x/3 = 14/21$

Step 1. Copy the problem: $\frac{2x}{3} = \frac{14}{21}$

Step 2. $\text{LCM}(3, 21) = 21$.

Step 3. Multiply both sides by 21: $(21)\left(\frac{2x}{3}\right) = (21)\left(\frac{14}{21}\right)$

Step 4. Simplify: $14x = 14$

Step 5. Divide both sides by 14: $\frac{14x}{14} = \frac{14}{14}$

Step 6. Simplify: $x = 1$

Exercise Set 7-7

Solve each of the following.

Part A

1. $(3/4)x = 12$
2. $2y/3 = 18$
3. $2y - 4/9 = 7/9$
4. $(2/3)x + 1/4 = 7/4$
5. $(1/9)x + 2/3 = (2/9)x - 1/3$
6. $-10 + (5/6)y = -1/12$
7. $0.7y - 0.3y = 0.2y + 2$
8. $(3/5)y + (2/15)y = 2$
9. $2/5 = -6 + 16/x$
10. $8y - 3/5 = 2y$

Part B

11. $2(x - 7) + 4 = 8$
12. $2x/3 = 6/9$
13. $6x/5 - 9/2 = 2/3$
14. $(3/4)x - 11/2 = (1/4)x + 1/2$
15. $3x/5 = 21/7$
16. $2x/5 + 7/10 = 9/10$

Answers

2. 27
4. 9/4
6. 119/10 or 11 9/10
8. 30/11 or 2 8/11
10. 1/10
12. 1
14. 12
16. 1/2

Chapter 8

Powers and Roots

When multiplying, we often find a number being multiplied by itself two or more times. In order to write such a product in a simple form, the use of exponents is helpful (see Sec. 1-6). For example, if we are interested in $11 \times 11 \times 11 \times 11$, we say that this product is the *fourth power of 11* and we use the exponent **4**, writing 11^4 (read *the fourth power of eleven* or *eleven to the fourth power*). To see how helpful such notation is, consider 17^{131}. If we were to write out this power of 17 in the equivalent multiplication form, it would probably require an entire page.

A second notation that we need to learn involves the reverse idea. Instead of asking what the nth power of a number is, we shall ask what number has a given nth power. If we were told that the cube of some number was 8, how could we write this? We could let x stand for the number and write $x^3 = 8$ but this would only be expressing x in an indirect way. The notation used in this case is $x = \sqrt[3]{8}$ (read *x equals the cube root of 8*). These two statements say the same thing but the latter expresses x directly.

8-1 POWERS

We have discussed powers sufficiently in the introduction; simply remember that the nth power of a number x is the product of x with itself a total of n times and is written x^n.

$$x^n = x \cdot x \cdot x \cdot \ldots \cdot x$$

x is used n times.

The exponents 2 and 3 are used so often that they have acquired special names. The second power of a number is known as its *square*, and the third power is known as its *cube*.

One caution: Remember that $x^n + y^n \neq (x + y)^n$. You might check this with some examples. For instance, $4^3 + 2^3 = 64 + 8 = 72$, but $(4 + 2)^3 = 6^3 = 216$. As practice with powers and in preparation for Sec. 8-2, we might list the first few powers of some whole numbers.

$2^1 = 2$	$3^1 = 3$	$5^1 = 5$	$10^1 = 10$
$2^2 = 4$	$3^2 = 9$	$5^2 = 25$	$10^2 = 100$
$2^3 = 8$	$3^3 = 27$	$5^3 = 125$	$10^3 = 1{,}000$
$2^4 = 16$	$3^4 = 81$	$5^4 = 625$	$10^4 = 10{,}000$
$2^5 = 32$	$3^5 = 243$		
$2^6 = 64$			
$2^7 = 128$			
$2^8 = 256$			

These powers of 2, 3, 5, and 10 occur frequently, and you should be familiar with them.

We can also take powers of fractional or decimal numbers in exactly the same way.

Example 8-1

$$\left(\frac{1}{2}\right)^3 = \frac{1}{2} \cdot \frac{1}{2} \cdot \frac{1}{2} = \frac{1}{8}$$

$$(1.2)^2 = (1.2) \cdot (1.2) = 1.44$$

Exercise Set 8-1

Part A

1. Find the indicated power of a whole number.
   (a) 6^2 (b) 4^3 (c) 10^8 (d) 9^3
   (e) 2^{10} (f) 10^2 (g) 9^5 (h) 2^3
2. Find the indicated power.
   (a) $(1/2)^4$ (b) $(1/2)^5$ (c) $(2.5)^2$ (d) $(0.5)^4$
   (e) $(2/5)^4$ (f) $(3/2)^4$ (g) $(1.1)^4$ (h) $(1.21)^2$

Part B

3. What is the area of the top of a metal block that measures 2.5 in. by 2.5 in. by 3.2 in. high?
4. What is the volume of a cube which is 0.5 in. on each side?
5. If a mechanic can tune five cars per day, how many cars can five mechanics tune in one 5-day week?
6. How many cubic inches of paper would you have if you had eight stacks, each 8 in. high, of 8 in. by 8 in. tablets?
7. The area of a circle is equal to πr^2, where $\pi \doteq 3.14$ and r is the radius of the circle. Find the area of circles with the following radii:

   (a) 1 in. (b) 3 ft (c) 2 mm (d) 11 cm
   (e) 6 in. (f) 2.1 cm (g) 0.2 in. (h) 7.3 mm

Part C

8. If a piston has a 1 1/4-in. radius, what is the area of its top surface? How would you calculate the displacement of such a piston? (See Prob. 7.)
9. What is the displacement of a piston with a 3.2-cm radius and a 9.1-cm stroke? (See Prob. 7.)
10. What is the total cross-sectional area of 16 pipes, half of which have a 2-in. radius and half of which have a 2-in. diameter? (See Prob. 7.)

Answers

2. (a) 1/16 (b) 1/32 (c) 6.25 (d) 0.0625 (1/16)
   (e) 16/625 (f) 81/16 (g) 1.4641 (h) 1.4641
4. 1/8 in.^{3}
6. 4096 in.^{3}
8. $25/16\pi \doteq 4.90625$ in.^{2}
10. 125.6 in.^{2}

8-2 ROOTS

In the introduction to this chapter we introduced the notation $\sqrt[3]{8}$ for the cube root of the number 8.

> *Definition:* The *nth root of the number x*, denoted $\sqrt[n]{x}$, is the number whose nth power is x.

The number n is called the *index* of the root and is written above the left edge of the root symbol.

$\sqrt[n]{x} = r$ means $x = r^n$
The index of the root of x is the exponent of r.

When writing a square root, with n = 2, the index is frequently omitted. Thus, $\sqrt{4}$ means $\sqrt[2]{4}$, which is 2. In order to find the nth root of a whole number, try writing the number as a product of prime factors (see Sec. 1-8). Then, arrange the prime factors in n *identical* groups, if possible. If this can be done, the product of the factors in any of these identical groups will be the nth root of the original number.

Example 8-2: To find $\sqrt[4]{1,296}$:

$$\begin{aligned} 1296 &= 2 \cdot 648 \\ &= 2 \cdot 2 \cdot 324 \\ &= 2 \cdot 2 \cdot 2 \cdot 162 \\ &= 2 \cdot 2 \cdot 2 \cdot 2 \cdot 81 \\ &= 2 \cdot 2 \cdot 2 \cdot 2 \cdot 3 \cdot 27 \\ &= 2 \cdot 2 \cdot 2 \cdot 2 \cdot 3 \cdot 3 \cdot 3 \cdot 3 \\ &= (2 \cdot 3)(2 \cdot 3)(2 \cdot 3)(2 \cdot 3) \\ &= 6 \cdot 6 \cdot 6 \cdot 6 \end{aligned}$$

So $\sqrt[4]{1,296} = 6$ because $1296 = 6^4$.

Example 8-3: To find $\sqrt[3]{27,000}$:

$$\begin{aligned} 27,000 &= 27 \cdot 1,000 \\ &= 3 \cdot 3 \cdot 3 \cdot 10 \cdot 10 \cdot 10 \\ &= 3 \cdot 3 \cdot 3 \cdot 2 \cdot 2 \cdot 2 \cdot 5 \cdot 5 \cdot 5 \\ &= (3 \cdot 2 \cdot 5)(3 \cdot 2 \cdot 5)(3 \cdot 2 \cdot 5) \end{aligned}$$

So $\sqrt[3]{27,000} = 30$ because $27,000 = 30^3$.

Example 8-4: To find $\sqrt[3]{216{,}000}$:

$$\begin{aligned}216{,}000 &= 1{,}000 \cdot 216\\ &= 2\cdot 2\cdot 2\cdot 5\cdot 5\cdot 5\cdot 2\cdot 2\cdot 2\cdot 27\\ &= 2\cdot 2\cdot 2\cdot 5\cdot 5\cdot 5\cdot 2\cdot 2\cdot 2\cdot 3\cdot 3\cdot 3\\ &= (2\cdot 5\cdot 2\cdot 3)(2\cdot 5\cdot 2\cdot 3)(2\cdot 5\cdot 2\cdot 3)\\ &= (60)^3\end{aligned}$$

So $\sqrt[3]{216{,}000} = 60$.

If we need a root of a common fraction such as $\sqrt[3]{8/27}$, we can rewrite this as $\sqrt[3]{8}/\sqrt[3]{27} = 2/3$. In a general setting, we can say that

$$\boxed{\sqrt[n]{\frac{a}{b}} = \frac{\sqrt[n]{a}}{\sqrt[n]{b}}}$$

Example 8-5

$$\sqrt{\frac{9}{16}} = \frac{\sqrt{9}}{\sqrt{16}} = \frac{3}{4}$$

Example 8-6

$$\sqrt[4]{\frac{1}{625}} = \frac{\sqrt[4]{1}}{\sqrt[4]{625}} = \frac{1}{5}$$

Example 8-7

$$\sqrt[3]{\frac{375}{81}} = \sqrt[3]{\frac{125}{27}} = \frac{\sqrt[3]{125}}{\sqrt[3]{27}} = \frac{\sqrt[3]{5\cdot 5\cdot 5}}{\sqrt[3]{3\cdot 3\cdot 3}} = \frac{5}{3}$$

Note that Example 8-7 required simplification of the fraction 375/81 before we actually started working on finding the cube root. The fraction was reduced by a common factor of 3.

The most common example of use of square roots involves right triangles. In a right triangle, there are two legs and a hypotenuse, which is longer than either leg and is located opposite the right angle (Fig. 8-1). If we denote the lengths of the two legs with the letters a and b and the length of the hypotenuse with the letter c, then the famous **Pythagorean theorem** says that $\mathbf{a^2 + b^2 = c^2}$.

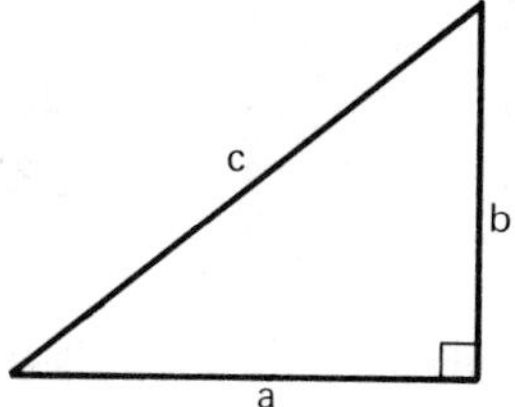

Figure 8-1. Right triangle.

Example 8-8: If the two legs of a right triangle have lengths of 5 in. and 12 in., we can find the length of the hypotenuse by solving

$$5^2 + 12^2 = c^2$$
$$25 + 144 = c^2$$
$$169 = c^2$$
$$c = \sqrt{169}$$
$$c = 13$$

Example 8-9: If one leg of a right triangle has length 6 in. and the hypotenuse has length 10 in., we can find the length of the other leg by solving

$$a^2 + 6^2 = 10^2$$
$$a^2 + 36 = 100$$
$$a^2 = 100 - 36$$
$$a^2 = 64$$
$$a = \sqrt{64}$$
$$a = 8$$

There are many physical problems that can be solved with the Pythagorean theorem, a few of which are included in the exercises. You might think about some situations in your field of interest that can be evaluated with the help of this technique. We shall see more examples and discussion of the Pythagorean theorem and its practical uses in Chapter 15.

Exercise Set 8-2

Part A

1. Find the indicated root by factoring.

   (a) $\sqrt{225}$ (b) $\sqrt{2{,}025}$ (c) $\sqrt[3]{512}$ (d) $\sqrt[3]{216}$

   (e) $\sqrt{1{,}600}$ (f) $\sqrt[3]{74{,}088}$ (g) $\sqrt[3]{1{,}728}$ (h) $\sqrt[4]{256}$

2. Find the indicated root by factoring.

(a) $\sqrt{25/4}$ (b) $\sqrt{48/75}$ (c) $\sqrt[3]{27/64}$ (d) $\sqrt[4]{1/81}$
(e) $\sqrt[3]{512/216}$ (f) $\sqrt[4]{162/512}$ (g) $\sqrt{98/72}$ (h) $\sqrt{75/147}$

Part B

3. In order to cut square pieces of paper with area of 64 in.^{2}, how long should the sides be?
4. If the two legs of a right triangle are 3 in. and 4 in. long, how long is the hypotenuse?
5. Find the center-to-center distance between the two holes shown in Fig. 8.2.

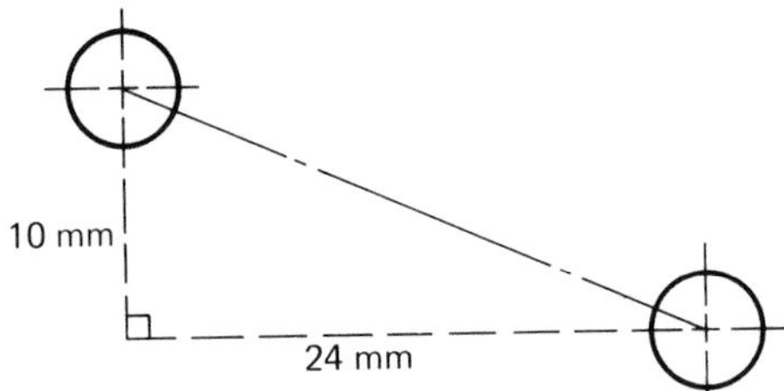

Figure 8-2

6. Find the length of the sides of a cube that has the same volume as the total volume of three cubes, one with 1-in. sides, another with 6-in. sides, and a third with 8-in. sides.

Part C

7. What is the length of the diagonal on a rectangle that is 9 cm by 12 cm?
8. What is the length of the sides of a cube that has a volume of 3 3/8 in.^{3}?
9. If an A-frame roof is 24 ft wide and the peak of the roof is 5 ft higher than its edges, how long should the rafters be?

Answers

2. (a) 5/2 (b) 4/5 (c) 3/4 (d) 1/3
   (e) 4/3 (f) 3/4 (g) 7/6 (h) 5/7
4. 5″
6. 9″
8. 3/2″ or 1 1/2″

8-3 ROOTS BY APPROXIMATION

If factoring does not work or if we need a root of a decimal fraction, we might try an approximation method. First, let's look at a quick method for obtaining the correct number of digits for an estimated root of a number. For instance, does

$\sqrt{400}$ have one, two, or three digits to the left of the decimal point? We know that $\sqrt{4} = 2$. So, is $\sqrt{400}$ equal to 2, 20, or 200?

Since $(2)^2 = 4$, $\sqrt{400} \neq 2$. ($\neq$ means is not equal to.)
Since $(20)^2 = 400$, $\sqrt{400} = 20$.
Since $(200)^2 = 40{,}000$, $\sqrt{400} \neq 200$.

Example 8-10: What whole number is closest to $\sqrt[3]{400}$?

$1^3 = 1$ $\quad$ $5^3 = 125$
$2^3 = 8$ $\quad$ $6^3 = 216$
$3^3 = 27$ $\quad$ $7^3 = 343$
$4^3 = 64$ $\quad$ $8^3 = 512$

We see that 7 appears to be the closest whole number to $\sqrt[3]{400}$ because 7^3 is closest to 400. To be certain, we check 7.5. Since $(7.5)^3 = 421.88$ is more than 400, the closest whole number to $\sqrt[3]{400}$ is 7.

Example 8-11: What whole number multiple of 10 is closest to $\sqrt{3478}$?

$10^2 = 100$ $\quad$ $40^2 = 1600$
$20^2 = 400$ $\quad$ $50^2 = 2500$
$30^2 = 900$ $\quad$ $60^2 = 3600$

We see that 60 appears to be the closest whole number multiple of 10 to $\sqrt{3478}$. Since $(55)^2 = 3025$, $\sqrt{3478}$ is between 55 and 60. So 60 is the closest multiple of 10 to $\sqrt{3478}$.

Many students now use calculators to do much of the arithmetic once done by hand. Suppose you were trying to find $\sqrt{342.75}$, and your friend gave you the digits 185135 as the answer from a calculator. However, your friend did not indicate the location of the decimal point. Now is the correct answer for $\sqrt{342.75}$ about 1.85135, 18.5135 or 185.135? Since $10^2 = 100$ and $20^2 = 400$, the answer is between 10 and 20. So 18.5135 must be the approximate answer.

The following are some practice problems. Place the decimal point in the proper location.

$\sqrt{4.37} \doteq 2\ 0\ 9\ 0\ 4\ 5$ $\quad$ $\sqrt{0.924} \doteq 0\ 9\ 6\ 1\ 2\ 5$
$\sqrt{43.7} \doteq 6\ 6\ 1\ 0\ 6$ $\quad$ $\sqrt{58.62} \doteq 7\ 6\ 5\ 6$
$\sqrt{0.0042} \doteq 0\ 0\ 6\ 4\ 8$ $\quad$ $\sqrt{82106.45} \doteq 2\ 8\ 6\ 5\ 4$

The problems above were done by using a calculator with a square root key. The following procedure may be used if you do not have a calculator or if your calculator does not have a square root key. Suppose we want to find $\sqrt[3]{20}$. We could first try factoring 20 as $2 \cdot 2 \cdot 5$, but we cannot arrange the factors in three identical groups. Since $\sqrt[3]{20}$ is the unique number that has 20 as its cube, our approximation technique will have us examining a series of numbers that are selected so that the cubes of these numbers are getting closer and closer to 20.

Guess	$(\text{Guess})^3$	Conclusion
2	8	Since $20 > 8$, $\sqrt[3]{20} > 2$.
3	27	Since $20 < 27$, $\sqrt[3]{20} < 3$.
(At this point, we have $2 < \sqrt[3]{20} < 3$.)		
2.5	15.625	$\sqrt[3]{20} > 2.5$
2.7	19.683	$\sqrt[3]{20} > 2.7$
2.8	21.952	$\sqrt[3]{20} < 2.8$
(At this point, we have $2.7 < \sqrt[3]{20} < 2.8$.)		
2.75	20.796	$\sqrt[3]{20} < 2.75$
2.72	20.124	$\sqrt[3]{20} < 2.72$
2.71	19.903	$\sqrt[3]{20} > 2.71$
(At this point, we have $2.71 < \sqrt[3]{20} < 2.72$.)		
2.715	20.013	$\sqrt[3]{20} < 2.715$

At this point we can say $\sqrt[3]{20} \doteq 2.71$, accurate to two decimal places. If we really need a more exact value, we can continue this process as long as necessary; however, you have probably already noticed that the arithmetic becomes very tedious.

Example 8-12: Approximate $\sqrt{3}$.

Guess	$(\text{Guess})^2$	Conclusion	$\sqrt{3}$ is between
1	1	Too small	
2	4	Too big	1 and 2
1.5	2.25	Too small	1.5 and 2
1.7	2.89	Too small	1.7 and 2
1.8	3.24	Too big	1.7 and 1.8
1.75	3.0625	Too big	1.7 and 1.75

So, to one decimal place, $\sqrt{3} \doteq 1.7$.

Example 8-13: Approximate $\sqrt{0.81}$.

Guess	$(\text{Guess})^2$	Conclusion	$\sqrt{0.81}$ is between
1	1	Too big	
0.5	0.25	Too small	0.5 and 1
0.7	0.49	Too small	0.7 and 1
0.8	0.64	Too small	0.8 and 1
0.9	0.81	Exact	

So, $\sqrt{0.81} = 0.9$, exactly. Can you show a good reason for this fact?

8-4 USE OF TABLES

For anyone who needs to find powers or roots very often, a good table of powers and roots can be very helpful.

The tables for powers and roots are in Appendices A and B. Appendix A lists the squares and square roots of numbers from 1 to 10 by increments of one-tenth. Appendix B lists the cubes and cube roots of the same numbers. Both appendix tables include special columns such as $\sqrt{10n}$ that add greatly to the versatility of the tables.

Reading across Appendix A, the table of squares and square roots, we find four numbers in any row. Starting from the left, they are a number (N) between 1 and 10, its square (N^2), its square root ($\sqrt{N}$), and the square root of 10 times the number ($\sqrt{10N}$). For example, the row starting with N = 1.1 gives us the following information.

$$(1.1)^2 = 1.21$$

$$\sqrt{1.1} = 1.049$$

$$\sqrt{(10)(1.1)} = \sqrt{11} = 3.317$$

In Appendix B, the table of cubes and cube roots, we find five numbers in each row. Starting from the left, they are the number (N) between 1 and 10, its cube (N^3), its cube root ($\sqrt[3]{N}$), the cube root of 10 times the number ($\sqrt[3]{10N}$), and the cube root of 100 times the number ($\sqrt[3]{100N}$).

For example, the row starting with N = 3.5 gives us the following information.

$$(3.5)^3 = 42.875$$

$$\sqrt[3]{3.5} = 1.518$$

$$\sqrt[3]{(10)(3.5)} = \sqrt[3]{35} = 3.271$$

$$\sqrt[3]{(100)(3.5)} = \sqrt[3]{350} = 7.047$$

One helpful technique in using such a table is to watch for the effect of powers of 10. Appendices A and B do some of this for you, but it should be explained how powers of 10 play a helpful role in finding powers and roots. The theory behind this work is summed up in the following equalities.

$$(a \cdot b)^n = a^n \cdot b^n$$
$$\sqrt[n]{a \cdot b} = \sqrt[n]{a} \cdot \sqrt[n]{b}$$
$$\left(\frac{a}{b}\right)^n = \frac{a^n}{b^n}$$
$$\sqrt[n]{\frac{a}{b}} = \frac{\sqrt[n]{a}}{\sqrt[n]{b}}$$

Example 8-14

$$(62)^2 = (6.2 \times 10)^2 = (6.2)^2 \times 10^2 = 38.44 \times 100 = 3{,}844$$

Example 8-15: Going in the other direction, if we want $(0.15)^3$, we can write

$$(0.15)^3 = \left(\frac{1.5}{10}\right)^3 = \frac{(1.5)^3}{10^3} = \frac{3.375}{1{,}000} = 0.003375$$

Example 8-16: For an example with roots, consider

$$\sqrt[3]{80} = \sqrt[3]{8 \cdot 10} = (\sqrt[3]{8})(\sqrt[3]{10})$$
$$= (2) \cdot (2.154)$$
$$= 4.308$$

Example 8-17

$$\sqrt[3]{0.9} = \sqrt[3]{\frac{9}{10}} = \frac{\sqrt[3]{9}}{\sqrt[3]{10}} = \frac{2.080}{2.154} = 0.966$$

Exercise Set 8-4

Part A

1. For each problem indicate which number is closest to the correct answer.
   (a) $\sqrt{39}$: 4, 5, 6, 7
   (b) $\sqrt{150.2}$: 9, 12, 13, 15
   (c) $\sqrt{0.041}$: 0.2, 0.3, 0.02, 0.03

(d) $\sqrt{1{,}000{,}000}$: 10, 100, 1000, 10,000
(e) $\sqrt[3]{65}$: 2, 3, 4, 5, 6
(f) $\sqrt[3]{125}$: 2, 3, 4, 5, 6, 7
(g) $\sqrt[3]{2568}$: 10, 11, 12, 13, 14
(h) $\sqrt[3]{0.9}$: 0.8, 0.9, 0.08, 0.09, 1.0

2. Indicate the correct position for the decimal point in each answer.
   (a) $\sqrt{3.69}$: 1 9 2 0 9
   (b) $\sqrt[3]{75}$: 4 2 1 7 2
   (c) $\sqrt[3]{0.008}$: 0 0 0 2 0
   (d) $\sqrt{342.8}$: 1 8 5 1 5
   (e) $\sqrt[3]{342.8}$: 6 9 9 8 6
   (f) $\sqrt{59723}$: 2 4 4 3 8

Follow the instructions in Exercises 3 through 7 unless directed otherwise by your instructor.

3. Approximate by the guess method of approximation.
   (a) $\sqrt{2.5}$ to two decimal places
   (b) $\sqrt{10}$ to one decimal place
   (c) $\sqrt[3]{21}$ to one decimal place
   (d) $\sqrt[4]{350}$ to the nearest whole number
   (e) $\sqrt{2}$ to two decimal places
   (f) $\sqrt{0.7}$ to two decimal places
   (g) $\sqrt[3]{100}$ to one decimal place
   (h) $\sqrt{0.06}$ to two decimal places

4. Approximate, to at least three decimal places, the square roots of the following numbers.
   (a) 2.5 (b) 10 (c) 2 (d) 0.7
   (e) 0.06 (f) 130 (g) 43.56 (h) 775

5. Find the following powers from Appendices A and B, using powers of 10 if necessary.
   (a) $(1.4)^2$ (b) $(8.1)^2$ (c) $(3.2)^3$ (d) $(6.9)^3$
   (e) $(21)^2$ (f) $(0.45)^2$ (g) $(23)^3$ (h) $(0.98)^3$

6. Find the square root of the following numbers from Appendix A, using powers of 10 if necessary.
   (a) 2.8 (b) 35 (c) 6.6 (d) 0.2
   (e) 94 (f) 5.4 (g) 77

7. Find the cube root of the following numbers from Appendix B, using powers of 10 if necessary.
   (a) 3.7 (b) 45 (c) 740 (d) 5.7
   (e) 680 (f) 1.7 (g) 6,200

Part B

8. In order to cut paper into square pieces with area of 38 in.^{2}, how long should the sides be, to the nearest 1/16 of an inch?

9. Find the center-to-center distance, to two decimal places, of the two holes in Fig. 8-3. (See Sec. 8-2.)

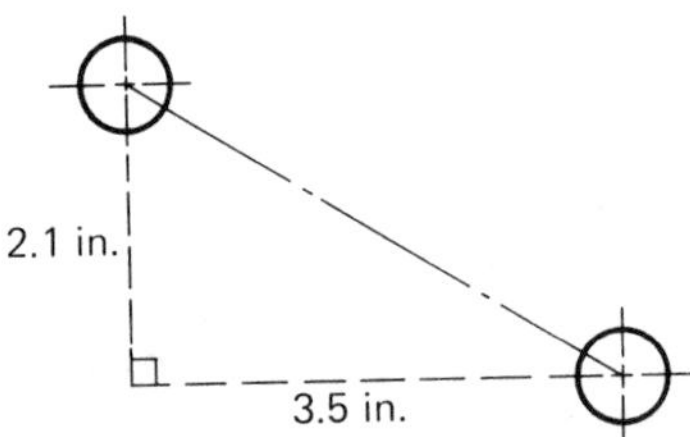

Figure 8-3

10. What is the length of the diagonal, to one decimal place, of a square that is 2 in. long on each side?

11. If we wanted the square root of 2.14, we would not be able to find it in our tables. We *could* just use the square root of 2.1, but that would not be very accurate. To find a better answer, we can use a method of approximation within the table known as *interpolation*. We notice that 2.14 is four-tenths of the way between 2.1 and 2.2 so the square root of 2.14 should be about four-tenths of the way between the square root of 2.1 and the square root of 2.2. Thus,

$$\begin{aligned}\sqrt{2.14} &\doteq \sqrt{2.1} + 0.4(\sqrt{2.2} - \sqrt{2.1})\\ &\doteq 1.449 + 0.4(1.483 - 1.449)\\ &= 1.449 + 0.4(0.034)\\ &= 1.449 + 0.0136\\ &= 1.4626\end{aligned}$$

So

$$\sqrt{2.14} \doteq 1.463$$

Use this method of interpolation in the tables to approximate

(a) $\sqrt{3.15}$ (b) $\sqrt{6.25}$ (c) $\sqrt{32.8}$
(d) $(1.01)^3$ (e) $\sqrt[3]{2.46}$ (f) $\sqrt[3]{39.7}$

Some of the exercises in the rest of this set and in the chapter review can be done best by using interpolation, but they can be done *roughly* by simply rounding off. Ask your instructor which way to do these.

12. Find the top area of a piston that has a 1.9 in. radius. (See Prob. 7, Exercise Set 8-1.)
13. What is the area of the triangle in Fig. 8-4?

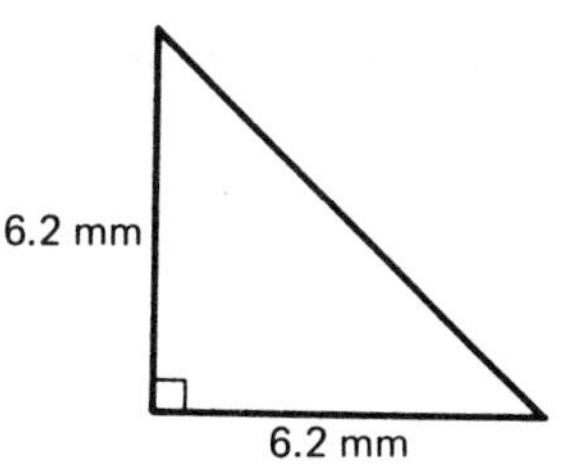

Figure 8-4

Part C

14. Two square heating ducts have sides of 10 in. and 12 in., respectively. Find the size of a single square duct that has the same area as the two given ducts.
15. The area of the top of a piston is 10 in.^{2} What is its diameter?
16. What amount of cross-sectional area is added if a pipe with a 1.9-in. radius is replaced by a pipe with a 2.2-in. radius?
17. What should be the radius of a semicircular fireplace if it must have an area of 1 m^2?
18. If you mistakenly milled a cube 9.2 in. on a side when the specifications called for 8.3 in. on a side, how much additional material (measured by its volume) must you take off?
19. What is the length of the hypotenuse of a right triangle with sides of lengths 12.6 in. and 10.2 in.?

Answers

2. (a) 1.9209 (b) 4.2172 (c) 0.2 (d) 18.515
   (e) 6.9986 (f) 244.38
4. (a) 1.581 (b) 3.162 (c) 1.414 (d) 0.8367
   (e) 0.2449 (f) 11.40175 (g) 6.600 (h) 27.8388
6. (a) 1.673 (b) 5.916 (c) 2.569 (d) 0.4472
   (e) 9.695 (f) 2.324 (g) 8.775
8. 6 3/16 in.
10. 2.8 in.
12. (a) 1.775 (b) 2.500 (c) 5.727 (d) 1.030
    (e) 1.351 (f) 3.411
14. 15.62 in.
16. 3.86 in.^{2}
18. 206.9 in.^{3}

8-5 SQUARE ROOTS BY CALCULATION

Another method for approximating square roots, but *only* square roots, is a method that resembles long division.

> Step 1. Starting at the decimal point, group digits by pairs in both directions from the decimal point. Each pair of digits will give one digit of the square root. If necessary, add a *zero* at the end of a decimal number to complete the last pair.

$$\sqrt{43.5\mathbf{0}}$$

> Step 2. Place the decimal point for the answer directly above the decimal point in the number.

$$\begin{array}{l} \quad\;\; . \\ \sqrt{43.50} \end{array}$$

> Step 3. Beginning at the left, determine the largest whole number whose square is less than or equal to the first pair.

$6^2 = 36$ and $7^2 = 49$, so 6 should be used.

> Step 4. Write this number above the pair and its square below the pair.

$$\begin{array}{l} \;\;\mathbf{6}. \\ \sqrt{43.50} \\ \;\mathbf{36} \end{array}$$

> Step 5. Subtract and bring down the next pair. This is now called the remainder.

$$\begin{array}{l} \;\;6. \\ \sqrt{43.50} \\ \;\underline{36} \\ \;\;\;7\,\mathbf{50} \end{array}$$

> Step 6. The trial divisor is determined by multiplying the partial answer by 20. This trial divisor is written to the left of the remainder.

```
      6.
    √43.50
      36
120|   7 50
```

> Step 7. Determine how many times the trial divisor will divide into the remainder.

120 goes into 750 six times.

> Step 8. Place that number above the pair brought down in Step 5 and also *add it to the trial divisor*.

```
      6. 6
    √43.50
   6| 36
 12Ø|  7 50
```

> Step 9. Multiply the *new* trial divisor by the last digit of the answer (the number you found in Step 7) and write this below the previous remainder. If it is larger, go back to Step 8 with the next smaller whole number.

```
      6. 6
    √43.50
      36
 126|  7 50
       7 56
```

> Step 10. If the desired accuracy has not been reached in the partial answer, continue by picking up at Step 5.

```
      6. 5
    √43.5000
   5| 36
 12ϕ|  7 50
       6 25
       1 2500
```

This method for finding a square root is sometimes called *square root by computation*. We shall finish the example:

```
                 6. 5 9 5 4 5
               √43.5000000000
                 36
          5 |
        12∅ |   7 50
                6 25
         9  |
       130∅ |   1 2500
                1 1781
          5 |
      1318∅ |       71900
                    65925
          4 |
     13190∅ |        597500
                     527616
          5 |
    131908∅ |         6888300
                      6595425
                       392975
```

So, to five decimal places,

$$\sqrt{43.5} \doteq 6.59545$$

Example 8-18: Approximate $\sqrt{4.64}$.

```
                 2. 1 5 4 0 6
               √04.6400000000
                 4
          1 |
        4∅  |   0 64
                  41
          5 |
        42∅ |     2300
                  2125
          4 |
       430∅ |      17500
                   17216
      43080 |        28400
                         0
          6 |
     43080∅ |        2840000
                     2584836
                      255164
```

So, to five decimal places,

$$\sqrt{4.64} \doteq 2.15406$$

Review Exercises

Part A

1. Find the following powers:
   (a) $(2.7)^3$ (b) $(9.81)^2$ (c) $(2/3)^4$ (d) $(0.81)^2$
2. Find the following roots by factoring:
   (a) $\sqrt[3]{24/81}$ (b) $\sqrt{144}$ (c) $\sqrt[4]{50{,}625}$ (d) $\sqrt{36/256}$
3. Find the following roots by approximation:
   (a) $\sqrt[3]{150}$ (b) $\sqrt{340}$ (c) $\sqrt{2.19}$ (d) $\sqrt{9.819}$
4. Find the following powers and roots in Appendices A and B:
   (a) $\sqrt[3]{11}$ (b) $\sqrt{0.21}$ (c) $\sqrt{219}$ (d) $(8.63)^2$

Part B

5. Write and solve two problems in your field of interest that require the use of powers or roots.
6. What is the center-to-center distance between the two holes in Fig. 8-5?

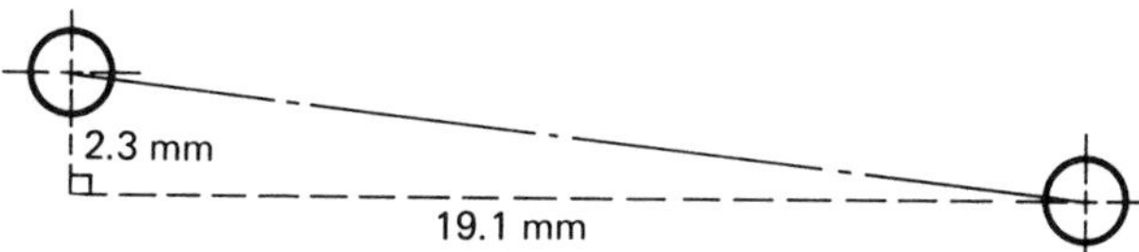

Figure 8-5

7. To the nearest hundredth of an inch, what is the length of the sides of a cube that has a volume of 3.65 in.^{3}?

Part C

8. If a square poster, 16.2 in. on a side, has a 2.7-in. margin on all four sides, what is the area of the printed matter? How much area is left blank for margins?
9. What volume of metal would a 1/2-in. drill remove from a 2-in.-thick metal plate if you drilled a hole completely through the plate?
10. In Fig. 8-6, a piston is connected to the crankshaft with a 10-in. connecting rod. When the connecting rod is at its largest angle (see figure below), the piston pin is 9.3 in. from the center line of the crankshaft. What is the radius (r) of the crankshaft throw?

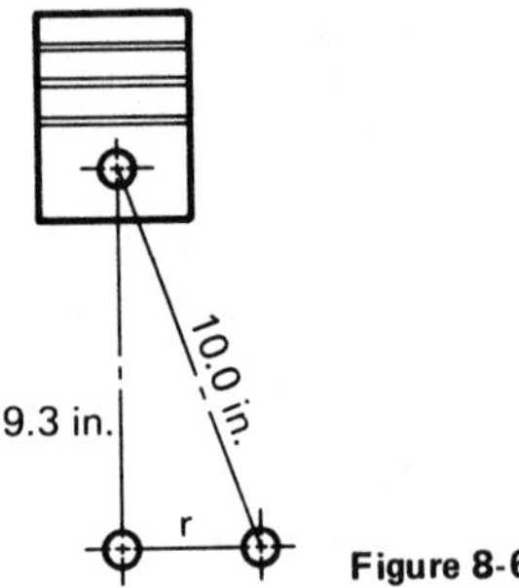

Figure 8-6

Answers

2. (a) 2/3 (b) 12 (c) 15 (d) 3/8
4. (a) 2.224 (b) 0.4583 (c) 14.7986 (d) 74.4769
6. Accurate as 19.2 mm
8. 116.64 in.^{2} printed; 145.8 in.^{2} in waste
10. 3.68 in.

Chapter 9

Algebra with One and Two Variables

A working knowledge of geometry and basic algebra is essential for solving many practical problems. In previous chapters, we discussed some basic concepts about algebra. In Chapter 7 we solved linear equations that had rational solutions. In this chapter we shall continue our study of linear equations and also work with second-degree equations. A second-degree equation is called a *quadratic*.

Some basic concepts that you should familiarize yourself with are discussed below. In Chapter 7 we illustrated and gave examples of terms and expressions. In the expression $ax + bx^2 + dx^3 + c$, the terms are ax, bx^2, dx^3, and c. In a term such as bx^2, the constant multiplier b is called a *coefficient*.

Now, consider the quadratic expression $ax^2 + bx + c$. The terms are ax^2, bx, and c. The coefficient of the term ax^2 is a, the coefficient of bx is b, and the constant c is called the *constant term* (or *constant coefficient* or *constant*). Remember that a, b, and c are real constant coefficients and that $a \neq 0$.

9-1 COMBINING GENERAL TERMS

In Chapter 7, we collected linear terms. Now we shall collect terms of linear and nonlinear expressions. [*Note:* Like terms must have the same letter(s) and same exponents(s), for example, $3x^2y^3$ and $4x^2y^3$.]

Example 9-1

$$5ax^2 + 3ax^2 - 2ax^2$$
$$8ax^2 - 2ax^2$$
$$6ax^2$$

Illustration of Example 9-1: Let $d = ax^2$ in $5ax^2 + 3ax^2 - 2ax^2$.

Replace ax^2 by d: $5(d) + 3(d) - 2(d)$

$8d - 2d$

$6d$

Now replace d by ax^2: $6(ax^2)$ or $6ax^2$

Example 9-2

$$6(2y^2 + x^2) - 4x^2$$

$$12y^2 + 6x^2 - 4x^2$$

$$12y^2 + 2x^2$$

The terms $12y^2$ and $2x^2$ may not be added because they represent two different types of things (unlike terms).

Exercise Set 9-1

Collect like terms.

Part A

1. $9x^2 - 5x^2$
2. $3(a^2 + b^2) - 7(a^2 - b^2)$
3. $(4/3)a^2y - (7/3)a^2y$
4. $3(d^2 + 2) - 2(d^2 - 4)$
5. $5(ax^2 + bx + c) - 3(ax^2 - 3bx + 2c)$
6. $2ab + (1/2)(b + c) + (3/2)ac$
7. $10d^2y^2 + 2d(dy^2 - 2d) + 5d^2$

Part B

Note: π, pi, is used in finding the circumference and area of a circle. π should be treated as any other symbol.

8. $5\pi r^2 - 6\pi(r^2 + 2) + 12\pi$
9. $(4/3)\pi r^3 + (8/3)\pi r^3 - 2\pi r^3$
10. $10(\pi r^3 - ba^2) - 4(\pi r^3 - 3ba^2)$
11. $8(\pi d + 3\pi r^2) - 4(\pi d - 2\pi r^2)$

Answers

2. $-4a^2 + 10b^2$
4. $d^2 + 14$
6. $(2\ 1/2)ab + 2ac$
8. $-\pi r^2$
10. $6\pi r^3 + 22ba^2$

9-2 LINEAR EQUATIONS

In Chapter 7, we examined simple linear equations. Now we are presenting more linear equations and equations with variables in the denominator.

Example 9-3: Solve for x:

$$\frac{x-1}{5} + \frac{2}{5} = 9$$

Step 1. Copy the problem: $\frac{x-1}{5} + \frac{2}{5} = 9$

Step 2. $\text{LCD}\left(\frac{x-1}{5}, \frac{2}{5}\right) = \text{LCM}(5, 5) = 5.$

Step 3. Multiply both sides by 5: $\left(\frac{x-1}{5}\right)(5) + \frac{2}{5}(5) = 9(5)$

Step 4. Simplify: $(x-1)(1) + 2 = +45$

Step 5. $x - 1 + 2 = 45$

Step 6. $x + 1 = 45$

Step 7. $x = 44$

Check: Replace x by 44.

$$\frac{44-1}{5} + \frac{2}{5} \stackrel{?}{=} 9$$

$$\frac{43}{5} + \frac{2}{5} \stackrel{?}{=} 9$$

$$\frac{45}{5} \stackrel{?}{=} 9$$

$$9 \stackrel{\checkmark}{=} 9$$

Example 9-4: Solve for x. $3/x + 1 = 7/x$

Step 1. Copy the problem: $$\frac{3}{x} + 1 = \frac{7}{x}$$

Step 2. LCD = x.

Step 3. Multiply by x: $$\left(\frac{3}{x}\right)(x) + 1(x) = \left(\frac{7}{x}\right)(x)$$

Step 4. Simplify: $$3 + 1x = 7$$

Step 5. $$x = 4$$

Check: Replace x by 4.

$$\frac{3}{4} + 1 \stackrel{?}{=} \frac{7}{4}$$

$$\frac{3}{4} + \frac{4}{4} \stackrel{?}{=} \frac{7}{4}$$

$$\frac{7}{4} \stackrel{\checkmark}{=} \frac{7}{4}$$

Therefore, x = 4 is a solution.

Example 9-5: Find the solution of

$$\frac{4x + 6}{x - 2} - 3 = \frac{14}{x - 2}$$

Step 1. Copy the problem: $$\frac{4x + 6}{x - 2} - 3 = \frac{14}{x - 2}$$

Step 2. LCD = x - 2

Step 3. Multiply by (x - 2): $$\left(\frac{4x + 6}{x - 2}\right)(x - 2) - 3(x - 2) = \left(\frac{14}{x - 2}\right)(x - 2)$$

Step 4. Simplify: $$\left(\frac{4x + 6}{x - 2}\right)\left(\frac{x - 2}{1}\right) - 3(x - 2) = \left(\frac{14}{x - 2}\right)\left(\frac{x - 2}{1}\right)$$

Step 5. $$(4x + 6)(1) - 3(x - 2) = 14(1)$$

Step 6. $$4x + 6 - 3x + 6 = 14$$

Step 7. $$x + 12 = 14$$

Step 8. $$x = 2$$

Now check x = 2 to see if it can be used as the solution for the problem.

Check: Replace x by 2.

$$\frac{4(2)+6}{(2)-2} - 3 = \frac{14}{(2)-2}$$

$$\frac{8+6}{0} - 3 = \frac{14}{0}$$

But

$$\frac{14}{2-2} = \frac{14}{0}$$

is undefined. The number 2 is the only possible solution that we have for the equation. Therefore, no number will work in the equation. When a replacement is made for a variable and it does not check to be an equality, then that number must be rejected as a solution for the equation.

> If one or more denominators become zero when the unknown is replaced, then the problem cannot possibly check for that replacement and *there is no solution*.

Exercise Set 9-2

Solve the following equations.

1. $\frac{x-1}{5} - \frac{2}{5} = 3$
2. $\frac{y-8}{3} - \frac{y}{6} = -1$
3. $6/x - 1/2 = 1/4$
4. $\frac{7}{x+2} + 4 = \frac{7}{x+2}$
5. $5/3x - 7/12 = 1/4x$
6. $2/x + 3 = 7/x$
7. $2x/3 - 1/5 = 3/4$
8. $\frac{x-1}{5} + \frac{2(x-3)}{5} = 7$
9. $8/(x+5) = -7 + -3/(x+5)$
10. $2(x-6)/3 + 7/8 = 1$

Answers

2. 10
4. No solution
6. 5/3
8. 14
10. 99/16

9-3 SECOND-DEGREE EQUATIONS

To realize the usefulness—and the limitations—of guesswork, the student should try to find the solution for each of the following equations:

1. $x^2 + 3x + 2 = 0$
2. $x^2 + 2x - 3 = 0$
3. $2x^2 - 9x - 5 = 0$
4. $6x^2 + x - 1 = 0$
5. $x^2 + 5x + 3 = 0$

Solution

1. $(x + 1)(x + 2)$	$-1, -2$
2. $(x + 3)(x - 1)$	$1, -3$
3. $(2x + 1)(x - 5)$	$5, -1/2$
4. $(3x - 1)(2x + 1)$	$1/3, -1/2$
5. Prime	$(-5 \pm \sqrt{13})/2$

You probably had considerable difficulty in guessing the solutions to some of the preceding equations, and for others you probably could not guess a solution at all. It is commendable to be able to look at a problem and make a good estimate of your solution by inspection.

As we saw in the problems above, solving some equations by inspection is at best a clumsy method.

The quadratic formula, which we shall discuss later in this chapter, is a very useful method to solve a quadratic equation because it always works; however, it is not always the quickest or easiest way. For this reason we shall discuss some of the other methods of solving a quadratic.

First we shall discuss factoring and then relate it to solving quadratic equations.

9-4 FACTORING BINOMIALS

In this section we shall discuss a further use of the distributive law as it relates to factoring. First, let's look at an expression of two terms. An expression of two terms is called a *binomial*. $5a + 3$ is a binomial and $2x + 10$ is a binomial. *To factor the expression* $2x + 10$ *means to write a multiplication expression that has the product* $2x + 10$.

Example 9-6: $2x + 10 = 2(x + 5)$ where 2 and $(x + 5)$ are factors of $2x + 10$. The factor 2 is called a *common factor* because it is a factor of the first term and also of the second term.

> *Definition: When* 1 *or* -1 *are the only factors in common in an expression, the expression is said to be a prime expression.*

Example 9-7: $5a + 3 = 1(5a + 3)$ or $5a + 3 = -1(-5a - 3)$. This expression cannot be factored in any other way; therefore, we say this expression is *prime*.

Example 9-8: Factor $7x^3 - 2x$.

$$7x^3 - 2x = x(7x^2 - 2)$$

Therefore, x and $7x^2 - 2$ are factors of $7x^3 - 2x$.

Exercise Set 9-4

Part A

Factor the following expressions.

1. $6x - 4$
2. $x^2 - x$
3. $5m^2 + 10m$
4. $8a^3 - 16a^2$
5. $2a^2 + 6a^2y$

Part B

Factor the following expressions, if possible.

6. $3x + 12$
7. $x^2 - 7x$
8. $x^3 + 9x$
9. $7x^2 + 9$
10. $4x^3 - 2x^2$

Answers

2. $x(x - 1)$
4. $8a^2(a - 2)$

6. $3(x + 4)$
8. $x(x^2 + 9)$
10. $2x^2(2x - 1)$

9-5 FACTORING QUADRATICS

Factoring expressions with three or more terms can be done in a manner similar to that used to factor expressions with two terms. For example, the "trial and error method" is commonly used; however, expressions with three or more terms may be factored using other methods. We shall take the liberty to explain one such method and leave it to your instructor to explain other methods if he or she desires.

Example 9-9: Factor $2x^2 + 11x + 5$ by use of the distributive law. First, select the coefficient of the first term (2) and the constant (5) and take their product $(2 \cdot 5 = 10)$. Now, find the factors of the product (10) that have a sum equal to the coefficient of the middle term (11).

$$10 = (2)(5)$$

$$(2) + (5) = 7$$

Since $2 + 5 \neq 11$, we must try two other factors of 10.

$$10 = (1)(10)$$

$$(1) + (10) = 11$$

The factors of 10 that add up to 11 are 1 and 10. Therefore, the quadratic can be expressed as follows:

$2x^2 + 11x + 5$	or	$2x^2 + 11x + 5$
$2x^2 + (10x + 1 \cdot x) + 5$		$2x^2 + (1 \cdot x + 10x) + 5$
$(2x^2 + 10x) + (1 \cdot x + 5)$		$(2x^2 + 1 \cdot x) + (10x + 5)$
$2x(\mathbf{x + 5}) + 1(\mathbf{x + 5})$		$x(\mathbf{2x + 1}) + 5(\mathbf{2x + 1})$
$(\mathbf{x + 5})(2x + 1)$		$(\mathbf{2x + 1})(x + 5)$

So

$$2x^2 + 11x + 5 = (x + 5)(2x + 1) = (2x + 1)(x + 5)$$

Example 9-10: Factor $6x^2 - 7x - 3$ by use of the distributive law. First, select the coefficient 6 and the constant -3. Multiply $6 \cdot (-3) = -18$. Now, find the factors of -18 that add up to -7.

Factors	Sum of Factors = −7?	Yes or No
$-18 = (-1)(18)$	$(-1) + (18) = 17$	No
$= (-2)(9)$	$(-2) + (9) = 7$	No
$= (-3)(6)$	$(-3) + (6) = 3$	No
$= (1)(-18)$	$(1) + (-18) = -17$	No
$= (2)(-9)$	$(2) + (-9) = -7$	Yes

(With experience and luck, you might not have to check this many factors.) Therefore,

$$\begin{aligned} 6x^2 - 7x - 3 &= 6x^2 + (2x - 9x) - 3 \\ &= (6x^2 + 2x) + (-9x - 3) \\ &= 2x(3x + 1) - 3(3x + 1) \\ &= (3x + 1)(2x - 3) \end{aligned}$$

The middle term could have been represented as $-9x + 2x$, which would give us

$$\begin{aligned} 6x^2 - 7x - 3 &= 6x^2 - 9x + 2x - 3 \\ &= (6x^2 - 9x) + (2x - 3) \\ &= 3x(2x - 3) + 1(2x - 3) \\ &= (2x - 3)(3x + 1) \end{aligned}$$

Exercise Set 9-5

Factor each of the following completely if it is not prime.

Part A

1. $x^2 - x - 2$
2. $x^2 - 3x - 10$
3. $x^2 + 2x + 1$
4. $x^2 - 3x + 12$
5. $x^2 + 3x + 7$

Part B

6. $5x^2 - 9x - 2$
7. $2x^2 + 5x - 12$
8. $4x^2 + 12x + 9$
9. $2x^2 - x - 15$

Part C

10. $4x^2 - 9$
11. $2x^2 - 10x + 12$
12. $6x^2 - 24x + 18$
13. $3x^2 + 9x - 30$
14. $6x^3 + 22x^2 - 8x$
15. $3x^2 - 30x + 63$

Answers

2. $(x - 5)(x + 2)$
4. Prime
6. $(5x + 1)(x - 2)$
8. $(2x + 3)(2x + 3)$
10. $(2x + 3)(2x - 3)$
12. $6(x - 1)(x - 3)$
14. $2x(3x - 1)(x + 4)$

9-6 SOLVING A QUADRATIC EQUATION BY FACTORING

When solving quadratic equations, we may use the methods of factoring discussed in Sec. 9-5. Also we may make use of the fact that *if the product of two or more factors is zero, at least one of the factors must be equal to zero.*

Example 9-11: Solve $x^2 - 3x - 4 = 0$. By factoring.

$$x^2 - 3x - 4 = (x - 4)(x + 1)$$

So

$$(x - 4)(x + 1) = 0$$

If the product of two numbers is zero, either the first (or the second) factor is zero or both factors are zero. So all we have to do is set each of the factors equal to zero and solve. Hence

$$x - 4 = 0 \quad \text{and} \quad x + 1 = 0$$

$$x = 4 \qquad\qquad x = -1$$

Check:

$$x^2 - 3x - 4 = 0 \qquad x^2 - 3x - 4 = 0$$

Replacement of x with 4

$$(4)^2 - 3(4) - 4 = 0$$
$$16 - 12 - 4 = 0$$
$$16 - 16 = 0$$
$$0 = 0$$

Replacement of x with -1

$$(-1)^2 - 3(-1) - 4 = 0$$
$$1 + 3 - 4 = 0$$
$$0 = 0$$

Hence 4 and -1 are solutions for the quadratic equation. Even though -1 is a mathematical solution, it doesn't have to have a practical application.

Example 9-12: Solve $5x^2 + 7x - 12 = 2x - 2$.

$$5x^2 + 5x - 10 = 0$$
$$5(x^2 + x - 2) = 0$$
$$5(x + 2)(x - 1) = 0$$

There are three factors whose product is zero. To find the mathematical solutions we set each factor equal to zero.

$5 = 0$	$x + 2 = 0$	$x - 1 = 0$
No solution	$x = -2$	$x = 1$

The two mathematical solutions are $x = -2$ and $x = 1$.

Exercise Set 9-6

Solve each of the following quadratics by factoring.

Part A

1. $x^2 + 3x - 10 = 0$
2. $x^2 + 5x + 6 = 0$
3. $2x^2 + 2x - 4 = 0$
4. $x^2 + 2x + 1 = 0$
5. $x^2 + 8x + 7 = 0$

Part B

6. $4x^2 - 9 = 0$
7. $5x^2 - 9x - 2 = 0$
8. $2x^2 + 5x - 12 = 0$

9. $2x^2 - x - 15 = 0$
10. $4x^2 + 12x + 9 = 0$

Part C

11. $2x^2 - 10x + 12 = 0$
12. $6x^2 - 24x = -18$
13. $3x^2 = -9x + 30$
14. $8x^2 + 4x = 0$
15. $3x^2 + 63 = 30x$
16. The cross-sectional area of the angle iron in Fig. 9-1 is 15 in.[2] If the dimensions are 7 in. and 9 in., find the value of x in this exercise.

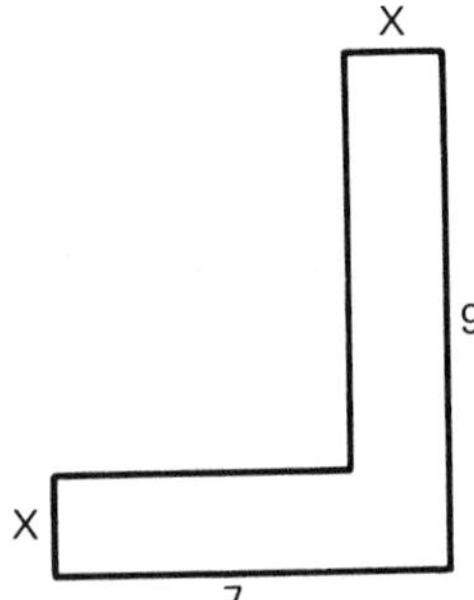

Figure 9-1

Answers

2. $-3, -2$
4. $-1, -1$
6. $3/2, -3/2$
8. $-4, 3/2$
10. $-3/2, -3/2$
12. $1, 3$
14. $0, -1/2$
16. 1"

9-7 SOLVING QUADRATIC EQUATIONS USING THE QUADRATIC FORMULA

Now let us consider the general form of the quadratic and relate this to the quadratic formula. The *standard form of the quadratic* is

$$ax^2 + bx + c = 0$$

where **a, b,** and **c** are real numbers and **a** is not zero. The solutions of the standard form of the quadratic are given by the *quadratic formula*, which says

$$x = \frac{-b \pm \sqrt{b^2 - 4ac}}{2a}$$

where a, b, and c are taken from the standard form of the quadratic.

Example 9-13: $2x^2 + 3x - 9 = 0$. The a, b, and c from the standard form of the quadratic are identified in this example by

$$a = 2 \qquad b = 3 \qquad c = -9$$

Now replace a, b, and c in the formula

$$x = \frac{-b \pm \sqrt{b^2 - 4ac}}{2a}$$

$$x = \frac{-(3) \pm \sqrt{(3)^2 - 4(2)(-9)}}{2(2)}$$

$$x = \frac{-3 \pm \sqrt{9 + 72}}{4}$$

$$x = \frac{-3 \pm \sqrt{81}}{4}$$

$$x = \frac{-3 \pm 9}{4}$$

Now we must make use of the fact that 9 is to be added to obtain one solution and subtracted to obtain the other solution.

$$x = \frac{-3 + 9}{4} \qquad x = \frac{-3 - 9}{4}$$

$$x = \frac{6}{4} \qquad x = \frac{-12}{4}$$

$$x = \frac{3}{2} \text{ or } 1\frac{1}{2} \qquad x = -3$$

The two solutions are 1 1/2 and −3.

The quadratic formula works only if the quadratic is in standard form.

Example 9-14: $3x^2 + 5x = 12$. We must write $3x^2 + 5x = 12$ in standard form.

$$3x^2 + 5x = 12$$

$$3x^2 + 5x - 12 = 0$$

Now proceed as you did in Example 9-13.

$$a = 3 \qquad b = 5 \qquad c = -12$$

$$x = \frac{-b \pm \sqrt{b^2 - 4ac}}{2a}$$

$$x = \frac{-(5) \pm \sqrt{(5)^2 - 4(3)(-12)}}{2(3)}$$

$$x = \frac{-5 \pm \sqrt{25 + 144}}{6}$$

$$x = \frac{-5 \pm \sqrt{169}}{6}$$

$$x = \frac{-5 \pm 13}{6}$$

$$x = \frac{-5 + 13}{6} \qquad x = \frac{-5 - 13}{6}$$

$$x = \frac{8}{6} \qquad x = \frac{-18}{6}$$

$$x = \frac{4}{3} \text{ or } 1\frac{1}{3} \qquad x = -3$$

Therefore, the solutions are -3 and $4/3$.

Exercise Set 9-7

Use the quadratic formula to find the solutions for each of the following.

Part A

1. $3x^2 + 1x - 2 = 0$
2. $2x^2 + 5x - 3 = 0$
3. $x^2 + 3x - 4 = 0$
4. $2x^2 - x - 6 = 0$
5. $2x^2 - 11x + 5 = 0$

Part B

6. $2x^2 + 5x = 3$
7. $2x^2 + x = 6$
8. $2x^2 = 3x + 9$
9. $2x^2 + 5 = 11x$
10. $x^2 = 7x + 8$

Part C

11. $x^2 + 4x - 1 = 0$
12. $2x^2 + 3x - 3 = 0$
13. $x^2 + 9x + 5 = 0$
14. $3x^2 + 5x = 1$
15. A rectangle is 8 ft longer than it is wide and its area is 273 ft.[2] How long is the rectangle?
16. Let the approximate distance (h) above the ground of a free-falling object be given by the following formula: $h = -16t^2 + v_0t + c$, where v_0 is the initial velocity (speed) and c is the initial height. Find the time (t) when a rock will be 40 ft high if it was thrown upward with an initial speed of 128 ft per sec from the top of a tower that is 360 ft high.

Answers

2. $-3, 1/2$
4. $-3/2, 2$
6. $1/2, -3$
8. $-3/2, 3$
10. $8, -1$
12. $(-3 \pm \sqrt{33})/4$
14. $(-5 \pm \sqrt{37})/6$
16. 10 sec

9-8 SYSTEMS OF LINEAR EQUATIONS

In this section we shall use some of the algebraic methods of solving simultaneous equations. In an equation such as $x - 2y = 6$ there are two variables, x and y. If $x = 4$, then it can be shown that $y = -1$. The values $x = 4$ and $y = -1$ may be represented as an ordered pair solution $(4, -1)$ where the first member is the value of x and the second member is the value of y.

> *Illustration 1:* Find the solution of the equation $x - 2y = 6$ when one variable's value is known. If $y = 3$, find x.

$$x - 2y = 6$$
$$x - 2(3) = 6$$
$$x - 6 = 6$$
$$x = 12$$

Therefore, the solution becomes the ordered pair (12, 3).

Illustration 2: If $x = 8$, find y in the equation $x - 2y = 6$.

$$x - 2y = 6$$
$$(8) - 2y = 6$$
$$-2y = -2$$
$$y = 1$$

Therefore, the solution becomes the ordered pair (8, 1).

> If in an equation like $x - 2y = 6$ one value is chosen, then the second value can be determined by solving the equation. The ordered pair (x, y) is called a solution set for the equation. For example, (12, 3) is a solution set for the above equation (see Illustration 1).

If the equations $3x - 2y = 6$ and $x + 2y = 10$ have the same solution set, then this solution set is called a *common solution*. When a solution set exists for both equations at the same time it is usually referred to as a *simultnaeous solution* set (see Illustration 3).

Illustration 3: Check the ordered pair (4, 3) in each of the above equations to see if (4, 3) is a solution for both.

$$3x - 2y = 6$$
$$3(4) - 2(3) \stackrel{?}{=} 6$$
$$12 - 6 \stackrel{?}{=} 6$$
$$6 \stackrel{\checkmark}{=} 6$$

(4, 3) is a solution that checks in the above equation.

$$x + 2y = 10$$
$$(4) + 2(3) \stackrel{?}{=} 10$$
$$4 + 6 \stackrel{?}{=} 10$$
$$10 \stackrel{\checkmark}{=} 10$$

(4, 3) is a solution that checks in the above equation.

Therefore, (4, 3) is a common solution for the system of simultaneous equations.

> In order to solve a system of simultaneous equations by algebraic method, we need to find some way to combine the two equations so that one equation with one unknown is left. This process is often referred to as the elimination method. After one of the variables is found by the elimination method, then the other variable may be found by substitution (see Illustration 1).

In the following examples, we shall illustrate the elimination method of solving a system of simultaneous equations.

Example 9-15: Solve for x and y.

$$\begin{cases} x + 2y = 7 \\ 3x - 2y = 5 \end{cases}$$

First we perform elimination by addition.

Step 1. Copy problem: $\begin{cases} x + 2y = 7 \\ 3x - 2y = 5 \end{cases}$

Note:

$$x + 3x = 4x$$

$$2y - 2y = 0$$

$$7 + 5 = 12$$

Step 2. $4x \quad = 12$

Step 3. $\dfrac{4x}{4} = \dfrac{12}{4}$

Step 4. $x = 3$

Now to find the value of y, substitute the value of x obtained by elimination into either of the original equations.

Step 5. Copy the first equation: $x + 2y = 7$

Step 6. Substitute x = 3: $(3) + 2y = 7$

Step 7. Subtract 3 from both sides: $2y = 4$

Step 8. Divide by 2: $\frac{2y}{2} = \frac{4}{2}$

Step 9. $y = 2$

Therefore, the common solution is (3, 2).

A check of (3, 2) can be made by replacing x by 3 and y by 2 in each of the equations.

$$x + 2y = 7$$
$$(3) + 2(2) \stackrel{?}{=} 7$$
$$3 + 4 \stackrel{?}{=} 7$$
$$7 \stackrel{\checkmark}{=} 7$$

Therefore, (3, 2) is a solution of the above equation.

$$3x - 2y = 5$$
$$3(3) - 2(2) \stackrel{?}{=} 5$$
$$9 - 4 \stackrel{?}{=} 5$$
$$5 \stackrel{\checkmark}{=} 5$$

Therefore, (3, 2) is a solution of the above equation.

Since (3, 2) checked in both of the equations, it is the common solution.

> *Note:* If all members of an equation are multiplied or divided by the same non-zero number, the solution of the equation remains unchanged.

Illustration 4: Show that the solution (3, 2) will check in the original equation, $2x - y = 4$, and the new equation after multiplying by 5.

Part A: Check (3, 2) in the original.

$$2x - y = 4$$
$$2(3) - (2) \stackrel{?}{=} 4$$
$$6 - 2 \stackrel{?}{=} 4$$
$$4 \stackrel{\checkmark}{=} 4$$

(3, 2) is a solution of the original.

Part B: Check (3, 2) in the equation multiplied by 5.

$$10x - 5y = 20$$
$$10(3) - 5(2) \stackrel{?}{=} 20$$
$$30 - 10 \stackrel{?}{=} 20$$
$$20 \stackrel{\checkmark}{=} 20$$

Therefore, (3, 2) is a solution of the original and the original multiplied by 5.

Illustration 5: If we know that (2, 3) is the common solution of the system

$$\begin{cases} 3x + 2y = 12 \\ 2x - 3y = -5 \end{cases}$$

then it can be shown that it is a common solution for the system

$$\begin{cases} 9x + 6y = 36 \\ 4x - 6y = -10 \end{cases}$$

where each equation has been multiplied by a nonzero number. One way to show that (2, 3) is a common solution for the system

$$\begin{cases} 9x + 6y = 36 \\ 4x - 6y = -10 \end{cases}$$

is to solve the system by elimination as in Example 9-15.

Step 1. Copy problem: $\begin{cases} 9x + 6y = 36 \\ 4x - 6y = -10 \end{cases}$

Step 2. By addition: $13x = 26$

Step 3. Divide by 13: $x = 2$

Step 4. Substitute $x = 2$: $9x + 6y = 36$
$9(2) + 6y = 36$

Step 5. Multiply: $18 + 6y = 36$

Step 6. Subtract 18: $6y = 18$

Step 7. Divide by 6: $y = 3$

Therefore, (2, 3) is a common solution for the system

$$\begin{cases} 9x + 6y = 36 \\ 4x - 6y = -10 \end{cases}$$

and also for the original system

$$\begin{cases} 3x + 2y = 12 \\ 2x - 3y = -5 \end{cases}$$

> *Steps to Solve a Linear System by Addition or Subtraction*
>
> Step 1. If necessary multiply one or both equations by a non-zero number, or by a set of non-zero numbers that will make one of the unknowns either add out or subtract out.
>
> Step 2. Add or subtract depending on whether the signs are different or the same.
>
> Step 3. Solve the equation for the value of the unknown that remains.
>
> Step 4. Substitute the result that you get in Step 3 for that unknown in one of the original equations.
>
> Step 5. Solve the equation that you substituted the value into for the other unknown.
>
> Step 6. The results that you have are the values for the common solution for the system of equations.

Example 9-16: Find the common solution for

$$\begin{cases} 3x - y = 3 \\ 2x + 3y = 2 \end{cases}$$

Step 1. Multiply $3x - y = 3$ by 3: $9x - 3y = 9$

Step 2. Add the second equation: $2x + 3y = 2$

$$11x = 11$$

Step 3. Divide by 11 to solve for x: $x = 1$

Step 4. Substitute $x = 1$ into the first equation in its original form: $3(1) - y = 3$

Step 5. Solve for y:

$$3 - y = 3$$
$$-1y = 0$$
$$y = 0$$

Step 6. Therefore, (1, 0) is the common solution.

Example 9-17: Find the common solution for

$$\begin{cases} -5x + 2y = 7 \\ 2x + 3y = 1 \end{cases}$$

Step 1. Multiply $-5x + 2y = 7$ by 2: $-10x + 4y = 14$
Multiply $2x + 3y = 1$ by 5: $10x + 15y = 5$

Step 2. Add the new equations: $19y = 19$

Step 3. Divide by 19 to solve for y: $y = 1$

Step 4. Substitute y = 1 into the first equation in its original form: $-5x + 2(1) = 7$

Step 5. Solve for x:

$$-5x + 2 = 7$$
$$-5x = 5$$
$$x = -1$$

Step 6. Therefore, (−1, 1) is the common solution.

Exercise Set 9-8

Solve each of the following linear systems for the common solution by addition or subtraction.

Part A

1. $\begin{cases} 2x + y = 2 \\ x + y = 2 \end{cases}$
2. $\begin{cases} -3x + y = 9 \\ x + y = 1 \end{cases}$
3. $\begin{cases} 2x - y = -3 \\ -x + y = -1 \end{cases}$
4. $\begin{cases} -x + 5y = 8 \\ x + 2y = 13 \end{cases}$
5. $\begin{cases} 3x - y = -1 \\ -x + y = -1 \end{cases}$
6. $\begin{cases} x - 3y = -3 \\ 2x + 3y = 12 \end{cases}$
7. $\begin{cases} 2x + y = 6 \\ -12x + y = -8 \end{cases}$

Part B

8. $\begin{cases} 2x + 3y = 9 \\ 5x - 2y = 13 \end{cases}$
9. $\begin{cases} -2x + y = -7 \\ x - 18y = 21 \end{cases}$
10. $\begin{cases} y - x = 8 \\ 5y + 2x = -2 \end{cases}$
11. $\begin{cases} 2x - y = 5 \\ 3x + 4y = 2 \end{cases}$
12. $\begin{cases} x + 6y = 18 \\ -x + 5y = 4 \end{cases}$
13. $\begin{cases} 2x - 3y = 13 \\ 3x + 2y = 13 \end{cases}$

Part C

14. $\begin{cases} 3x - 5y = -9 \\ 9x + 4y = 11 \end{cases}$
15. $\begin{cases} 5x - 3y = 12 \\ 2x + 2y = 14 \end{cases}$
16. $\begin{cases} 3x + 2y = 14 \\ 2x + 3y = 11 \end{cases}$

Answers

2. (−2, 3)
4. (7, 3)
6. (3, 2)
8. (3, 1)
10. (−6, 2)
12. (6, 2)
14. (1/3, 2)
16. (4, 1)

9-9 CONSISTENT, INCONSISTENT, AND DEPENDENT SYSTEMS

In the previous sections, we solved equations that had a single common solution. A system of equations with a single solution is called *consistent* equations. If a system has no common solution, then the system is known to be *inconsistent*. There is no ordered pair (x, y) which is a solution for both equations. A system of equations that has all solutions in common is called *dependent* equations. The following examples are selected to illustrate the three different types of systems of linear equations and to demonstrate what probably will take place in each case as you try to solve each system for the *common solution* (if it exists) by elimination (addition or subtraction).

Example 9-18

$$\begin{cases} 3x + 2y = 14 \\ 2x + 3y = 11 \end{cases}$$

$$\begin{cases} 6x + 4y = 28 \\ -6x - 9y = -33 \end{cases}$$

$$0x - 5y = -5$$

$$\begin{aligned} -5y &= -5 \\ y &= 1 \\ 3x + 2(1) &= 14 \\ 3x + 2 &= 14 \\ 3x &= 12 \\ x &= 4 \end{aligned}$$

The ordered pair (4, 1) is the common solution. Therefore, the system is *consistent*.

Example 9-19

$$\begin{cases} 4x - 2y = 7 \\ 2x - y = 9 \end{cases}$$

$$\begin{cases} 4x - 2y = 7 \\ -4x + 2y = -18 \end{cases}$$

$$0x + 0y = -11$$

Note: Both of the variables x and y were eliminated at the same time, and the nonvariable side (constant term) is not zero. There are no values of x and y which could ever make 0x + 0y equal to -11. Therefore, the system is *inconsistent*.

Example 9-20

$$\begin{cases} 4x - 2y = 6 \\ 2x - y = 3 \end{cases}$$

$$\begin{cases} 4x - 2y = 6 \\ -4x + 2y = -6 \end{cases}$$

$$0x + 0y = 0$$

Note: Both of the variables x and y were eliminated and the nonvariable side (constant term) is zero. Any values of x and y would make 0x + 0y equal to 0. Therefore, the system is *dependent*.

Example 9-21: Find the common solution, if it exists, for

$$\begin{aligned} 10x - 6y &= 8 \\ -5x + 3y &= -6 \end{aligned}$$

Step 1. Multiply second equation by 2: $-10x + 6y = -12$

Step 2. Add the first equation: $10x + 6y = 8$

$$0x + 0y = -4$$

Step 3. Solve: 0( ) + 0( ) = -4

(second blank, for y): Any real number replacement for y will give a product of zero.

(first blank, for x): Any real number replacement for x will give a product of zero.

Hence, $0 + 0 \neq -4$. Therefore, no ordered pair will satisfy as a solution, and this system is *inconsistent*.

Example 9-22: Find the common solution, if it exists, for

$$\begin{cases} 10x - 6y = 8 \\ -5x + 3y = -4 \end{cases}$$

Step 1. Multiply second equation by 2: $-10x + 6y = -8$

Step 2. Add the first equation: $10x - 6y = 8$

$$0x + 0y = 0$$

Step 3. Solve: $0(\ ) + 0(\ ) = 0$

- Any real number replacement for y will give a product of zero.
- Any real number replacement for x will give a product of zero.

Hence, $0 + 0 = 0$, and this system is *dependent*.

In a dependent system of equations, any ordered pair of real numbers that are a solution for one equation will be a solution of the other equation (see Illustration 6).

Illustration 6: If $x = 2$ in the first equation, then

$$10x - 6y = 8$$
$$10(2) - 6y = 8$$
$$20 - 6y = 8$$
$$-6y = -12$$
$$y = 2$$

The common solution is (2, 2). If $x = 2$ in the second equation, then

$$-5x + 3y = -4$$
$$-5(2) + 3y = -4$$
$$-10 + 3y = -4$$
$$3y = 6$$
$$y = 2$$

The common solution is (2, 2). Therefore, the ordered pair (2, 2) was a solution of the first equation, and since the system is dependent it will be a solution of the second equation. There are other ordered pairs which are common solutions. For example, you can check that the ordered pairs (-1, -3) and (5, 7) are also common solutions to this dependent system.

> *Summary of the Three Types of Systems of Equations*
> In the process of solving for a solution by elimination, one of three possibilities will develop.
>
> *Consistent System:* If one and only one variable is eliminated, then the system has a single common solution.
>
> *Inconsistent System:* If both of the variables are eliminated and the non-variable (constant) term is not eliminated, then the system has no common solution.
>
> *Dependent System:* If both of the variables and the non-variable (constant) term are eliminated, then the system has all solutions in common. (One of the many solutions may be obtained by selecting a value for one variable and solving either of the original equations for the other variable.)

Exercise Set 9-9

Classify and solve the following linear systems when a single common solution exists.

1. $\begin{cases} 5x - y = 20 \\ 2x + y = 1 \end{cases}$

2. $\begin{cases} 5x - 5y = 13 \\ x - y = 2 \end{cases}$

3. $\begin{cases} 3x - 3y = 9 \\ x - y = 3 \end{cases}$

4. $\begin{cases} x + 2y = 8 \\ 3x + 6y = 27 \end{cases}$

5. $\begin{cases} 3x - 2y = 10 \\ 5x + 6y = -2 \end{cases}$

6. $\begin{cases} 3x - 2y = -2 \\ 6x - 4y = -4 \end{cases}$

7. $\begin{cases} 3x - 2y = 11 \\ 3x + 2y = 19 \end{cases}$

8. $\begin{cases} 2x + 3y = -1 \\ 3x + 5y = -2 \end{cases}$

9. $\begin{cases} 3x - y = 5 \\ -9x + 3y = -15 \end{cases}$

10. $\begin{cases} 3x + 2y = -9 \\ x + 3y = 25 \end{cases}$

11. $\begin{cases} -11x - 21y = -3 \\ 5x + 3y = 3 \end{cases}$

12. $\begin{cases} 2x - 3y = -1 \\ 3x + 4y = 24 \end{cases}$

13. $\begin{cases} 3x + 2y = 0 \\ 2x - 5y = 19 \end{cases}$

14. $\begin{cases} 6x + 5y = -2 \\ 10x + 7y = -4 \end{cases}$

Answers

2. Inconsistent; { }
4. Inconsistent; { }
6. Dependent; {(0, 1), (-2, -2), . . .}
8. Consistent; {(1, -1)}
10. Consistent; {(-11, 12)}
12. Consistent; {(4, 3)}
14. Consistent; {(-3/4, 1/2)}

Chapter 10

Percents

No one living in the United States fails to have his or her life affected by percents. We hear that inflation is increasing by 9%. Employees at a certain industrial plant pay 33.1% of their salaries for F.I.C.A. (social security), state tax, and federal income tax. The local department store is having a 20% reduction sale. Your instructor tells you that only 10% of the class will receive A's. You get 8% higher octane by buying a certain brand of gas. The bank tells you that you may secure a loan at 12 3/4% annual interest. The examples are endless. Just read any newspaper any night and you are guaranteed to find percents lurking within its pages.

10-1 THE MEANING OF PERCENT

Before we actually do any work, let's look at the word *percent*. Webster's dictionary tells us the *per* means *for each* or *for every* and *cent* means *a hundred*. Combining these two facts, we reach the conclusion that percent (this may be written *percent* or *per cent*) means *for every hundred* or *for each hundred*. Thus 25% means 25 for every 100 or 25 out of every 100 or 25 parts per 100 parts.

When comparing two numbers using percents, we must be careful about the order of the two numbers. One of them will be the *base* figure, and it is the one to which the comparison will be made. If we are interested in comparing the number of A's in a class to the total number of students, we would take the total number of students as the base figure and set up a fraction using it as the denominator. This would look like

$$\frac{\text{number of A's}}{\text{total number of students}}$$

Finally, we would convert this fraction to a percent as explained in the following sections.

As a general rule, in *the ratio of* A *to* B, B is the base number. If there is no indication of which number should be the base, these two guidelines might be applied:

> *Guidelines for Choosing Base Number*
>
> Guideline 1. If the two numbers occurred at different times, take the first-time number as the base.
>
> Guideline 2. If one number represents a portion of the objects represented by the other, take the number reflecting the total as the base.

For example, Guideline 1 would give

$$\frac{\text{college graduates in 1875}}{\text{college graduates in 1835}}$$

because 1835 is the first time mentioned.

For example, Guideline 2 would give

$$\frac{\text{women mechanics}}{\text{all mechanics}}$$

because women mechanics are a portion of all mechanics.

10-2 CHANGING FRACTIONS AND DECIMALS TO PERCENTS

In the first three chapters, you reviewed work in arithmetic and the use of fractions and decimals. One of the most important skills that we need when working with percents is the ability to write fractions and decimals as percents.

Let's look at some simple examples that you already know. Consider the I beam pictured in Fig. 10-1. Some would say that 1/2 of the beam has been painted, while others might say that 50% of the beam has been painted. This discussion suggests that

$$50\% = \frac{1}{2}$$

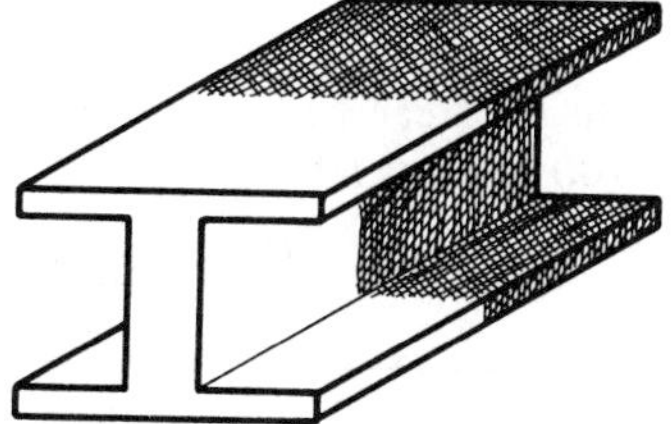

Figure 10-1. I beam.

According to the discussion in Sec. 10-1, 1% should mean 1 part for every 100 parts. The percent symbol % may be thought of as 1/100. For example,

$$27\% = 27 \times \frac{1}{100} = \frac{27}{100}$$

The equality between 50% and 1/2 may then be written

$$50\% = 50 \times \frac{1}{100} = \frac{50}{100} = \frac{\not{2} \cdot {}^{1}\not{5} \cdot \not{5}}{2 \cdot \not{2} \cdot \not{5} \cdot \not{5}} = \frac{1}{2}$$

Now suppose we want to change 7/20 to a percent. Since percent means parts per hundred, we write

$$\frac{7}{20} = \frac{\#}{100}$$

Since 20 times 5 is 100, we have

$$\frac{7}{20} = \frac{7 \cdot 5}{20 \cdot 5} = \frac{35}{100} = 35 \times \frac{1}{100} = 35\%$$

To summarize what we have done, suppose we have any fraction and we want to write it as a percent. Then we use the following rule.

> Rule. To change a given fraction to a percent, we write an equivalent fraction whose denominator is 100 and then write its numerator using our percent sign.

Fraction	Equivalent Fraction (denominator 100)	Percent
$\frac{1}{5}$	$\frac{1 \times 20}{5 \times 20} = \frac{20}{100}$	20%
$\frac{1}{8}$	$\frac{1 \times 12.5}{8 \times 12.5} = \frac{12.5}{100}$	12.5%
(Note that the 12.5 can be found by solving 1/8 = x/100. Thus, 8x = 100 and x = 100/8 = 12.5.)		
$\frac{2}{3}$	$\frac{2 \times 33\ 1/3}{3 \times 33\ 1/3} = \frac{66\ 2/3}{100}$	66 2/3 %

Changing a fraction to a percent seems easy, but as we shall soon find out, changing a decimal to a percent is even easier. When we have seen how to do this, we shall also learn another way to change a fraction to a percent.

We all know that 0.5 is the same as 50% because 0.5 = 1/2 = 50%. As with the fractions, we only need to write the decimal number as a fraction whose denominator is 100.

Decimal	Fraction	Equivalent Fraction (denominator 100)	Percent
0.3	$\frac{3}{10}$	$\frac{3 \times 10}{10 \times 10} = \frac{30}{100}$	30%
0.02	$\frac{2}{100}$	$\frac{2}{100}$	2%
2.13	$\frac{213}{100}$	$\frac{213}{100}$	213%
0.0025	$\frac{25}{10{,}000}$	$\frac{25 \div 100}{10{,}000 \div 100} = \frac{0.25}{100}$	0.25%
0.59	$\frac{59}{100}$	$\frac{59}{100}$	59%

Summarizing these results, we have

$$0.3 = 30\%$$
$$0.02 = 2\%$$
$$2.13 = 213\%$$
$$0.0025 = 0.25\%$$
$$0.59 = 59\%$$

> Rule. Write the fraction in decimal form, move the decimal point two places to the right, and insert the percent sign.

For example,

$$\frac{17}{40} = 0.425 = 42.5\%$$

Look at the following table. If you understand what was done going from one column to the next, then you are ready to work the problems. If not, then reread this section.

Fraction	Decimal	Percent
$\frac{2}{25}$	0.08	8%
$\frac{2}{9}$	$0.2\overline{2}$	$22.\overline{2}\%$, which is $22\frac{2}{9}\%$
$\frac{103}{100}$	1.03	103%
$\frac{25}{16}$	1.5625	156.25%

Exercise Set 10-2

Part A

1. Express each of the following fractions as percents.
   (a) 1/2 (b) 3/10 (c) 12/50 (d) 2/5 (e) 2/3
   (f) 7/25 (g) 9/75 (h) 11/11 (i) 5/2 (j) 30/150
   (k) 3/8 (l) 1/7
2. Write each of the following decimals as a percent.
   (a) 0.13 (b) 0.03 (c) 0.57 (d) $0.3\overline{3}$ (e) 0.27
   (f) 2.13 (g) 3.002 (h) 0.12 (i) 0.09 (j) 0.49
   (k) 0.85 (l) 1.0
3. If 9/25 of a person's wages are used to pay social security, taxes, and union dues, what percent of the person's pay goes for these items?
4. What percent of the square is shaded in Fig. 10-2?
5. In a shipment of 140 wheel rims, 7 were found to have pinholes and therefore are not suitable for tubeless tires. What percent of the rims were defective?

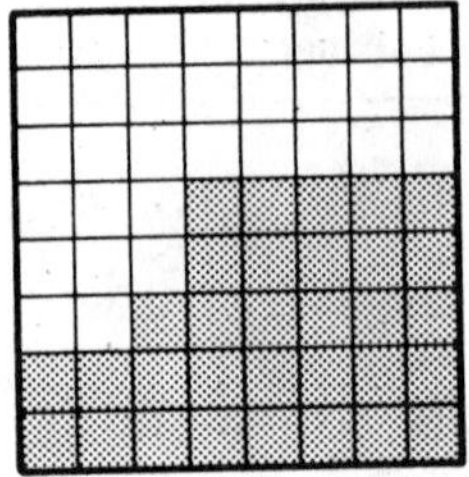

Figure 10-2

Part B

6. Express each of the following numbers as a percent.

(a) 1/7	(b) $0.6\overline{6}$	(c) 2 1/5	(d) 110/121
(e) 0.00023	(f) 1/8	(g) 1/16	(h) 0.75
(i) 1/4	(j) 0.909		

7. A study found that one out of every six cars sold during January, 1956, had a factory defect. What percentage of cars sold during this period had no defects?

8. A woman earning $4.95 per hour had her pay increased to $5.27 per hour. By what percent was her pay raised?

9. What percent of 12.75 is 1/4?

10. The tires used by some of the major bus companies have an average tread life of 60,000 mi. A bus equipped with these tires has traveled 45,000 mi. What percent of tread wear should its tires have left?

11. A new car cost $6,238.52. After 1 year it depreciated in value to $5,112.18. How much did it depreciate and what percent of its original value was it worth after 1 year?

12. An eight-cylinder engine has a single piston displacement of 44.625 in.^{3} If this engine is rebored to give it a total displacement of 380 in.^{3}, what will be the percent of increase in engine size?

13. A certain paint company supplies contractors with paint in 5-gal buckets. One contractor purchased 14 such buckets. He then painted four houses, averaging 12 gal per house. What percent of the paint did he have left? How many more houses could he paint if he maintained the same average number of gallons per house?

14. What percent of 1 mi is 1 km?

15. A machinist has a piece of sheet metal 8 ft by 3 ft. From this piece of sheet metal she stamps out 132 parts, each having an area of 24 in.^{2} What percent of metal is not used?

16. If a 1-in. margin should be left on the bottom and sides of an 8 in. by 11 in. page and a 2-in. margin should be left on the top, what percent of the page area is available for printed matter?

Answers

2.	(a) 13.0%	(b) 3.0%	(c) 57.0%	(d) 33.3%	(e) 27.0%
	(f) 213.0%	(g) 300.2%	(h) 12.0%	(i) 9.0%	(j) 49.0%
	(k) 85.0%	(l) 100.0%			
4.	50.0%				
6.	(a) 14.3%	(b) 66.7%	(c) 220.0%	(d) 90.9%	(e) 0.023%
	(f) 12.5%	(g) 6.3%	(h) 75.0%	(i) 25.0%	(j) 90.9%

8. 6.47%
10. 25%
12. 6.4%
14. 62.14%
16. 54.5%

10-3 CHANGING PERCENTS TO FRACTIONS AND DECIMALS

We have learned how to change fractions and decimals to percents, but probably more important for everyday use is the process of changing a percent to a fraction or a decimal.

We use percent (%) to give people a clearer understanding. Thus, rather than saying 30¢ off on a $1.00 toy, 75¢ off on a $2.50 hex wrench, $14.99 off on a $49.95 tool set, and so forth, we say 30% price reduction on all merchandise in the store. In order to find the exact amount off on any of the articles listed above, we should first change 30% to an equivalent fraction (or decimal) and multiply this by the original price.

> *Before doing any arithmetic involving a percent, the term should be changed to an equivalent fraction (or decimal).*

Making use of this statement is a very easy task. We simply reverse the process of changing a fraction or decimal to a percent.

> Percent to Fraction: To change a percent to a fraction, place the number over **100** as the denominator, delete the percent symbol, and reduce the fraction to lowest terms.
>
> Percent to Decimal: To change a percent to a decimal, move the decimal point two positions to the **left** and delete the percent symbol.

Example 10-1: Change 27% to a fraction.

$$27\% = 27 \times \frac{1}{100} = \frac{27}{100}$$

Example 10-2: Change 27% to a decimal number.

$$27\% = 0.27$$

Example 10-3: Change 4.2% to a fraction and a decimal.

$$4.2\% = 4.2 \times \frac{1}{100} = \frac{4.2}{100} = \frac{42}{1{,}000} = \frac{21}{500}$$

$$4.2\% = (4.2)(0.01) = 0.042 = \frac{42}{1{,}000} = \frac{21}{500}$$

$$4.2\% = 04.2\% = 0.042$$

$$4.2\% = (4.2)(0.01) = 0.042$$

$$4.2\% = 4\frac{1}{5}\% = \frac{21}{5}\% = \frac{21}{5} \times \frac{1}{100} = \frac{21}{500}$$

This example should show you that there may be several ways to solve a problem involving a percent.

Exercise Set 10-3

Part A

1. Change each of the percents to their equivalent decimal and fractional forms.
   (a) 25% (b) 32% (c) 18% (d) 105% (e) 23.2%
   (f) 1/2% (g) 0.7% (h) 2.5% (i) 0.25% (j) 1.05%
2. If 60% of all automatic transmissions in cars fail within the first 5 years, what fraction of these transmissions fail within this time limit?
3. On a certain test for which David had not studied, he made a 35%. If the test had 20 questions, how many did he do correctly?
4. Write 37 1/2% as a fraction. Is this fraction greater than or less than 4/9?
5. Change 100% to a fraction. What does 100% mean?

Part B

6. Explain why moving the decimal point two places to the right and adding the percent sign works when changing from a decimal to a percent.
7. If 25% of the nails a carpenter uses are 10-penny nails, 30% are 12-penny nails, and the rest are other weights and sizes, what fraction of each kind does she use?
8. Mr. Hairston and Mr. Lester, the two owners of a machine shop, took 25% and 37 1/2% of the profits, respectively, as personal income. The remaining profit was devoted to purchasing new equipment. What fraction of the profits did the two owners together take as income? What fraction was used for modernizing?
9. If 8.8% was changed to a fraction whose denominator was 250, what would be this fraction's numerator?
10. What is the denominator of the fraction, reduced to lowest terms, that is equivalent to $0.\overline{66}\%$.

Part C

11. Why do you think that we have to change a percent to a decimal or fraction before we can "mathematically" use it?
12. A river which flooded reached a height which was 208% of its normal height. What fraction above twice its normal height did the river reach?
13. If a printer chooses a page size where the width is 60% of the length, what fraction expresses the length in terms of the width?

Answers

2. 3/5
4. 3/8; less than
8. 5/8; 3/8
10. 3
12. 2/25

10-4 USE OF PERCENTS

We use percents in all phases of our lives, but basically there are only three types of uses. If we master these three types, then we should be able to handle any percent problems readily. In this section you will find a general statement and an example of each of the three types of problems.

Type I: You want to find a percent of a given quantity.

Example 10-4: The ad in Fig. 10-3 tells us that shirts are on sale for 20% below their usual selling price. If you want to buy one of these shirts, how much will it cost?

Figure 10-3

First we find the amount of the discount (20% of \$12.98). Second, we round this off to the nearest penny and subtract this amount from the usual selling price (\$12.98). That will be our answer. Here is the work:

20% of \$12.98 = 0.2 × \$12.98 (Note that we changed 20% to a decimal and replaced the word *of* by the multiplication sign.)

= \$2.596

≐ \$2.60 (Note that ≐ means *approximately equal to*.)

The sale price of the shirt is

\$12.98 − \$2.60 = \$10.38

selling price − discount = sale price

In most locations, we must also add a sales tax to this price, and this is again a percent problem. Suppose our sales tax is 4%. We find the actual price of the shirt by computing the tax and adding it to the sale price.

4% of \$10.38 = 0.04 × \$10.38

= \$0.4152

= \$0.42

The actual amount you pay is

\$10.38 + \$0.42 = \$10.80

Type II: You want to find what percent one quantity is of another quantity.

Example 10-5: A family buys a used car with a list price of \$2,114.75. The sales representative tells them that if they buy today they can buy it for \$1,950.00. What percent discount would they be getting?

First we find the actual amount saved by subtracting the sales representative's price from the list price. Second, we determine what percent this is of the list price. This percent is the percent discount. Here is the work:

\$2,114.75 − \$1,950.00 = \$164.75

$$\frac{\$164.75}{\$2{,}114.75} = \frac{16{,}475}{211{,}475} \quad \left(\text{multiplied by } \frac{100}{100} = 1\right)$$

$$= \frac{659}{8{,}459} \quad \text{(reducing)}$$

$$\doteq 0.0779 \quad \text{(rounding off)}$$

$$= 7.79\% \text{ discount}$$

Type III: You want to find a total quantity, given part of it, and the percent that the part represents.

Example 10-6: A tire and rubber company sold 50 aircraft tires to the U.S. Navy on a single day. If this represents 10% of their daily production of aircraft tires, how many tires do they produce daily?

We want to find a number such that 50 is 10% of that number. First, we change 10% to its equivalent decimal (or fraction); then we divide 50 by this result, and we have our answer.

Here is the work: Let T be the daily production of tires.

$$10\% \text{ of } T = 50$$

$$0.10 \times T = 50$$

$$T = 50 \div 0.10$$

$$T = 500 \text{ aircraft tires per day}$$

Let's look at additional examples, making use of simple equations.

Example 10-7: What number is 32% of 128? Let n be the desired number. Then the English sentence can be translated into mathematical form, giving us

$$n = 32\% \text{ of } 128$$

Now, we continue with

$$n = 0.32 \times 128$$

$$n = 40.96$$

Example 10-8: What percent of 45 is 15? Translating into mathematical form and letting x represent the desired percent, we find

$$x\% \text{ of } 45 = 15$$

$$x \cdot \frac{1}{100} \cdot 45 = 15$$

$$x \cdot \frac{45}{100} = 15$$

$$x = 15 \cdot \frac{100}{45}$$

$$x = \frac{1{,}500}{45}$$

$$x = 33\ 1/3$$

So 15 is 33 1/3% of 45.

Example 10-9: Twenty-seven is 15% of what number? Letting n represent the desired number and translating, we find

$$27 = 15\% \text{ of } n$$

$$27 = 0.15n$$

$$27 \div 0.15 = n$$

$$180 = n$$

Exercise Set 10-4

Part A

1. Find the missing quantity in each of the following.
   (a) 16% of 32 is what number?
   (b) What number is 3.5% of 240?
   (c) 23.1% of $182.00 is how much?
   (d) What percent of 100 is 16?
   (e) 12 is what percent of 48?
   (f) What percent of 200 is 2.5?
   (g) 30 is 19% of what number?
   (h) 20% of what number is 120?
   (i) $1.50 is 15% of how much?
2. A car getting 20 mi per gallon was tuned. The tune-up increased the gas mileage by approximately 20%. What was the new gas mileage? If the tune-up cost $30, how long do you think it would take to pay for itself, driving 20,000 mi per year?
3. During a flu epidemic a machine shop's employees produced 25% of normal production. If they produced 200 parts during this period, how much would they have produced if there had been no epidemic?
4. A woman earns $284.12 for a 40-hr week. If she saves 9% of this amount each week, how much does she save in 1 year?
5. Compute the cost of 100 bearings if a single bearing costs $11.50 and you are given a 10% discount when you purchase 100 or more.

Part B

6. In the late sixties, the gross amount of a man's paycheck was $212.52. He payed 4.15% F.I.C.A., 18% federal income tax, 2% state income tax, and 1.5% to the United Fund. How much was his net pay? How does this compare with the percents paid now?
7. A savings and loan association will pay 7 3/4% annual interest on your account by sending you a monthly interest check. How much would you have to have in your account at retirement age to receive a monthly check for $100?
8. When building a model of a machine, the machinist decreased the machine's size 70% by volume. If the model had a volume of 222.6 in.^{3}, what was the volume of the machine?

9. A general contractor buys oil in 50-gal drums, which saves him 12.5% of the wholesale cost of the same amount of oil purchased in cans. If a single quart can costs 82¢ wholesale, how much will 100 gal cost if two drums are bought and there is a 4% sales tax on the purchase?

10. If printing machinery depreciates 20% during the first year and 15% for each of the next 2 years, what is your printing machinery worth after 3 years if it cost you $250,000 originally?

Part C

11. A Virginia businessman saw the ad in Fig. 10-4 in a local newspaper. He then purchased one typewriter originally costing $143.50, six staplers originally $5.98 apiece, and six desk files whose original total price was $29.95. Including the 4% sales tax, how much did the man pay for all these items?

Figure 10-4

12. The cost of making a single plug tap for internal threading is $4.20. If the manufacturer wants to make a 25% profit above cost on sales less than 100 and a profit of 15% on sales of 100 or more, what prices will he charge?

13. In a single year a woman received $10,825.12 in dividends and interest. Stock dividends gave a return of 14.6% and accounted for 75% of the income. The average interest on her savings was 9.2% and accounted for the remaining part of this income. How much money did she have invested in stocks and how much in savings?

14. Burning 1 lb of wood produced 6,000 to 7,800 Btu of heat (Btu = British thermal unit); burning the same amount of anthracite (hard coal) produces 8,000 to 14,500 Btu. On the average, what percent increase in heat produced per pound do you have when you switch from wood to hard coal?

Answers

2. 24 mi per gallon; about 1 1/2 months (2,400 mi at $1.50/gal)
4. $1,329.68

6. \$158.01
8. 742 in.^{3}
10. \$125,000
12. \$5.25 for less than 100; \$4.83 for 100 or more
14. About 63%

Review Exercises

Part A

1. Complete the following table.

	Percent	Decimal	Fraction
(a)	9%		
(b)		0.021	
(c)			3/5
(d)	5 1/8%		
(e)		$0.0\overline{9}$	
(f)			7/6
(g)	140%		
(h)		0.00054	
(i)	8.9%		
(j)		0.103	

2. Willie bought a used car for \$225. He sold it 6 months later to a friend for \$300. What percent profit did he make?
3. Bonnie paid \$495 down on a new car. If this was a 7% down payment, how much did the car cost?
4. Fill in the blanks.
   (a) 13% of a number is 21.06. The number is ________.
   (b) 15 is ________% of 600.
   (c) 98% of \$1,215.20 is ________.
5. If we want to put money in a savings account, we have three basic choices: (a) a regular account paying 6% interest, (b) a regular account requiring a minimum balance of \$150 paying 6 1/4% interest, and (c) a special 90-day-notice account paying 8 3/4% interest. If you wanted to deposit \$275, find the amount of interest you would earn in each account if it is compounded annually.

Part B

6. The bead of a tire is the part that holds the tire to the rim. Its width is a major factor in curing the tire. Suppose for a certain tire a percent increase in bead width corresponds to the same percent increase in curing time. A tire with a 3.62-in. bead width takes 52 min to cure. How long (to the nearest minute) will it take to cure this tire if its bead width is increased to 3.80 in.?

7. If 3 lb of hamburger contains 38% fat, how much meat and how much fat would you have after cooking it if 18% of the fat is lost in cooking?

Part C

8. The formula for the volume of a sphere is $V = 4/3\pi r^3$. If you increase the radius by 10%, by what percent does the volume increase? See Fig. 10-5.

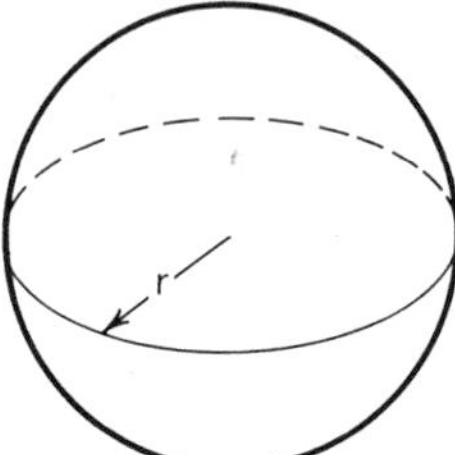

Figure 10-5

9. A metal alloy called Kelumese Shiney Glow used in heating wire contains 74% nickel, 20% chromium, and 1.9% manganese by weight. How much of each element is in 57 lb of wire? Account for the fact that your answers don't add up to 57 lb.

10. The outside diameter (OD) and shoulder width (SW) are two very important measurements on a tire (see Fig. 10-6). If each of these dimensions has a 2% tolerance, find the limits for each if the specifications are 25.75 in. for OD and 7.68 in. for SW.

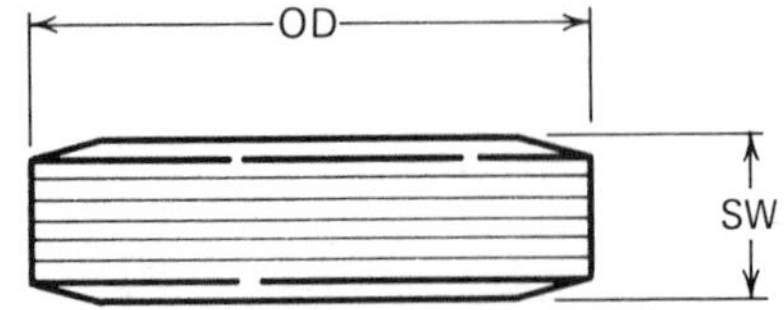

Figure 10-6

11. The volume of the rectangular block shown in Fig. 10-7 is found using the formula $V = l \cdot w \cdot h$. Suppose we mill the sides down until the area of the base has been decreased 15%. What will be the new volume? What percent of decrease in volume will this give us?

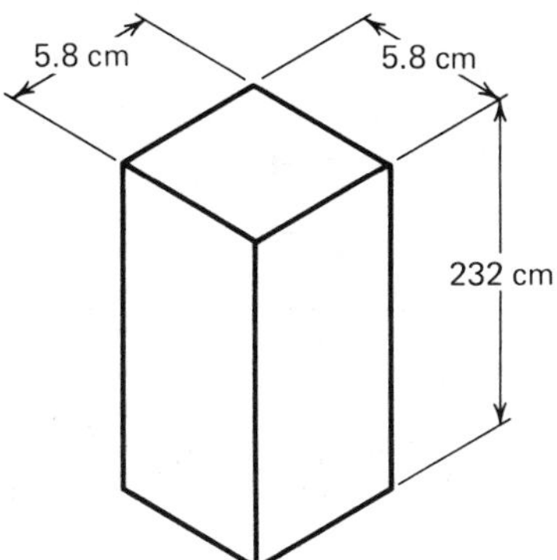

Figure 10-7

12. If the height of a note card should be between 50 and 60% of its width, what are the limits on the height of a card that is 4 in. wide?
13. Your shop bought a delivery truck for $15,000 and you are permitted to depreciate it at 30% for the first year and 20% for the second year. How much is the truck worth at the end of 2 years?
14. Write a practical problem involving percentages from your major field of interest.

Answers

2. 33 1/3%
4. (a) 162 (b) 2.5% (c) $1,190.90
6. 55 min
8. About 33%
10. 25.235″-26.265″ for OD; 7.5264-7.8336″ for SW
12. 2″ to 2.4″

Chapter 11

Ratio and Proportion

Two very useful concepts of mathematics are *ratio* and *proportion*. These occur in almost every field of work and in almost every home. When we cook, a recipe may call for 2 cups of flour for every 1 cup of sugar. We could say that the ratio of flour to sugar is two to one or that the ratio of sugar to flour is one to two. Either statement should tell us that we need twice as much flour as sugar.

We could make use of proportions by realizing that if we want to double our recipe, we would need 4 cups of flour and 2 cups of sugar. Carrying the idea one step further, suppose we had 7 cups of flour to use. Then a proportion would tell us that we need 3 1/2 cups of sugar.

To have another example of ratio, consider a car with a 4.44 gear ratio. This ratio means that for every tooth on the pinion gear there are 4.44 (correct to two decimal places) teeth on the ring gear. How could we have 0.44 of a tooth? Of course we can't. If our owner's repair manual tells us that the pinion gear has 9 teeth, we use a proportion to find that the ring gear has 40 teeth.

One of the most important ratios we use today was discovered in ancient times. This ratio is used to find the circumference and area of a circle. We have used *pi* (π) since elementary school, but most of us have not realized that it is a ratio. Pi is the ratio of the circumference of a circle (distance around the circle) to its diameter (distance across the circle), and $\pi = 3.1415926535$, approximately. We usually express π as 3.14, 3.1416, or 22/7, and these values generally give us answers that are accurate enough for everyday problems. It is an interesting fact that there is no exact decimal or fractional representation of pi.

11-1 RATIO

When we have two quantities, we may compare them using division.

> *Definition:* The *ratio* of two numbers is a comparison between them by division.

Suppose two gears are meshed together. In Fig. 11-1, the smaller gear has 14 teeth and the larger gear has 42 teeth.

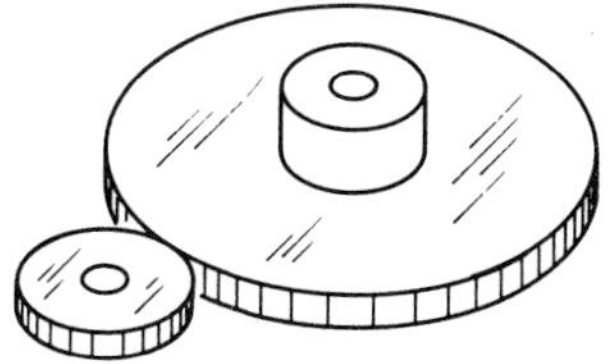

Figure 11-1

We may compare the number of teeth on the smaller gear to the number on the larger gear by means of a ratio. This may be done in either of the following ways:

1. $\dfrac{\text{number of teeth on smaller gear}}{\text{number of teeth on larger gear}} = \dfrac{14}{42} = \dfrac{1}{3}$

   We say their ratio is *one-third*.

2. $\begin{pmatrix}\text{number of teeth}\\ \text{on smaller gear}\end{pmatrix} : \begin{pmatrix}\text{number of teeth}\\ \text{on larger gear}\end{pmatrix}$

   $$14 : 42$$

   $$1 : 3$$

   We say their ratio is *one to three.*

Both of the ratios above tell us that for each tooth on the smaller gear there are three teeth on the larger gear. Knowing this ratio, we should be able to see that for every three turns of the small gear the large gear makes one turn.

In both of these ratios, notice that the original ratio has been reduced to lowest terms in the same manner that we would reduce a fraction. There are two forms that are considered to be *reduced*:

1. Reduced like a fraction. As examples, 40/9 and 2/5.
2. Reduced until one of the terms is one (1). As examples, 40/9 = 4.44/1 = 4.44 gear ratio; 2/5 = 1/2.5 = 1:2.5 = 1 to 2.5.

When using ratios we prefer to compare quantities that have the same units of measure. For example, we prefer not to have a ratio of yards to inches or quarts to gallons. This is illustrated by the following examples.

Example 11-1: What is the ratio of 6 yd to 240 in.? We change yards to inches by multiplying 6 yd times 36 in. per yard and obtain 216 in. Our ratio is $216/240 = 9/10$.

Example 11-2: What is the ratio of 127 cm to 100 in.? Since 2.54 cm = 1 in., we find that 100 in. = 100×2.54 cm = 254 cm, and our ratio is $127/254 = 1/2$.

Exercise Set 11-1

Part A

1. Express each of the following ratios in lowest terms.
   (a) 12:24 (b) 27:18 (c) 44:100
   (d) 26/39 (e) 72/144 (f) 612 to 468
2. Two gears have 84 teeth and 105 teeth, respectively. What is the ratio of the number of teeth in the smaller gear to the number of teeth in the larger gear?
3. The circumference of a circle is 704 cm. Its diameter is about 224 cm. Find the ratio of the circumference of this circle to its diameter. What do we usually call this value?
4. Express each of the following ratios in simplest terms.
   (a) 12 in. to 3 ft (b) 3 yd to 6 in. (c) 100 cm to 5 m
   (d) 6 oz to 1 lb (e) 2 mi to 2,640 yd (f) 18 qt to 7 gal
5. Find the ratio of 1,270 cm to 10 in.

Part B

6. Express the following ratios as decimal fractions correct to two decimal places.
   (a) 36/48 (b) 11.5:3.5 (c) 7 to 8
   (d) 1,464:100 (e) 30 to 78 (f) 60 to 66
7. A speed reducer has a speed ratio of 14:64. If the large gear is missing and the smaller gear has 28 teeth, how many teeth should a replacement gear have in order to maintain the ratio?
8. A garage owner sold 984 gal of gasoline and 16 qt of oil during 1 day of business. What is the ratio of the quantity of oil sold to the quantity of gasoline sold?
9. In a single hour, a typesetter set 18 pages; when they were checked by the

quality control person, 3 pages were found to have misprints. How many good pages did the typesetter produce for each misprint?

10. Find six pairs of integers whose ratio is 3 to 7.

Part C

11. Express each of the following ratios in two reduced forms.
    (a) 1/2:3/4 (b) 5/6 to 11/66 (c) 2 in.^{2} to 1 ft^2
    (d) 1 yd^3 to 9 ft^3 (e) 0.314:1.57 (f) 3 pt to 6.25 gal
12. The *pitch* of a screw is defined as the distance between a point on one thread and the corresponding point next to it, measured parallel to the axis. If one screw is threaded with 10 American National threads per inch and another is threaded with 14 American National threads per inch, what is the ratio of the pitch of the first screw to the pitch of the second?
13. A blueprint of a building is drawn to scale and has a scale 1/16 in. equals 1 ft. What is the ratio of the drawing to the corresponding distance in the building?
14. Two wheels have diameters of 22 in. and 13 in., respectively. What is the ratio of the larger wheel to the smaller wheel in terms of area? In terms of circumference?
15. The pinion gear in a certain car has 16 teeth, and the ring gear has 52 teeth. Find the gear ratio correct to two decimal places.

Answers

2. 4/5
4. (a) 1 to 3 (b) 18 to 1 (c) 1 to 5 (d) 3 to 8 (e) 4 to 3 (f) 9 to 14
6. (a) 0.75 (b) 3.29 (c) 0.88 (d) 14.64 (e) 0.38 (f) 0.91
8. 1/246
10. 3 to 7; 6 to 14; 9 to 21; 12 to 28; 15 to 35; 18 to 42
12. 7/5
14. 484/169; 22/13

11-2 MORE APPLICATIONS OF RATIOS

A mixture of antifreeze may have 3 parts alcohol to 5 parts water by volume. Someone wants to sell 20 gal of this mixture for $85. We know that pure alcohol can be bought for $3.59 a quart. Is his price fair? We can answer this question by use of ratios. First, each "unit" of the mixture must contain 8 parts (5 parts water and 3 parts alcohol). For convenience, let the 8 parts be 8 qt (5 qt water and 3 qt alcohol). Since we know the price of a quart of alcohol, we want to find out how many quarts of alcohol are in the mixture. We know 20 gal is 80 qt. Since each unit of the mixture has 8 qt, we divide 80 qt by 8 qt and find 10 units. Each unit has 3 qt of alcohol so we have 30 qt of alcohol all together. This would cost \$107.70, so we see that it is cheaper to buy the mixture than to make it ourselves.

Let's look at the steps we used to solve the problem. These same steps may be used to solve other problems of this type.

The mixture of antifreeze contained alcohol and water in the ratio 3:5. This meant that if the mixture was uniform, each unit of it contained 3 + 5 = 8 parts. We divided this into the total quantity to find out how many units there would be if the units were made up of quarts. We multiplied this by 3 to obtain the number of quarts of alcohol since each unit contained 3 qt of alcohol. Finally, we noted that 30 times \$3.59 is \$107.70, which is more than \$85, so the price was fair.

Example 11-3 will show us how to solve a problem involving a triple ratio.

Example 11-3: A dry mix for concrete is made of cement, sand, and gravel in the ratio 1:2:3, respectively, by weight. How much of each substance is there in 120 lb of this concrete? The ratio is 1:2:3; therefore, each unit has 1 + 2 + 3 = 6 parts, where each part is considered to be 1 lb. In 120 lb we have 120 ÷ 6 = 20 units. Each unit contains 1 lb cement, 2 lb sand, and 3 lb gravel. We have

$$
\begin{aligned}
1 \times 20 &= 20 \text{ lb cement} \\
2 \times 20 &= 40 \text{ lb sand} \\
3 \times 20 &= \underline{60} \text{ lb gravel} \\
& 120 \text{ lb concrete}
\end{aligned}
$$

Exercise Set 11-2

Part A

1. Divide each of the following numbers into parts having the indicated ratio.
   (a) 72, ratio 1:2 (b) 150, ratio 3:7 (c) 27, ratio 2:1
   (d) 2,412, ratio 4:5 (e) 126, ratio 8:1 (f) 28.8, ratio 7:5
2. If 2,000 lb of concrete contains cement, sand, and crushed stone in the ratio 1:2:3 by weight, how much cement is contained in the mixture?
3. The ratio of nickels to dimes in a vending machine is 4:3. If there is \$49.00 in the machine, all nickels and dimes, how many of each are there?
4. A windshield solvent is to be mixed with water in the ratio one to seven (solvent to water). How many cups of solvent do you need to prepare 1 gal of the mixture?
5. The average daily profit for your newspaper is \$591.50. The profit is divided among the three stockholders in the ratio 7:5:1. What is the average daily income for each of the stockholders from your newspaper?

Part B

6. Divide each of the following numbers into parts having the given ratio.
   (a) 12.1, ratio 1:4:6
   (b) 93,765, ratio 1:2:3:4:5
   (c) 0.125, ratio 4:1
   (d) 1, ratio 2:7:11
   (e) 1/2, ratio 1:2
   (f) 3/8, ratio 2:5:7
7. A customer at your paint store has asked you to mix 16 gal of dark green paint using one part semigloss black to seven parts of your darkest stock green. How many gallons of each should you use to produce the 16 gal?
8. Bill and Eddie contracted to paint a car for $175. They both worked 12 hr the first day; Bill worked 6 hr the second day to finish the job. If they both worked at the same rate of pay, how much did Bill receive?
9. A fuel mixture used for a small engine contains 6.25% oil and the rest gasoline. What is the ratio of oil to gasoline in this fuel?
10. The ratio of two gears is 4:1. If together they have 90 teeth, how many teeth are on each gear?

Part C

11. Three workers contracted to paint a roof for $767. (Paint was furnished by the owner.) One worker worked for 10 hr, another for 25 hr, and the third for 30 hr. If they divided the money in the same ratio as the number of hours worked, how much did each receive?
12. A dry mix of concrete is to be made of gravel, sand, and cement in the ratio 5:3:1 by volume. How many cubic yards of each must be used to make 162 yd^3 of concrete if there is 10% shrinkage on mixing?
13. During a year, a tool and die company found that the ratio of perfect dies to blemished dies to defective dies tooled was 16:3:1. They produced 20,000 dies. They sold 25% of the perfect dies to the automobile industry at a profit of 19¢ per die. What was the profit on the sale of dies to the automobile industry?
14. Find the quotient of all pairs of numbers that have the following ratios.
   (a) 5:2
   (b) 1:12
   (c) 5/9
15. A metal alloy called linotype metal, used in printing, contains lead, antimony, and tin in the ratio 79:16:5 by weight. How many pounds of each metal are needed to produce 1,500 lb of linotype metal?

Answers

2. 333 1/3
4. 2 cups

6. (a) 1.1:4.4:6.6 (b) 6,251:12,502:18,753:25,004:31,255
   (c) 0.100:0.025 (d) 0.1:0.35:0.55
   (e) 1/6:1/3 (f) 3/56:15/112:3/16
8. $105
10. 72 and 18
12. 100 gravel; 60 sand; 20 cement
14. (a) 2.5 (b) $0.08\overline{3}$ (c) $0.\overline{5}$

11-3 DIRECT PROPORTION

There are many problems that cannot be solved using a single ratio. Some of these can be solved by using a proportion.

> *Definition:* A *proportion* is a statement of equality between two or more ratios.

Proportions may be represented in two different forms, as shown below:

$$\frac{1}{2} = \frac{3}{6} \quad \text{or} \quad 1:2::3:6$$

Both forms of this proportion should be read *1 is to 2* as *3 is to 6*. The numbers in a proportion are called *terms*. In the proportion above, the numbers 1, 2, 3, and 6 are terms.

The middle terms of a proportion are called the *means*; the other two are called the *extremes*.

1:2::3:6 — 2 and 3: means; 1 and 6: extremes

$$\frac{1}{2} = \frac{3}{6}$$

1 and 6 are extremes

2 and 3 are means

> The product of the extremes equals the product of the means.

If 1:2::3:6, then $1 \times 6 = 3 \times 2$. If $1/2 = 3/6$, then $1 \times 6 = 2 \times 3$. If a, b, c, and d are numbers, and $a/b = c/d$, then $ad = bc$.

Example 11-4: Find the value of n that makes $5/8 = 15/n$.

Since the product of the extremes equals the product of the means

$$5n = 8 \cdot 15$$

$$n = \frac{120}{5}$$

$$n = 24$$

So, 5:8::15:24.

In the remaining part of this section we shall see how to apply some of these facts to problems involving direct proportions. We shall not give a mathematical definition for a direct proportion but shall show what we mean by a direct proportion by using examples.

The cost of articles purchased is directly proportional to the number of articles. (If we buy more articles, our cost increases.) The distance covered by a vehicle moving at a constant speed is directly proportional to the time it travels. The interest earned by a savings account is directly proportional to the amount of money in the account. The tuition income of a college is directly proportional to the number of students enrolled.

There are many more examples but these should give you the idea that a direct proportion between two quantities requires a proportional increase (decrease) in one quantity for any increase (decrease) in the other quantity.

> We form a direct proportion by setting a first-quantity ratio equal to a second-quantity ratio.

Example 11-5: If 6 screw-type visual battery caps cost $1.98, how much will 48 cost?

Using a proportion to solve this problem, we let c represent the cost of 48 caps. If we increase the number of caps, the cost also increases, so we need a direct proportion. By establishing the two ratios, 6 caps to 48 caps and $1.98 to c dollars,

$$\frac{6 \text{ caps}}{48 \text{ caps}} = \frac{\$1.98}{c}$$

$$\frac{6}{48} = \frac{1.98}{c}$$

$$6c = (48)(1.98) \qquad \text{(Product of extremes equals product of means.)}$$

$$c = \frac{95.04}{6}$$

$$c = 15.84$$

So 48 caps would cost $15.84.

Example 11-6: The ratio between the length of a bearing and the diameter of the shaft it supports is 3:1. For instance, a 3-in. bearing supports a 1-in. shaft. The diameter of another shaft is 4.5 in. What is the length of its bearing? Letting n represent the length of the bearing, we set up our direct proportion:

$$\frac{\text{3-unit length}}{\text{n-unit length}} = \frac{\text{1-unit diameter}}{\text{4.5-unit diameter}}$$

$$3(4.5) = 1 \cdot n$$

$$13.5 = n$$

Exercise Set 11-3

Part A

1. Find the missing term in each of the following proportions.
   (a) 1/2 = n/24 (b) 3/7 = 36/x (c) 11/a = 352/96
   (d) x/16 = 10/160 (e) 12:75::d:25 (f) m:69::6:18
2. If 1 doz stabilizer bushings cost $11.68, how much will 100 of these bushings cost?
3. A picture 3 1/2 in. by 5 in. is to be enlarged so its height is 15 in. How many inches will its width measure?
4. A machinist can produce three key pins in a 48-hr week. If he maintains this rate, how many key pins will he produce in a year? (Assume he works 50 weeks in a year.)
5. How many templates can be stamped from a copper strip 10 ft long if 24 can be stamped from an 8-ft strip?

Part B

6. The weight of 5 ft of copper pipe with an outside diameter of 3/4 in. is 2 lb. How much will 75 ft of this pipe weigh?
7. An aluminum machine part weighing 5 lb 6 oz is to be replaced with the same part made of an alloy that weighs 0.13 lb per cubic inch. What is the weight of the new part? (*Hint:* Aluminum weighs 0.09 lb per cubic inch.)
8. Charles' law states *at constant pressure, the volume occupied by a given weight of gas is directly proportional to its absolute temperature.* If the volume of a certain weight of gas is 4 ℓ at 300°K (Kelvin or absolute) and the pressure is held constant, what will be the volume at a temperature of 450°K?
9. A printing press prints 620 pages in 10 min. How many hours will it take to print 30,256 pages?

Part C

10. Areas of similar figures are directly proportional to the squares of corresponding dimensions. A drawing of a rectangular trapdoor uses a scale of 1:4. One side of the drawing is 8 in. long, and the area is 96 in.^{2} What is the area of the actual trapdoor?

11. The scale used for the drawing in Fig. 11-2 is 1:120. Using a ruler or other measuring device, find the rise, run, and span in feet or meters.

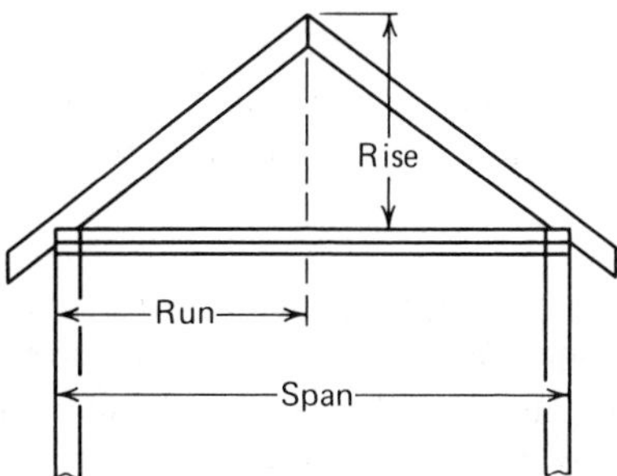

Figure 11-2

12. A glider released at 12,000 ft fell to 8,000 ft over a distance of 82 mi in 40 min. Assuming that conditions remain the same, how much longer could the pilot remain aloft and how far would she have glided?

13. Figures in the ratio 89:144 **(golden ratio)** are supposed to be the most pleasing to the human eye. If a living room meets this requirement and is 16 ft wide, what is its area?

14. How long should you cut a card that needs to be 3 in. wide if you want it to approximate the golden ratio?

Answers

2. \$97.33
4. 150
6. 30 lb
8. 6 ℓ
10. 1536 in.^{2}
12. 80 min; 246 mi
14. about 5 in.

11-4 INVERSE PROPORTION

The inverse proportion has as many applications as the direct proportion. Again, as in the previous section, we will not actually define inverse proportion but rather gain a working knowledge of what it means with examples.

First, let's list some situations where we have inverse proportions:

The time required to complete a job is inversely proportional to the number of people employed. (A job gets done faster when more people are working.)

Speed is inversely proportional to time over a constant distance. (If we increase the speed, we decrease the time.)

When two gears are meshed, the revolutions per minute (rpm) are inversely proportional to the number of teeth. (If you increase the number of teeth on one gear, you decrease its rpm.)

When two pulleys are connected by a belt, the rpm are inversely proportional to the diameters. (If you increase the diameter of one pulley, you decrease its rpm.)

As illustrated in the examples above, an *inverse proportion* **(or indirect proportion)** between two quantities requires a proportional decrease **(increase)** in one quantity for any increase **(decrease)** in the other quantity.

Remember that the quantities in a direct proportion change in the same direction but the quantities in an inverse proportion change in opposite directions.

> To establish an inverse proportion, we set a first-quantity ratio equal to the reciprocal (inverse) of a second-quantity ratio.

Now let's look at some examples that will show us how to apply inverse proportions.

Example 11-7: The large gear in Fig. 11-3 is turning at 64 rpm. How many rpm is the small gear turning?

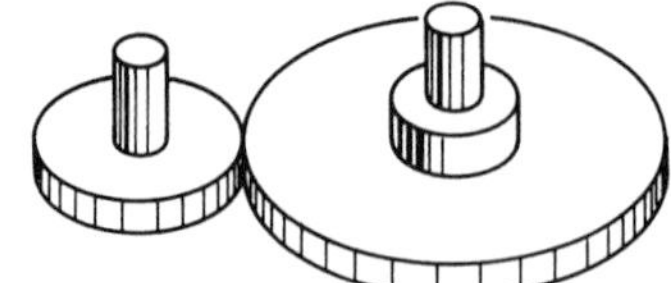

Figure 11-3

Since this is an inverse proportion, we set up the two ratios:

$$\frac{\text{rpm of large gear}}{\text{rpm of small gear}} \qquad \frac{\text{no. of teeth on large gear}}{\text{no. of teeth on small gear}}$$

Now, the inverse proportion is

$$\frac{\text{rpm of large gear}}{\text{rpm of small gear}} = \frac{\text{no. of teeth on small gear}}{\text{no. of teeth on large gear}}$$

(Notice that the second ratio has been "flipped" to form its reciprocal.) The large gear has 24 teeth and the small gear has 12 teeth, giving us the proportion

$$\frac{64}{n} = \frac{12}{24}$$

Reducing the ratio on the right, we have

$$\frac{64}{n} = \frac{1}{2}$$

$$128 = n$$

Therefore, the small gear is turning 128 rpm. What would happen if the small gear had 4 more teeth? The small gear now has 16 teeth so we set up the ratios

$$\frac{64 \text{ rpm}}{t \text{ rpm}} \quad \text{and} \quad \frac{24 \text{ teeth}}{16 \text{ teeth}}$$

Invert the second ratio and establish the inverse proportion.

$$\frac{64}{t} = \frac{16}{24}$$

$$\frac{64}{t} = \frac{2}{3}$$

$$192 = 2t$$

$$96 = t$$

In this case the small gear would be turning at 96 rpm. Notice that by increasing the teeth we decreased the rpm.

Example 11-8: A car covers 180 mi in 4 hr 30 min. What average speed would be required to cover this distance in just 4 hr?

This is also an inverse proportion so we have

$$\frac{\text{original average speed}}{\text{new average speed}} = \frac{\text{new time}}{\text{original time}}$$

In order to find the original speed, we divide the distance by the time (4 hr 30 min = 4.5 hr).

$$180 \div 4.5 = 40 \text{ mph}$$

Now we set up our inverse proportion

$$\frac{40}{s} = \frac{4}{4.5}$$

$$180 = 4s$$

$$45 \text{ mph} = s$$

Example 11-9: Three auto mechanics can give 15 cars a complete tune-up in 10 hr. How many mechanics would be needed to complete the job in 4 hr?

Setting up our inverse proportion we have

$$\frac{\text{no. of auto mechanics working 10 hr}}{\text{no. of auto mechanics working 4 hr}} = \frac{\text{new time (4)}}{\text{original time (10)}}$$

Note that the number of cars is constant, so we don't need that to solve our problem. We have

$$\frac{3}{m} = \frac{4}{10}$$

$$30 = 4m$$

$$7\frac{1}{2} = m$$

We know it is impossible to have 1/2 a mechanic, so the conclusion is that 8 mechanics could do the job in 4 hr.

As you work these problems, you should keep in mind that an inverse proportion is needed when an increase in one quantity implies a decrease in a related quantity.

Exercise Set 11-4

Part A

1. Solve each of the following proportions.
   (a) $x/36 = 12/27$ (b) $24/40 = n/60$ (c) $17/x = 51/9$
   (d) $16/m = m/4$ (e) $2\ 1/2 : 3\ 1/4 :: 7\ 1/2 : x$ (f) $b : 5/12 :: 5/3 : b$
2. The forces exerted by two weights balanced on a lever are inversely proportional to their distances from the fulcrum. If a 60-lb weight 3 ft from the fulcrum balances an 80-lb weight, how far from the fulcrum is the 80-lb weight? See Fig. 11-4.

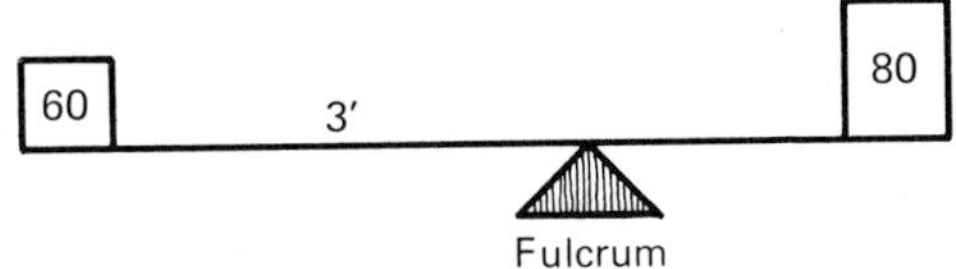

Figure 11-4

3. A rectangle is 3.6 m high and 2.4 m wide. If we keep the area the same and make the height 0.3 m, what will the new width be?
4. A gear with 18 teeth is turning at 250 rpm. This gear drives a larger gear at 100 rpm. How many teeth does the larger gear have?
5. A 9-in. pulley turning at 700 rpm drives a 12-in. pulley. What is the rpm of the driven pulley? See Fig. 11-5.

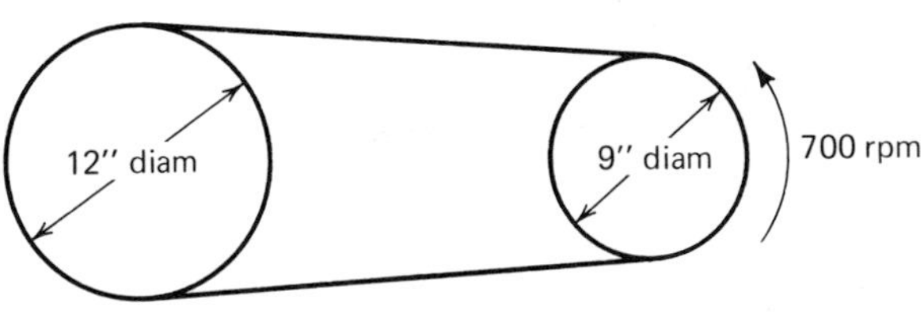

Figure 11-5

Part B

6. A pulley turning at 420 rpm drives an 8-in. pulley at a rate of speed 20% slower. What is the diameter of the drive pulley?
7. The pull of gravity at the earth's surface is 32.2 ft per second squared. As an object moves away from the earth's surface, this force becomes smaller and is inversely proportional to the square of the distance from the center of the earth. The earth's radius is approximately 4,000 mi. What is the force of the earth's gravity on a space capsule 1,000 mi from the earth?
8. At a constant temperature, the volume of a given quantity of gas is inversely proportional to the pressure. *Standard pressure* is the pressure exerted by a column of mercury exactly 760 mm high. Two liters of nitrogen gas occupy 2 liters at a pressure of 73.5 mm of mercury. If the temperature remains constant, what will be the volume at standard pressure?
9. In each of the diagrams in Fig. 11-6 the gear on the left is the drive gear. Find the number of teeth and rpm of all the gears.

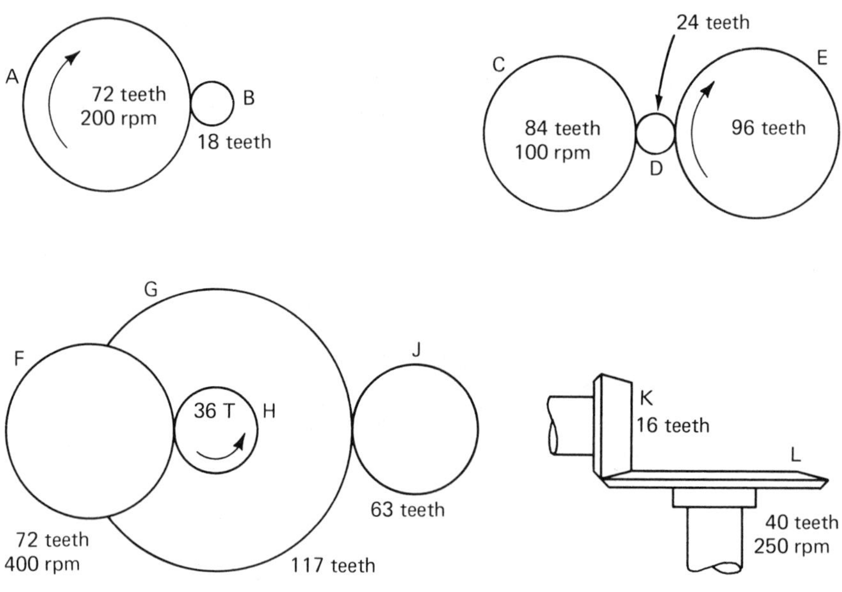

Figure 11-6

10. If 15 people working 8 hr per day complete a job in 22 days, how many people would it take to complete the job in fourteen 10-hr days?

Part C

11. A *dyne* is a unit of force. It is the force required to accelerate one gram of mass at a rate of one centimeter per second squared. The gravitational attraction between two bodies is inversely proportional to the square of the distance between them. If the attraction is 0.0042 dyne when the distance is 100 cm, find the attraction when the distance is 50 cm.

12. An auto parts factory employs 572 people. This work force produces 5,673 parts per 40-hr week. If the number of employees is cut by 30%, how many hours will the remaining employees have to work to produce the same number of parts each week?

13. A motor turns a 5-in. pulley at 1,760 rpm. What size pulley would you belt to the drive pulley to operate a machine requiring 1,000 rpm?

14. Shaft A in Fig. 11-7 is being driven at 400 rpm. Compute the speed of shafts B and C.

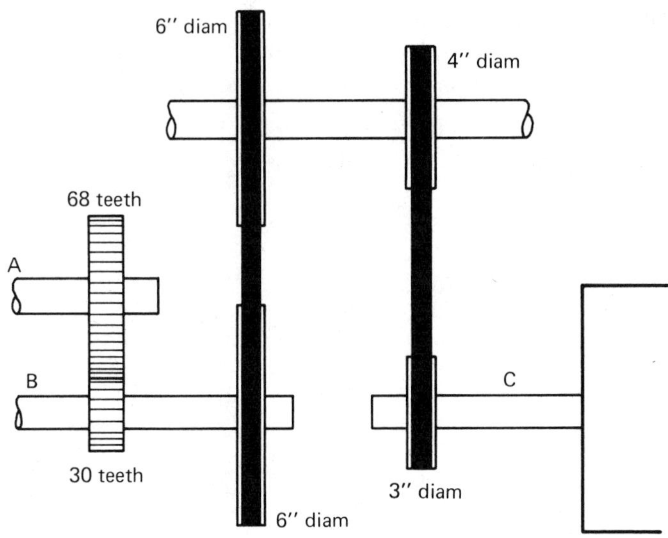

Figure 11-7

15. M is directly proportional to V^2 and inversely proportional to R. If M = 540 when V = 90 and R = 5, find M when V = 120 and R = 4.

Answers

2. 2.25 ft
4. 45 teeth
6. 6.4″
8. About 0.19 liter
10. 19 men
12. 57.2 hr
14. 906 2/3 rpm for B; 1,208 8/9 rpm for C

Review Exercises

Part A

1. Express each of the following ratios in simplest terms.

(a) 63 : 49	(b) 36/108	(c) 75 to 15
(d) 6 qt to 3 gal	(e) 5,280 ft to 440 yd	(f) 1 m to 5,000 cm

2. A penny is composed of 19 parts copper to 1 part tin, by weight. Find the weight of each metal in 380 lb of pennies.

3. Find the missing term in each of the following proportions.
   (a) Y/12 = 4/6 (b) 3/N = N/27 (c) 65/A = 5/2
   (d) 81/729 = 729/X (e) 4/112 = X/56 (f) 1:7::8:X
4. What is the gear ratio of the gears pictured in Fig. 11-8? If the smaller gear is turning at 50 rpm, what is the speed of the large gear?

Figure 11-8

5. A map is drawn to the scale 2.25 in. = 10 mi. Find the distance between two points 18 in. apart on the map.

Part B

6. The weight of 1 doz small screws is 10 1/2 oz. Find the weight of 300 screws of the same size.
7. The amount of water discharged by a pipe under a fixed pressure is directly proportional to the square of the inside radius of the pipe. A pipe with a 6-in. inside radius discharges water at 19.8 gal per second. What size pipe would you use to replace this pipe to obtain a flow of 24 gal per second?
8. Soft solder is composed of tin and lead in the ratio 3:2 by weight. How much of each metal would you need to produce 400 rolls if each roll holds 1.75 lb of solder?
9. If the air-fuel mixture in the cylinder of an engine is compressed from a volume of 450 cc to 50.4 cc, what is the compression ratio?

Part C

10. Compute all the speeds at which the machine shaft in Fig. 11-9 may be operated.

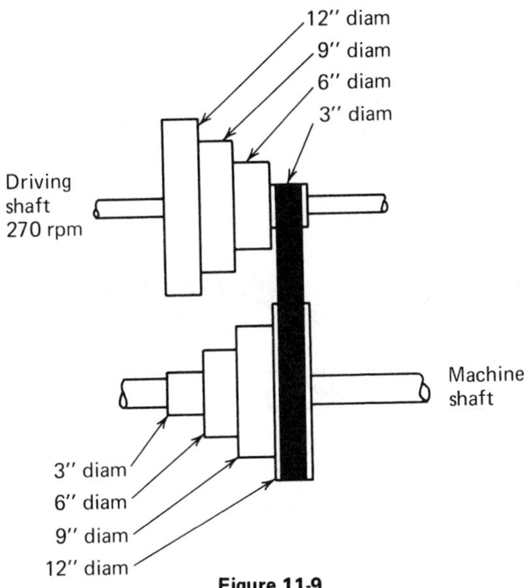

Figure 11-9

11. Mrs. Roberts, Mr. Motley, and Miss Sanderson invest in the purchase of a gas station in the ratio 2:3:7, respectively. At the end of the first year their net profit is $13,884.60. If they split the profit in the same ratio as they invested, how much will each receive?
12. Write an equation relating Q, M, and N, if Q is directly proportional to M and inversely proportional to N.
13. The period of vibration of a pendulum is directly proportional to the square root of its length. If the period of a 5.29-cm pendulum is 0.81 sec, what is the period of a 9-cm pendulum?
14. Pewter is composed of tin and lead in the ratio 4:1 by weight. If there is a 7% loss of tin and a 3% loss of lead in the production of pewter, how much of each metal would you need to produce 600 lb of pewter?
15. A staff member suggested that you change the page size of your newspaper to be more pleasing to the eye. She recommended the golden ratio, and you want to keep a 15-in. width. What length should you change to?

Answers

2. 19 lb of tin; 361 lb of copper
4. 2:1; 25 rpm
6. 262.5 oz
8. 420 lb of tin; 280 lb of lead
10. 67.5 rpm; 180 rpm; 405 rpm; 1,080 rpm
12. Q = km/N; k is constant
14. 511.7 lb of tin; 127.9 lb of lead

Chapter 12

Perimeters, Areas, and Volumes

Frequent use of mathematics is made in problems involving a perimeter, an area, or a volume. A simple use of perimeters would occur in finding how many feet of fencing you might need to enclose your yard, garden, or any other area. In this case you would simply measure the distance around the area you wanted to enclose.

One of the most common problems involving area might arise when you want to buy some carpeting for your home. In this case you would have to find the area of the floor, perhaps in square yards, and then multiply this number by the price per square yard to find the cost.

A frequent problem involving volumes occurs when you want to see how much liquid a container will hold. An example of this would be finding the amount of water needed to fill a swimming pool.

In this chapter we shall learn to find areas of figures with simple or complicated shapes and to find volumes of right prisms.

12-1 PERIMETER AND AREA OF PLANE FIGURES

There are special formulas for the perimeters of many figures, and you will find some of these in Table 12-1. Knowing these special formulas is not nearly so important as understanding the concept of perimeter.

> The *perimeter* of any figure is simply the distance around the outside of the figure.

For example, we may find the perimeter of Fig. 12-1 by starting at A, "walking" to B, to C, to D, to E, and then back to A. The perimeter would then be the total distance we "walked," in this case 148 ft.

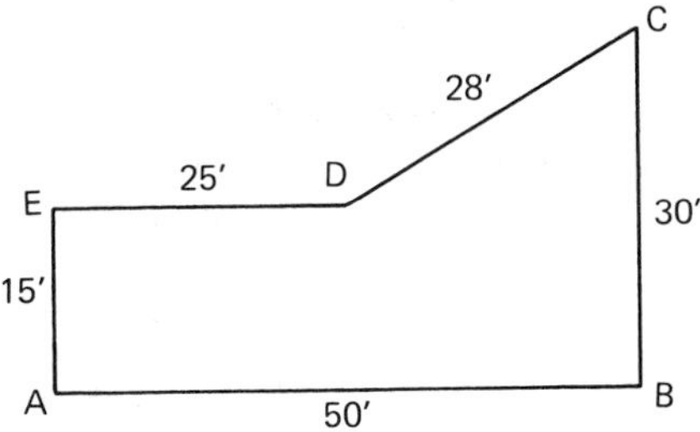

Figure 12-1

The most common plane figures we come in contact with are the rectangle, square, triangle, circle, and parallelogram. These are pictured in Fig. 12-2. Because these figures are used often, you will be expected to know and use the formulas when finding their perimeters.

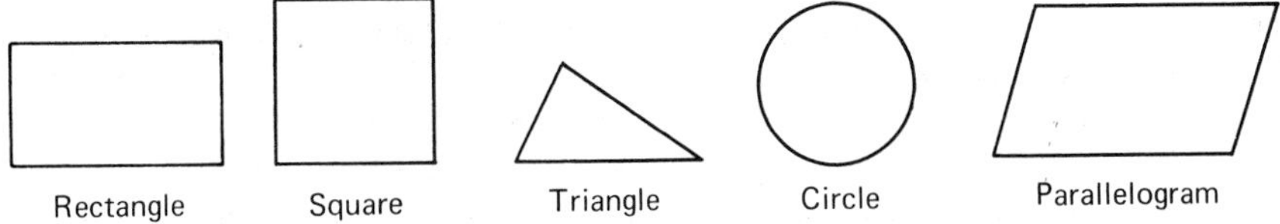

Figure 12-2. Common plane figures.

These same five common figures will form a basis for our discussion of area. If we know how to find the area of each of these, we shall be able to find the area of figures that are combinations of these by dividing the figure into several parts, finding the area of each part, and adding these areas together. For example, Fig. 12-3 may be subdivided as shown in Fig. 12-4.

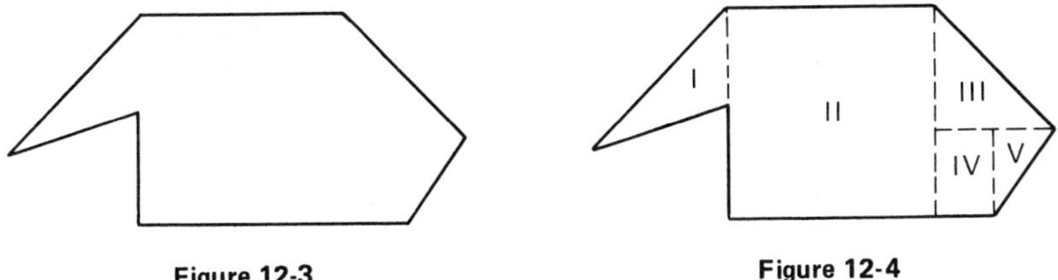

Figure 12-3 **Figure 12-4**

Then the area of the whole figure would be equal to the sum of the areas of sections I, II, III, IV, and V.

We are now ready to present an intuitive approach to finding the area of the five common figures shown in Fig. 12-2. Consider a rectangle with length 12 units and width 4 units (see Fig. 12-5).

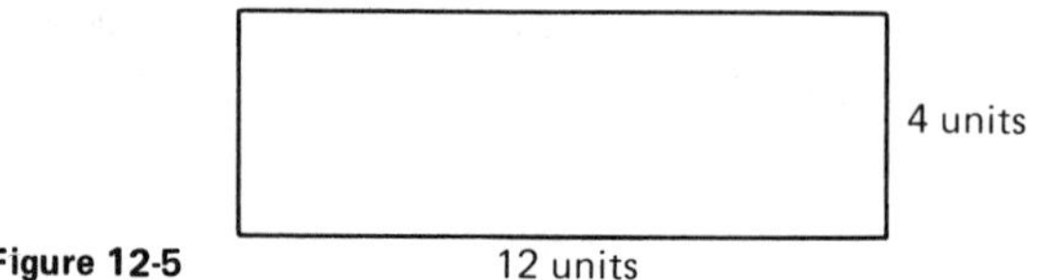

Figure 12-5

We define the *area of a rectangle* as the number of squares, 1 unit on a side, that completely fill the spaces inside the rectangle. By counting the squares in Fig. 12-6, we find that in this case the area is 48 *square units*. Many of you realize that we

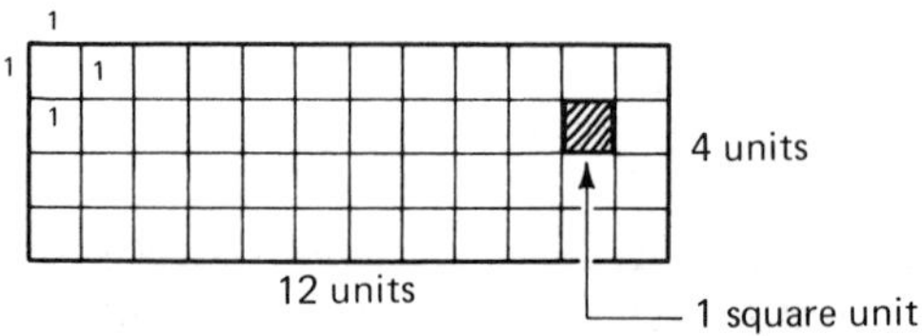

Figure 12-6. Forty-eight square units.

could have found the same result by multiplying 4 times 12. We may then write a formula for the area of a rectangle as $\mathbf{A} = \boldsymbol{l} \cdot \mathbf{w}$ where $\boldsymbol{l}$ represents the length of the rectangle and $\mathbf{w}$ is its width.

The *square* is a rectangle whose length is equal to its width. Therefore, the formula for the area of a square with side $\mathbf{s}$ is given by $\mathbf{A} = \mathbf{s} \cdot \mathbf{s} = \mathbf{s}^2$.

The area of a triangle is found by using the formula $\mathbf{A} = \mathbf{1/2bh}$ where $\mathbf{b}$ is the base of the triangle and $\mathbf{h}$ is the height or altitude to the base. Identify one side of the triangle as the *base*. The *height* is defined as the perpendicular distance between the base and the vertex opposite the base.

The formula for the area of a circle comes from the ancient Egyptians and is given by the formula $\mathbf{A} = \pi \mathbf{r}^2$ where $\mathbf{r}$ is the radius (distance from center to any point on the circle) and π is a constant approximately equal to 22/7 or 3.1416.

Figure 12-7 shows a parallelogram. The *height* of a parallelogram is defined to be the perpendicular distance between the base and the side parallel to the base. We shall develop a formula for its area by using a formula we already have. A triangle can be cut off the end of the parallelogram by cutting along h. If this triangle is placed on the other end of the parallelogram, we obtain Fig. 12-8.

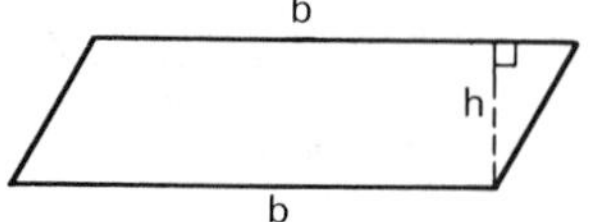

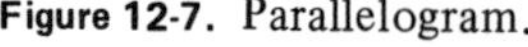

Figure 12-7. Parallelogram.

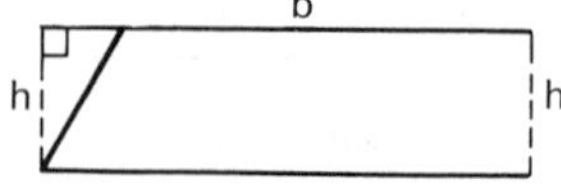

Figure 12-8. Rectangle.

This is a rectangle with length b and height h so the formula for the area of a parallelogram may be given by $\mathbf{A} = \mathbf{bh}$ where $\mathbf{b}$ is the base and $\mathbf{h}$ is the height measured perpendicular to the base. Now that we can find the area of the simple figures, we are ready to find the area of complicated figures.

Example 12-1: Find the area and perimeter of the figure pictured in Fig. 12-9. We divide the area into separate areas as shown and fill in the needed dimensions to find the individual areas, as shown in Fig. 12-10.

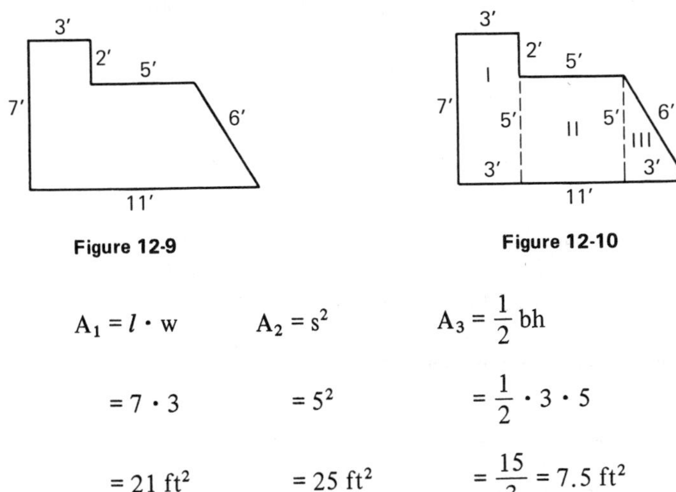

Figure 12-9

Figure 12-10

$$A_1 = l \cdot w \qquad A_2 = s^2 \qquad A_3 = \frac{1}{2}bh$$

$$= 7 \cdot 3 \qquad = 5^2 \qquad = \frac{1}{2} \cdot 3 \cdot 5$$

$$= 21 \text{ ft}^2 \qquad = 25 \text{ ft}^2 \qquad = \frac{15}{3} = 7.5 \text{ ft}^2$$

Then the total area is $= 21 + 25 + 7.5 = 53.5 \text{ ft}^2$. The perimeter or distance around is found by adding the lengths of each of the sides.

$$P = 11 + 7 + 3 + 2 + 5 + 6 = 34 \text{ ft}$$

Example 12-2: Find the area of Fig. 12-11.

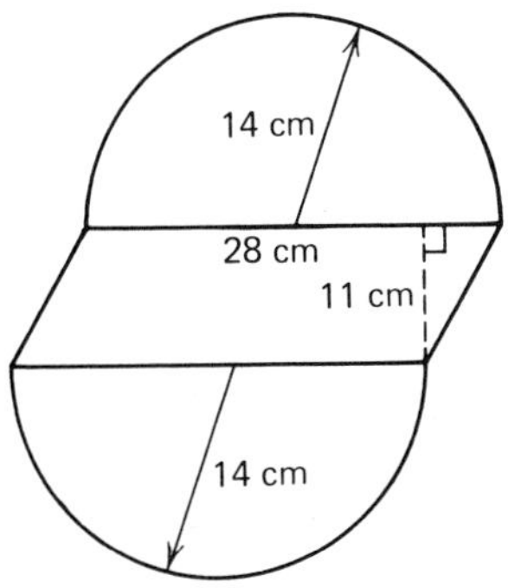

Figure 12-11

Divide the figure into a parallelogram and two semicircles. For the parallelogram:

$$\begin{aligned} A &= bh \\ &= (28)(11) \\ &= 308 \end{aligned}$$

The two semicircles could be put together to form one circle of radius 14 cm. The area of the circle is

$$\begin{aligned} A &= \pi r^2 \\ &\doteq \frac{22}{7}(14)^2 = \frac{22}{7}(196) = 22(28) = 616 \end{aligned}$$

So the total area of the figure is

$$308\text{ cm}^2 + 616\text{ cm}^2 = 924\text{ cm}^2$$

We summarize this section with Table 12-1.

TABLE 12-1

	Figure	Area	Perimeter
w, l	Rectangle	$A = l \cdot w$	$P = 2l + 2w$
s	Square	$A = s^2$	$P = 4s$
a, h, c, b	Triangle	$A = \frac{1}{2}bh$	$P = a + b + c$
d, r	Circle	$A = \pi r^2$	$C = 2\pi r = \pi d$ (*Note:* The perimeter of a circle is usually called the *circumference*.
b, a, h, a, b	Parallelogram	$A = bh$	$P = 2b + 2a$

Exercise Set 12-1

Part A

1. Find the area and perimeter (if possible) of each part in Fig. 12-12 (use $\pi \doteq 22/7$).

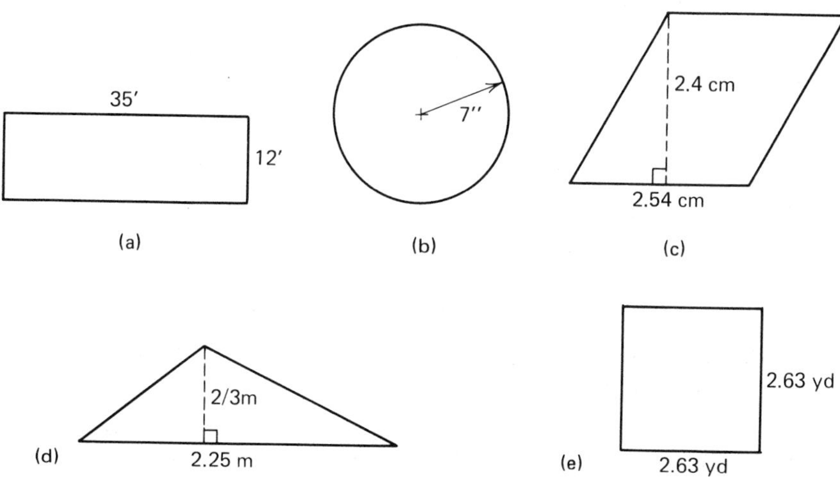

Figure 12-12

2. Find the area and perimeter (if possible) of each part in Fig. 12-13. In (c) and (d), find the area of the shaded portion and perimeters inside and outside (use $\pi \doteq 3.14$).

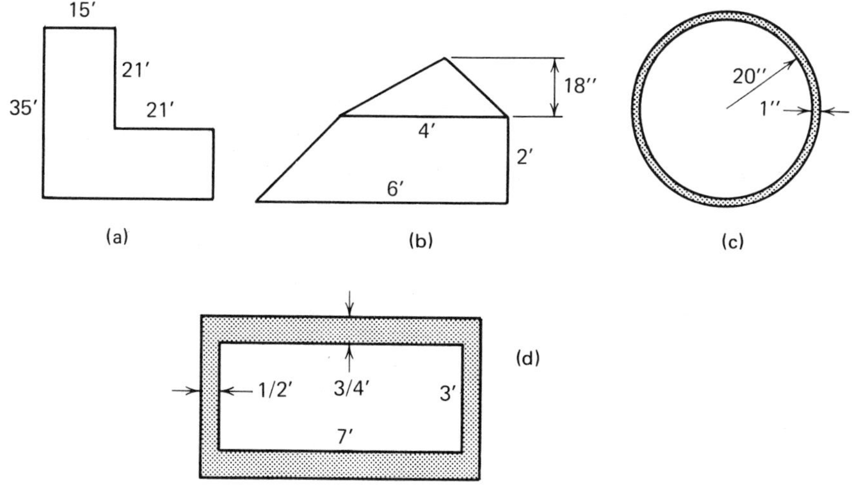

Figure 12-13

3. The diameter of a telescope is 200 in. What is the circumference of the telescope?
4. Find the area of the basketball key shown in Fig. 12-14. What is its perimeter?
5. Find the area and perimeter of the bracket pictured in Fig. 12-15.

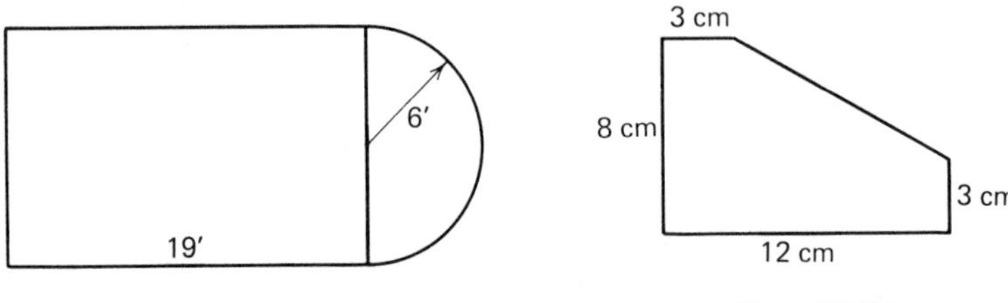

Figure 12-14

Figure 12-15

Part B

6. A warehouse will rent space at a cost of 12¢ per square foot per month. Find the minimum cost of storing 12 cartons 8 ft by 4 ft by 3 ft for 1 year. Assume the cartons cannot be stacked.
7. A circular disc 32 in. in diameter is to be cut from a square piece of metal 36 in. on a side. How many pieces of metal will be needed to cut 10 such discs? How much scrap metal will be left over?
8. Fig. 12-16 shows the cross section of a swimming pool. Find the area.

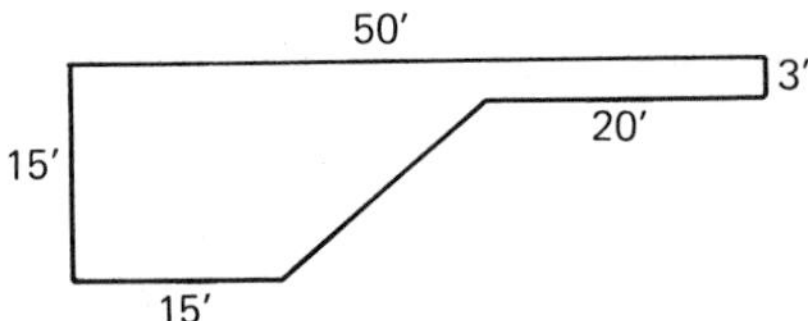

Figure 12-16

9. The floor plan of a house is given in Fig. 12-17. The owner wants to put wall-to-wall carpeting costing $15.99 per square yard in the living room, a hall runner costing $6.99 per square yard that will leave *6 in. of flooring* exposed

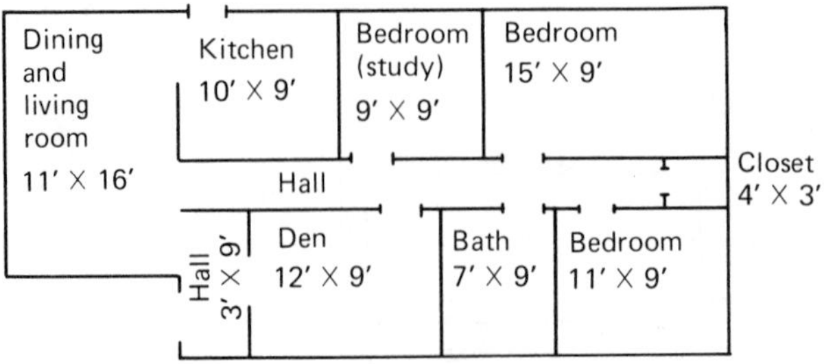

Figure 12-17

on each side, and wall-to-wall carpeting in the master bedroom costing \$12.50 per square yard. Find the cost of each type of carpet and the cost of the total job.

10. Figure 12-18 shows an external dovetail. Find the total area of this cross-sectional view.

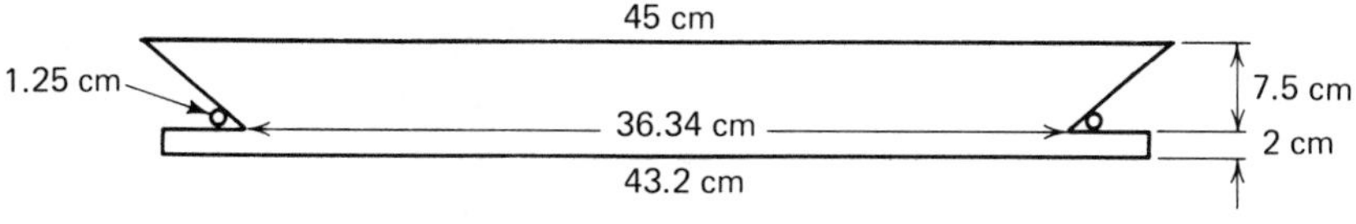

Figure 12-18

Part C

11. A machinist wants to plate the brace shown in Fig. 12-19. If the plating costs \$1.47 per square foot, what will it cost to plate this brace? The holes have a 2-in. radius. Neglect thickness.

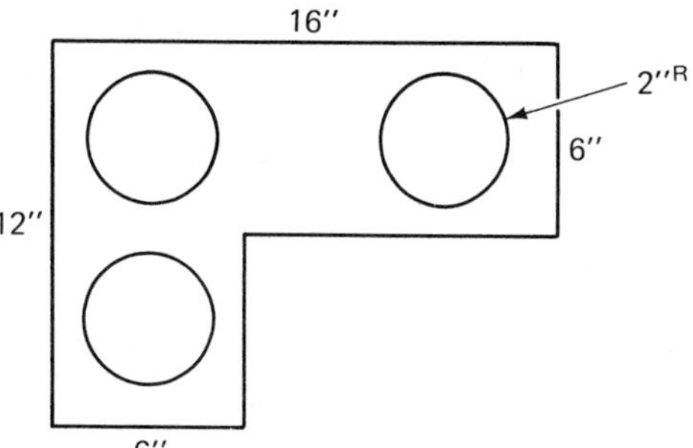

Figure 12-19

12. A racetrack owner wants to asphalt her 2-mi track. The track is 105 ft wide, and the outside edges of the ends of the two straightaways are semicircles formed by an inside radius of 140 yd. How many square yards of asphalt does she need? Use $\pi \doteq 22/7$.

13. A trapezoid is a quadrilateral with two parallel sides. For the trapezoid in Fig. 12-20, write a formula for its area. Will your formula work for all trapezoids? Explain.

Figure 12-20

14. Calculate the area of one side of the snap ring in Fig. 12-21 whose width is 2.75 cm. Use $\pi \doteq 3.1416$.

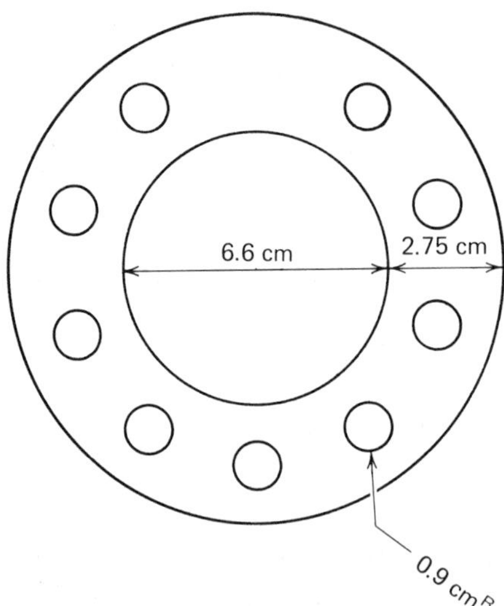

Figure 12-21

15. A machinist has an order to make 1,000 washers with an inside diameter of 1 in. and an outside diameter of 3 in. The metal for the washers costs \$3.00 per linear foot and the sheets are 10 ft wide. Allowing 0.1 in. for cutting, how much will it cost him to fill the order?

Answers

2. (a) P = 142 ft; A = 819 ft^2
   (b) P = not possible; A = 13 ft^2
   (c) IC = 125.6 in.; OC = 131.88 in; A = 128.74 $in.^2$
   (d) IP = 20 ft; OP = 25 ft; A = 15 ft^2
4. A = 284.52 ft^2; P = 68.84 ft
6. \$207.36
8. 420 ft^2
10. 401.2375 cm^2
12. 127,050 yd^2
14. 57.8761 cm^2

12-2 VOLUMES

The *volume* of an object is the amount of space it occupies. We know that space has three dimensions, namely, length, width, and height. It seems logical that volume would be measured in some three-dimensional unit, such as cubic inches ($in.^3$), cubic feet (ft^3), or cubic meters (m^3).

What are some common objects whose volume we may need to find? One example is the rectangular box pictured in Fig. 12-22. We divide this box into cubes, each of which measures 1 unit on an edge. The volume of the box will then be the

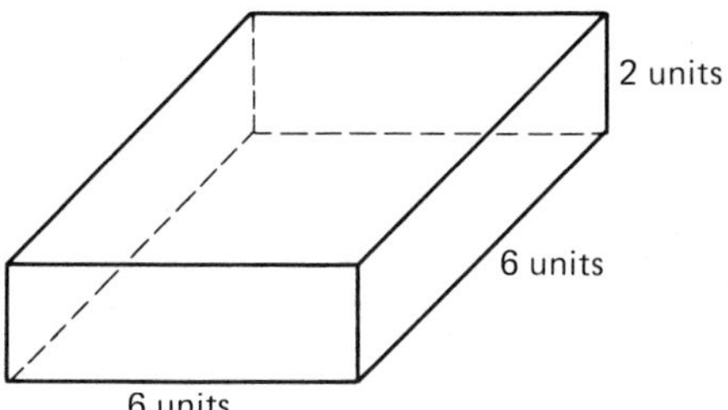

Figure 12-22. Rectangular box.

number of cubes, 1 unit by 1 unit by 1 unit, that the box could hold, as shown in Fig. 12-23.

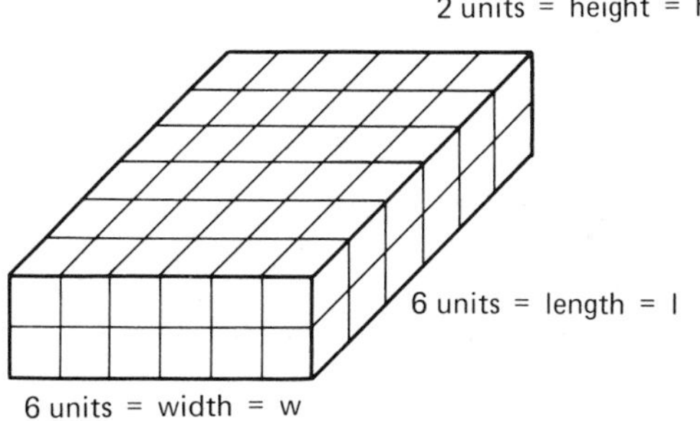

Figure 12-23. The volume of the rectangular box is determined by the number of 1-unit cubes.

In a physical situation, you could actually count the number of cubes. Most of you probably already know that we can find the number of such cubes by multiplying the length times the width times the height. (Of course, each of these dimensions must be expressed in the same units.) If we let V equal the volume, then the formula for the volume of a rectangular box is

$$V = lwh$$

We note here that since wh is equal to the area of one end of the box, we may say that we can find the volume of a rectangular box by multiplying the area of one of the ends times its length.

Finding the volume of a rectangular box does not seem difficult, but what about the objects pictured in Fig. 12-24?

> *Definition:* A right prism is a solid whose ends are parallel and whose sides are perpendicular to both ends.

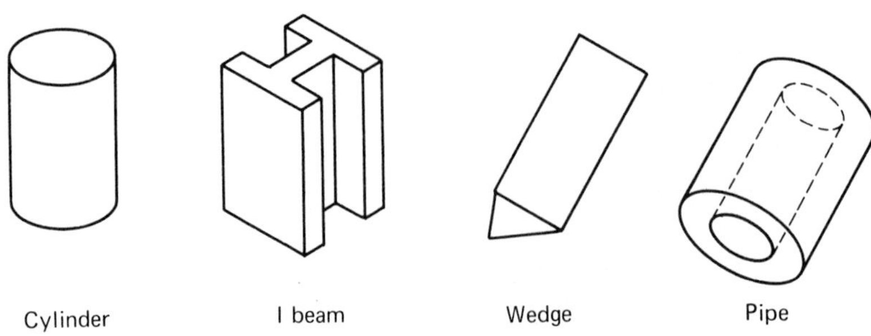

Figure 12-24. Right prisms.

It is a relatively easy task to find the volume of a right prism. As in the case of a rectangular box, we may find the area of one end and multiply it times the length of the right prism.

> The volume V of any right prism is given by the formula
>
> $$V = A \cdot l$$
>
> where A is the area of one end and l is the length (or height) of the right prism.

Example 12-3: Find the volume of concrete contained in the section of sewer pipe pictured in Fig. 12-25.

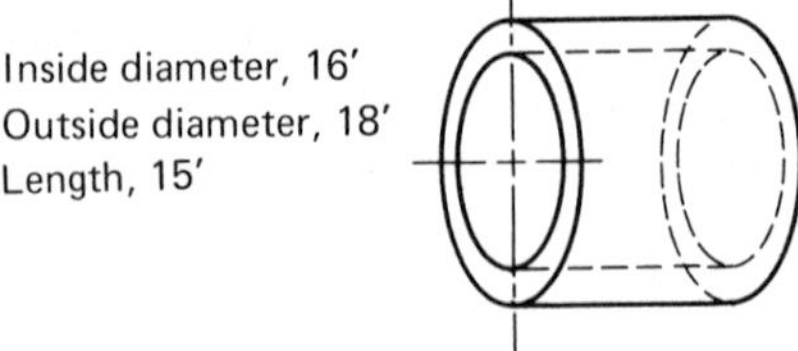

Figure 12-25

First we look at a cross section (Fig. 12-26) and find its area.

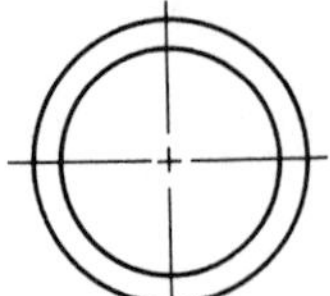

Figure 12-26

Note that the cross-sectional area can be found by finding the difference between the areas of the two circles.

Outside Circle	Inside Circle
$A = \pi r^2$	$A = \pi r^2$
$= \pi(9)^2$	$= \pi(8)^2$
$= 81\pi$	$= 64\pi$

The cross-sectional area is $81\pi - 64\pi = 17\pi$. Now we multiply this cross-sectional area times the length to find the volume.

$$V = A \times l$$
$$= 17\pi \cdot 15$$
$$= 255\pi \doteq 801$$

In addition to right prisms there are other common objects whose volumes we may need to find. Figures 12-27 and 12-28 show two of these objects, their names, and the formulas for their volumes.

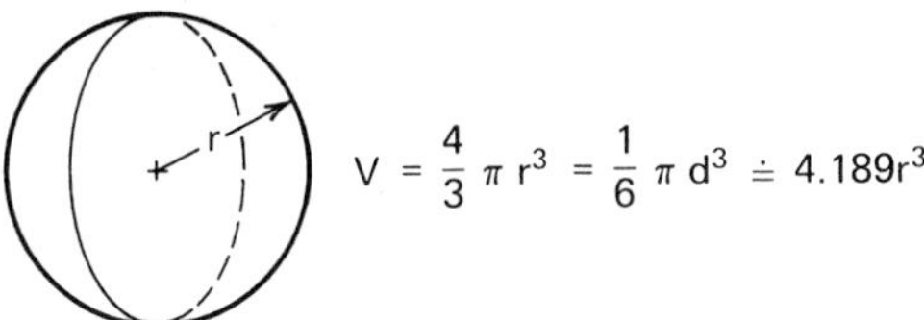

Figure 12-27. The radius of the sphere is r.

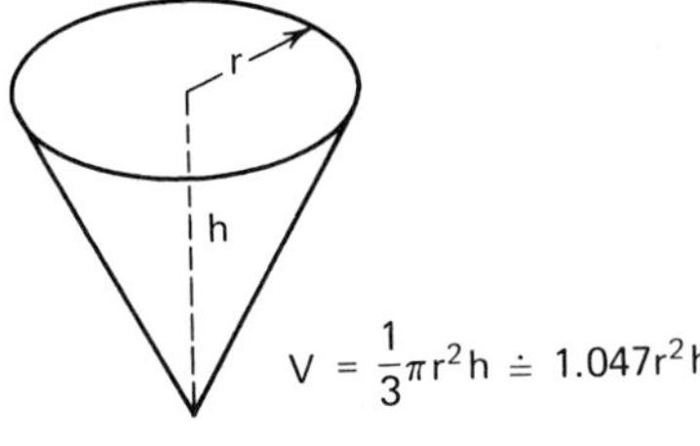

Figure 12-28. The radius of the base of the cone is r and the height of the cone is h.

Exercise Set 12-2

Part A

1. Find the volume of each of the objects pictured in Fig. 12-29. Use $\pi \doteq 22/7$.

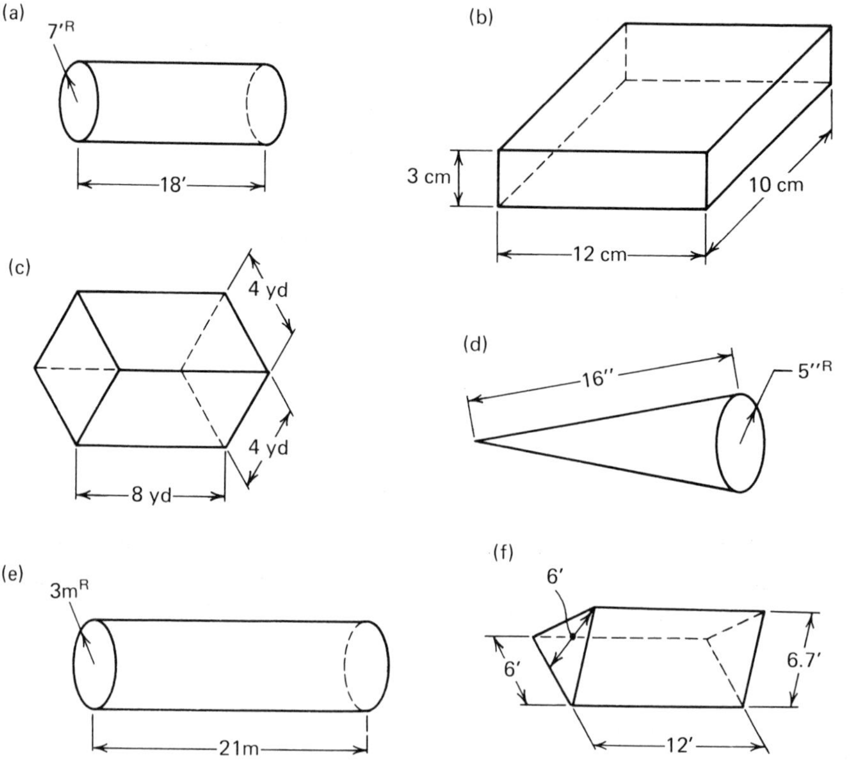

(a) 7′R 18′

(b) 3 cm 12 cm 10 cm

(c) 4 yd 4 yd 8 yd

(d) 16″ 5″R

(e) 3mR 21m

(f) 6′ 6′ 6.7′ 12′

Figure 12-29

2. Find the volume, in cubic feet, of insulation needed to fill the side walls and attic of the building shown in Fig. 12-30.

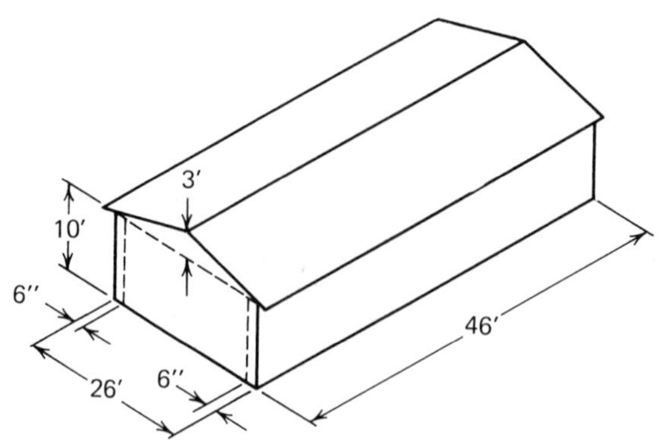

Figure 12-30

3. Find the volume of steel in the beam pictured in Fig. 12-31.

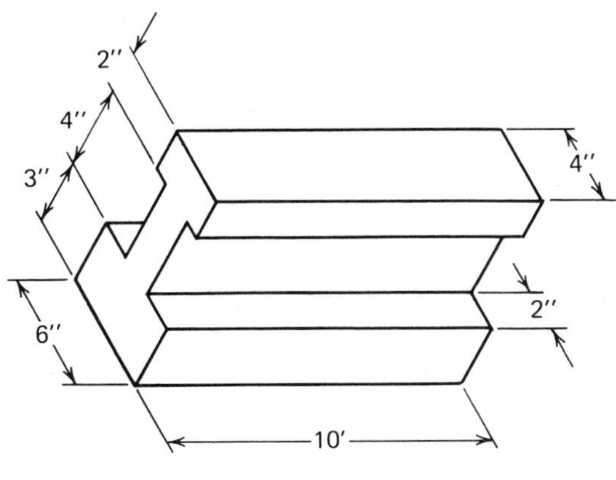

Figure 12-31

4. The diagram in Fig. 12-32 shows a square keyway cut through a steel disc. What is the volume of the disc after the cut has been made?

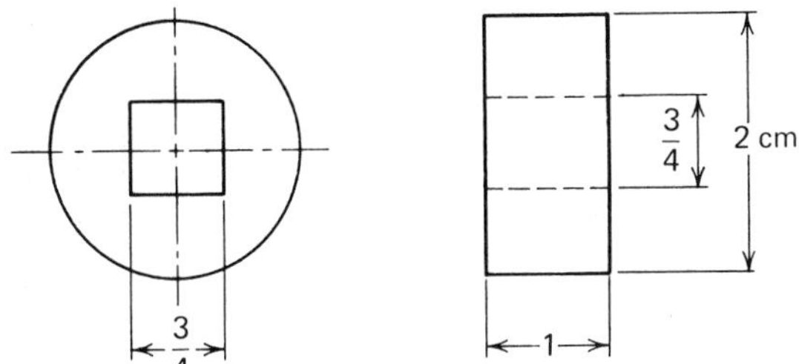

Figure 12-32

5. A gas storage tank in the shape of a sphere has an inside radius of 21 ft. What volume of gas will it hold?

Part B

6. A homeowner wants to build a concrete patio 15 ft by 12 ft and 4 in. thick. Since the homeowner will do all the work, the only expense will be the cost of the concrete. If the concrete costs $2 per cubic foot, how much will the patio cost?

7. A cylindrical pump 7 in. in diameter with a stroke length of 8 in. pumps 100 strokes per minute. If each stroke pumps the cylinder full, what volume of water will it pump in 10 min.?

8. The side view of a swimming pool is pictured in Fig. 12-33. The pool is 15 ft wide. If 1 gal of water occupies 231 in.^{3}, how many gallons of water would it take to fill the pool?

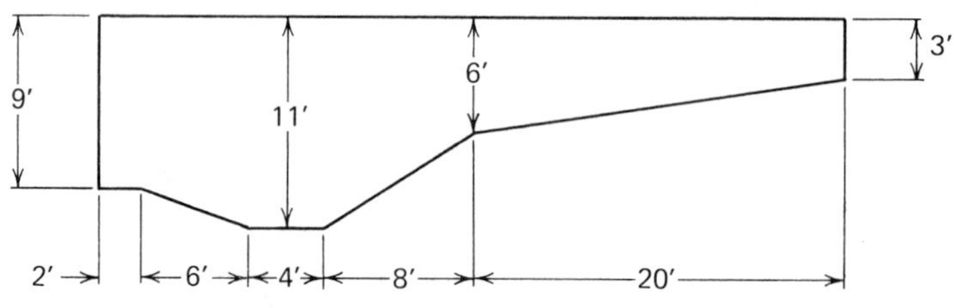

Figure 12-33

9. After consulting a contractor about the installation of a septic tank, a homeowner found that a ditch 36 ft long, 3 ft deep, and 1 1/2 ft wide and a rectangular hole 12 ft by 12 ft and 10 ft deep had to be dug. How many cubic yards of earth will be excavated?

10. Find the volume of the figures pictured in Fig. 12-34.

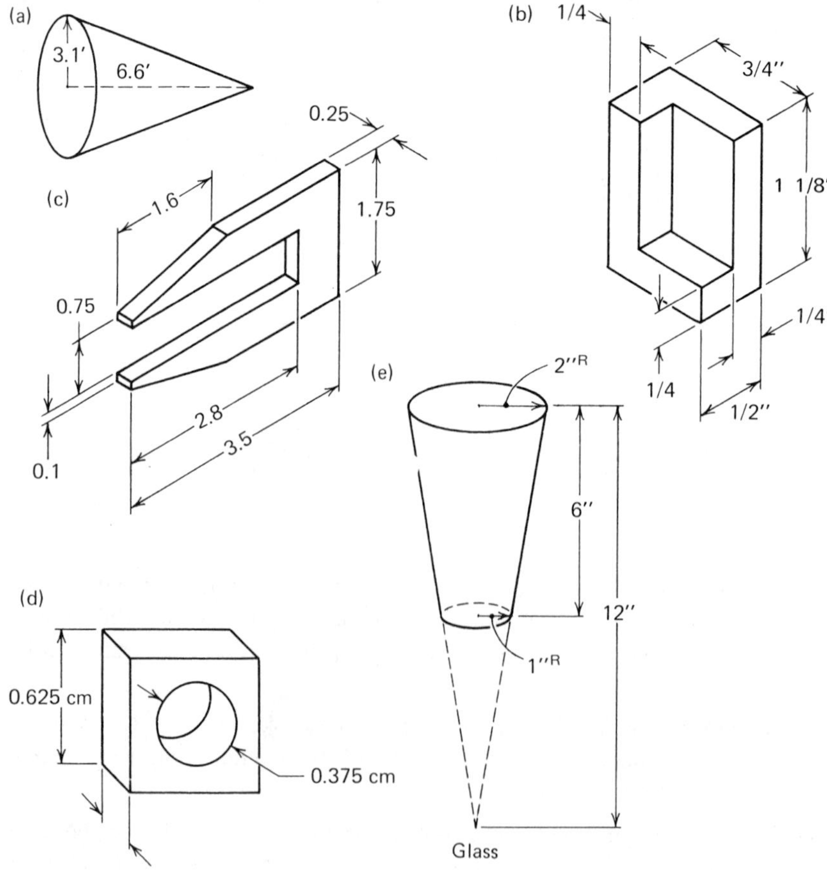

(a) 3.1′ 6.6′

(b) 1/4 3/4″ 1 1/8″ 1/4″ 1/4 1/2″

(c) 0.25 1.6 1.75 0.75 0.1 2.8 3.5

(e) 2″R 6″ 12″ 1″R Glass

(d) 0.625 cm 0.375 cm

Figure 12-34

Part C

11. A driveway 100 ft long and 9 ft wide is to be covered with a layer of crushed rock that will be spread to an average depth of 6 in. If this rock costs $11.80 per cubic yard, how much will it cost to cover the driveway?
12. Two pistons have equal stroke lengths. The radius of one is twice the radius of the other. How do their displacements compare?
13. The unit to measure liquid capacity in the metric system is the liter. In volume, 1 ℓ occupies 1,000 cc (cubic centimeters). We know from Exercise 8 that 1 gal occupies 231 in.^{3} How many liters and how many gallons will a spherical tank with a 21-ft diameter hold?
14. A shipper wants to send 200 cylindrical containers 7 ft high with 6 ft diameter and 300 cartons 4 ft by 6 ft by 5 ft. Find the total volume that he intends to ship.
15. Find the volumes of each of the figures in Fig. 12-35.

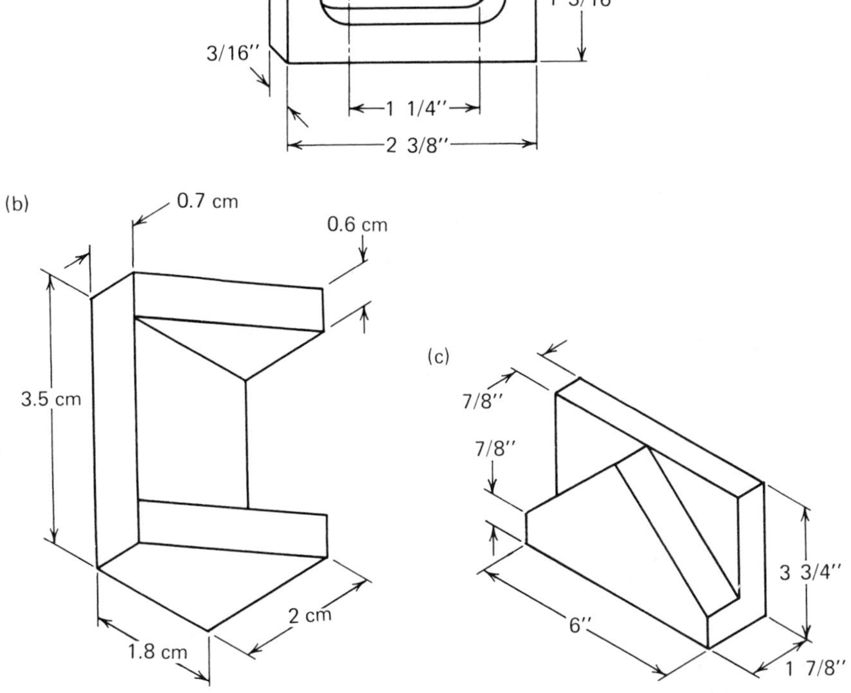

Figure 12-35

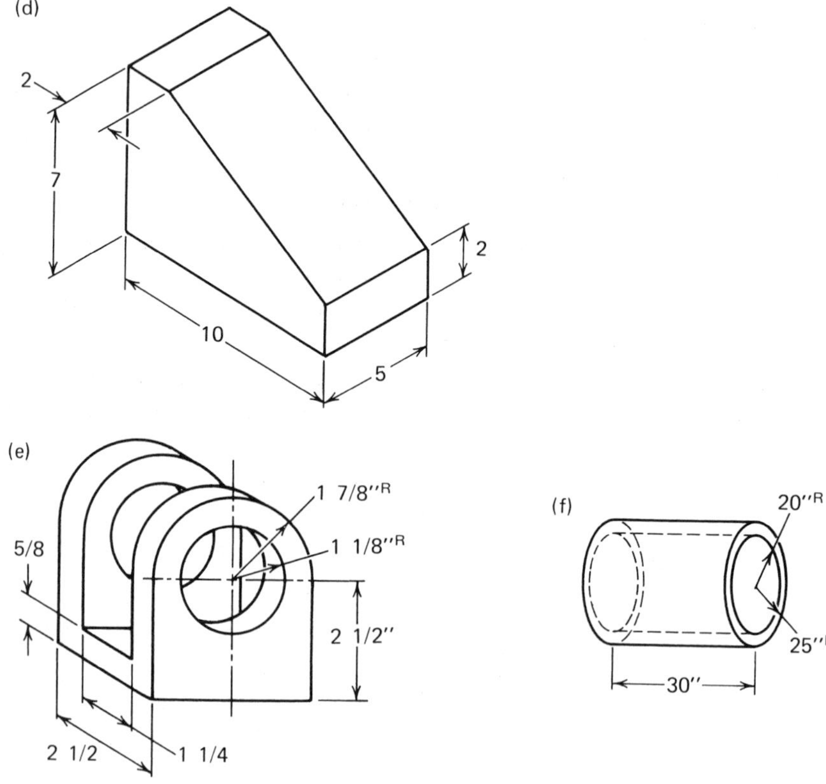

Figure 12-35. continued.

Answers

2. 2,254 ft^3
4. 2.58 cm^3
6. $120
8. 31,418.18 gal
10. (a) 66.4197 ft^3 (b) 5/16 in.^{3}
    (c) 0.8463 cubic units (d) 43.96 in.^{3} (e) 0.0876 cm^3
12. One is 4 times the other.
14. 75,564 ft^3

12-3 SURFACE AREAS

So far we have learned how to find the area of various two-dimensional figures and to find the volume of various three-dimensional objects. We are now ready to find the surface area of some three-dimensional objects.

There are two types of surface area.

> *Lateral surface area* is the sum of the areas of each of the sides.
>
> *Total surface area* is the sum of the lateral surface area and the areas of the bases.

You may ask, why do we have two types of surface area? The answer to this question may best be given by an example.

Suppose a food manufacturer wants to produce a can that will hold 10 oz of peaches. The cost of making and labeling such a can must be determined. The amount of tin needed to make the can itself could be determined by finding the can's total surface area; the amount of paper needed for the label would be equal to the lateral surface area of the can. (The tin of a given thickness costs a certain amount per square unit.) To help the manufacturer solve the problem, we need to find the lateral surface area (denoted S_L) and the total surface area (denoted S_T) for the can illustrated in Fig. 12-36.

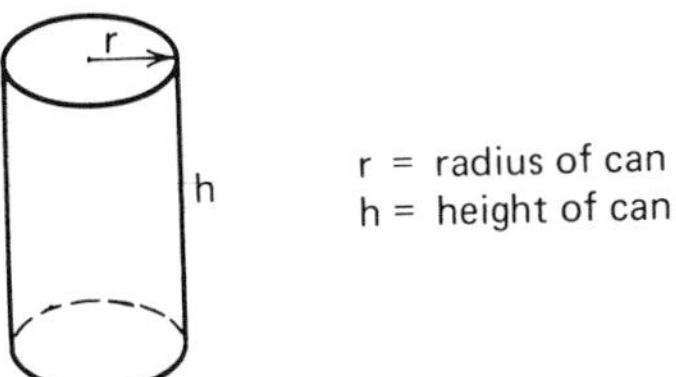

Figure 12-36. Can.

The area of the label is the lateral surface area of the can. Figure 12-37 shows the label before it is wrapped around the can. Note that the length of this rectangle must be equal to the circumference of the top of the can. (The label wraps around exactly once.)

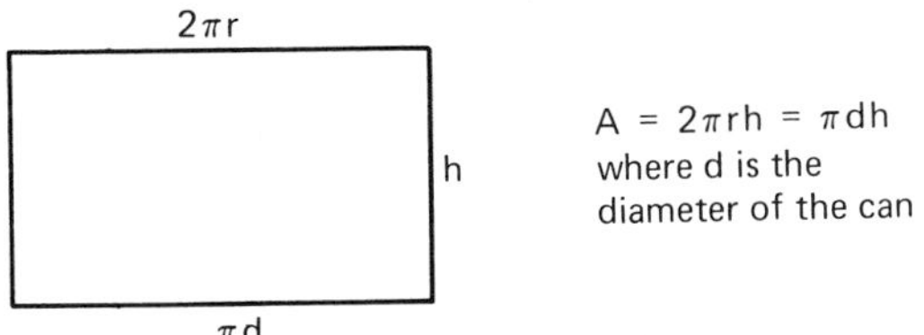

Figure 12-37. Label.

From Fig. 12-37, we can see that

$$S_L = 2\pi rh = \text{area of label}$$

The area of tin needed to produce the can will be the total surface area. To find S_T, we need to add S_L and the areas of the two bases (top and bottom). Each base is a circle of radius r, so each has area πr^2. The total surface area is

$$S_T = 2\pi rh + 2\pi r^2$$

which is the area of the tin in the can.

Exercise Set 12-3

Part A

1. Find the lateral surface area of each part in Fig. 12-38.

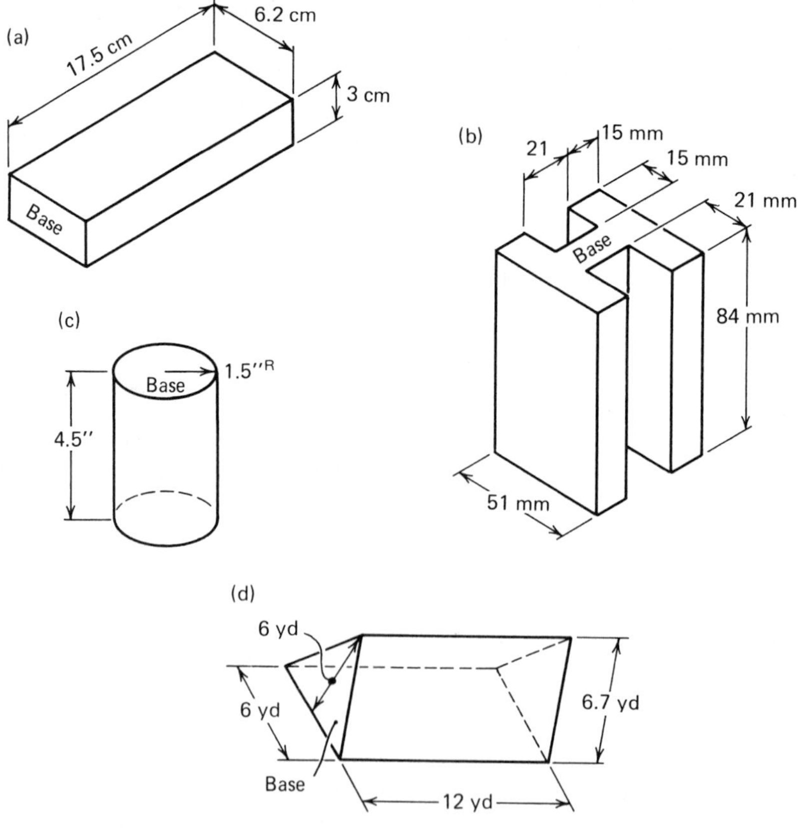

Figure 12-38

2. Find the total surface area of each of the figures in Exercise 1.
3. A room is 14 ft by 12 ft and 8 ft high. One wall has a door 6 ft 8 in. by 3 ft, and there are two windows 3 ft by 4 ft in one of the other walls. How many

square feet of wallpaper would be needed to paper the room, excluding waste?

4. A town wants to paint the outside of its water tank. The tank is shaped like a cylinder and is 100 ft high with a 70-ft diameter. If the top and bottom are included in the painting, how many square feet need to be painted?

5. Find the lateral surface area of the hexagonal (six-sided) prism pictured in Fig. 12-39.

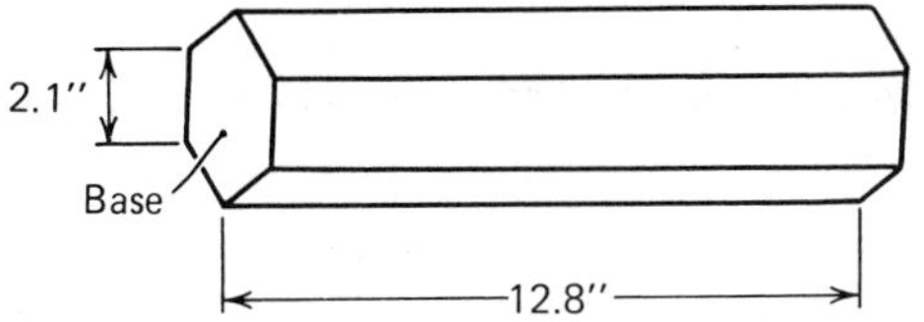

Figure 12-39

Part B

6. The pipe pictured in Fig. 12-40 is to be zinc-plated at a cost of $6.25 per square foot. What will be the total cost? Use $\pi \doteq 3.14$.

7. Find the area of the grinding surface shown in Fig. 12-41.

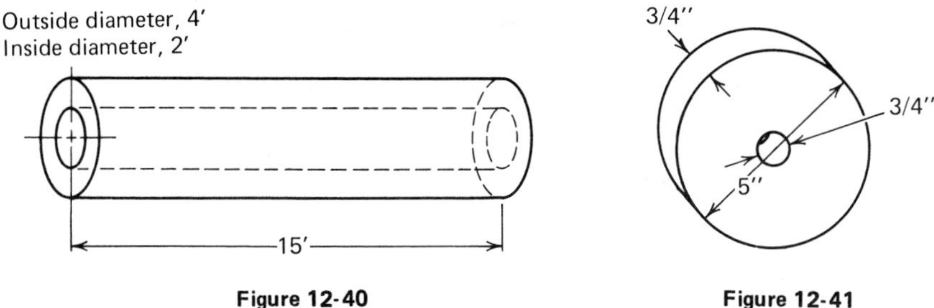

Figure 12-40

Figure 12-41

8. Find the lateral surface area and the total surface area of the solid pictured in Fig. 12-42.

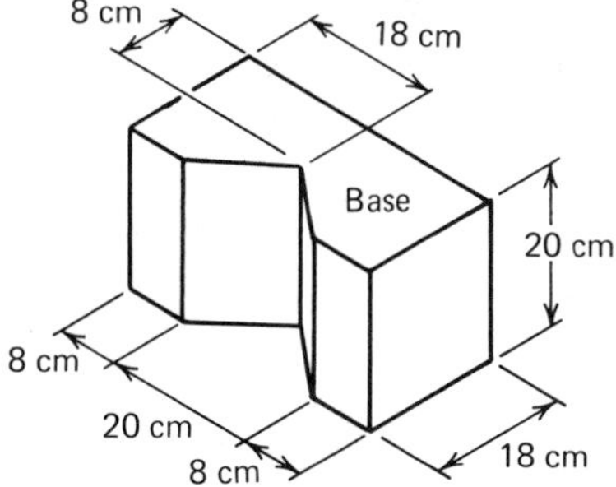

Figure 12-42

9. (a) What would be the minimum amount of paper needed to cover a shoe box 13.8 in. by 6.21 in. by 5 in.?
   (b) If paper with the store's name on it cost 15¢ per square foot, how much would it cost to cover 6,000 such boxes?

10. If the formula for the surface area of a sphere is given by $S = 4\pi r^2$ (r is the radius of the sphere), then find the amount of leather needed to make a ball with a 30.12-cm diameter.

Part C

11. If r is the radius of the ends of a cylinder and h is the height, write formulas for its total surface area S_T and its lateral surface area S_L.

12. If a pint of stain cost \$3.30 plus 4% tax and covers 20 ft², how much will it cost to stain six bookshelves like the one pictured in Fig. 12-43?

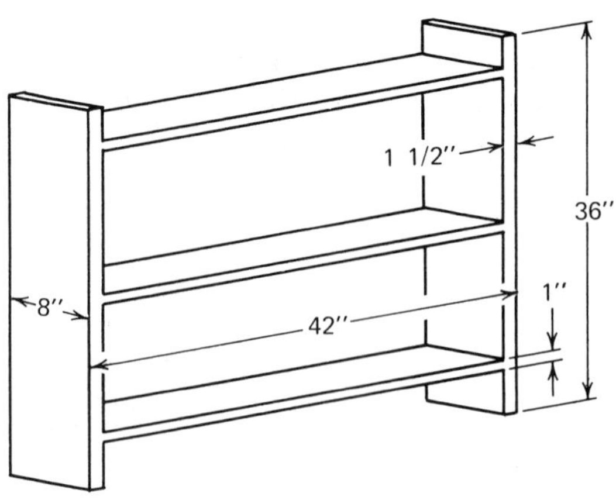

Figure 12-43

13. Find the total surface area exposed to air of a steel plate 8 ft by 6 ft by 3 ft after four 6-in. diameter holes have been bored in it.

14. Derive a formula for the total surface area of a wedge with equilateral triangle ends. Use the diagram in Fig. 12-44.

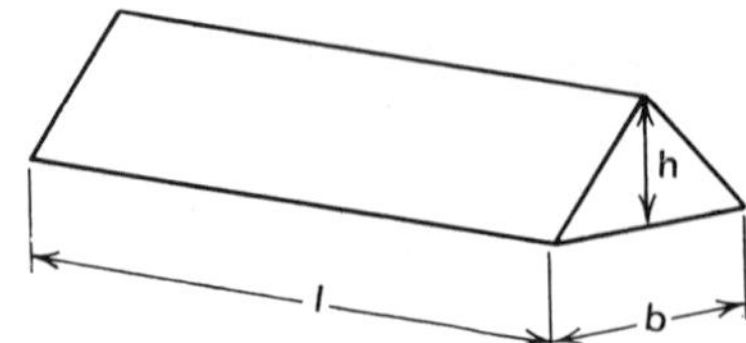

Figure 12-44

15. Find the formula for the volume of the frustrum of a cone. (See Fig. 12-45.)

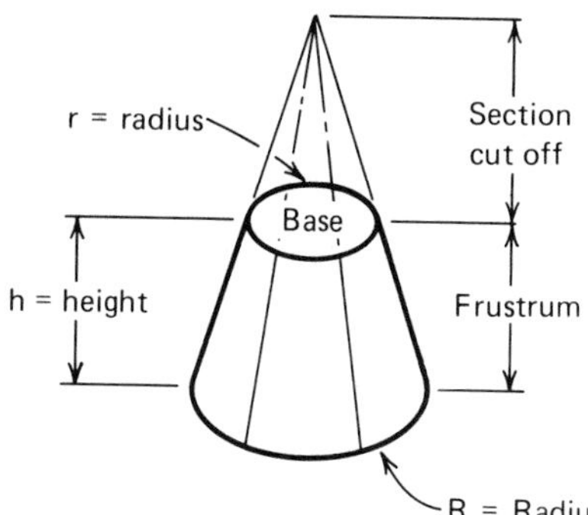

Figure 12-45

Answers

2. (a) 359.2 cm^2 (b) 26,874 mm^2 (268.74 cm^2)
   (c) 56.52 $in.^2$ (d) 268.8 yd^2
4. 29.673 ft^2
6. \$1,884.00
8. $S_L = 2{,}320$ cm^2; $S_T = 3{,}416$ cm^2
10. 2,850 cm^2
12. Must buy 8 pints for \$27.46
14. $S_T = b(3l + h)$

Review Exercises

Part A

1. Find the area and perimeter of each of the diagrams in Fig. 12-46. (Use $\pi \doteq 3.14$.)
2. Figure 12-47 is a diagram of a tennis court. Find the area and perimeter of the following:
   (a) Doubles court
   (b) Singles court
   (c) One service area

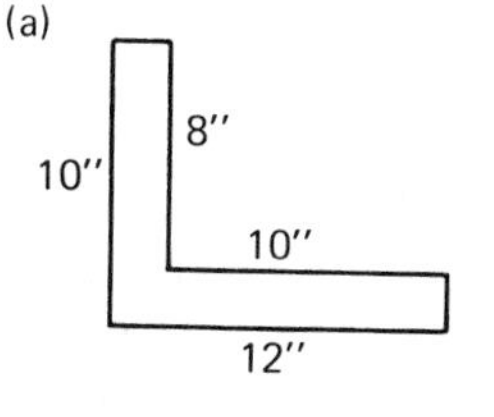

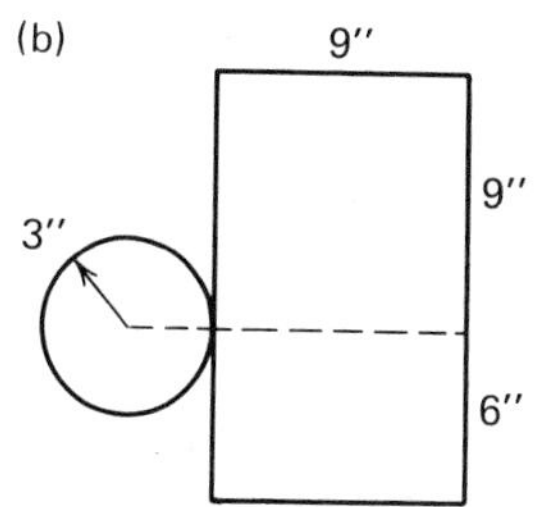

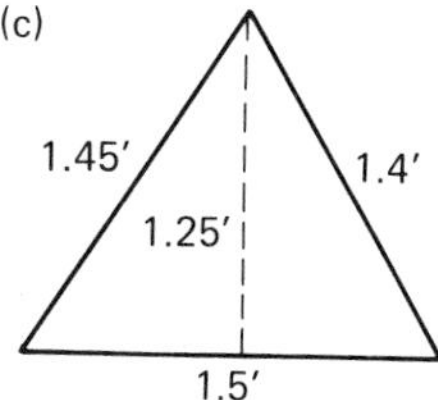

Figure 12-46

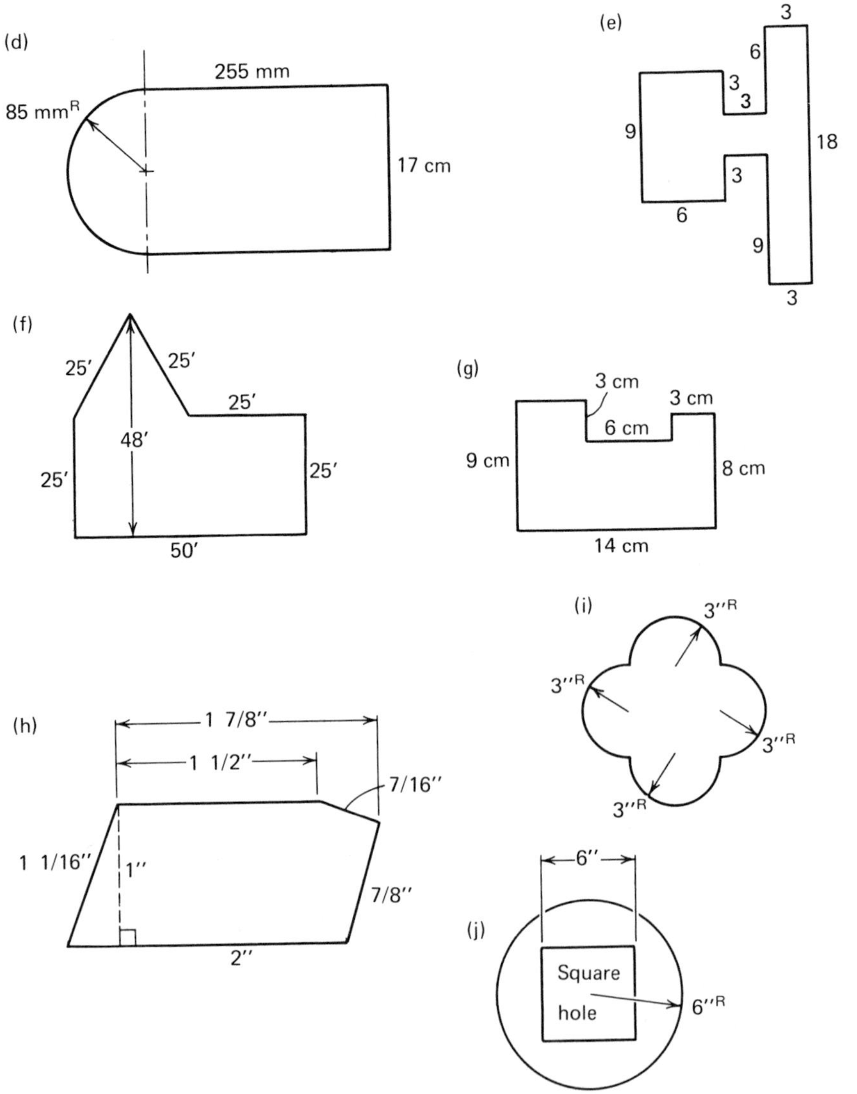

(d)

255 mm

85 mm^R

17 cm

(e)

3
6
3
3
9
18
3
6
9
3

(f)

25′
25′
25′
48′
25′
25′
50′

(g)

3 cm
3 cm
6 cm
9 cm
8 cm
14 cm

(h)

1 7/8″
1 1/2″
7/16″
1 1/16″
1″
7/8″
2″

(i)

3″^{R}
3″^{R}
3″^{R}
3″^{R}

(j)

6″
Square hole
6″^{R}

Figure 12-46. continued.

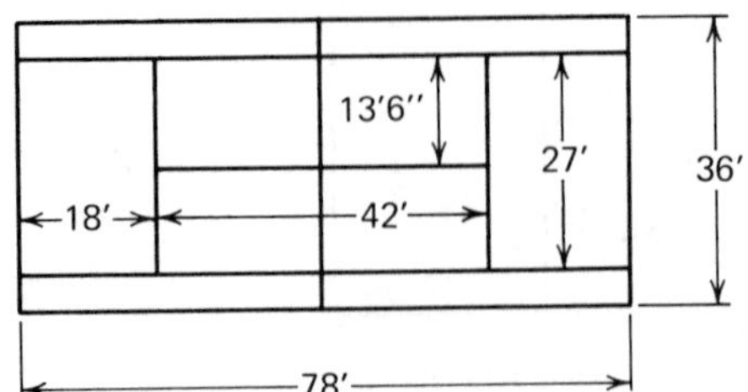

Figure 12-47

3. Find the volume of the objects pictured in Fig. 12-48 (use $\pi \doteq 3.14$).

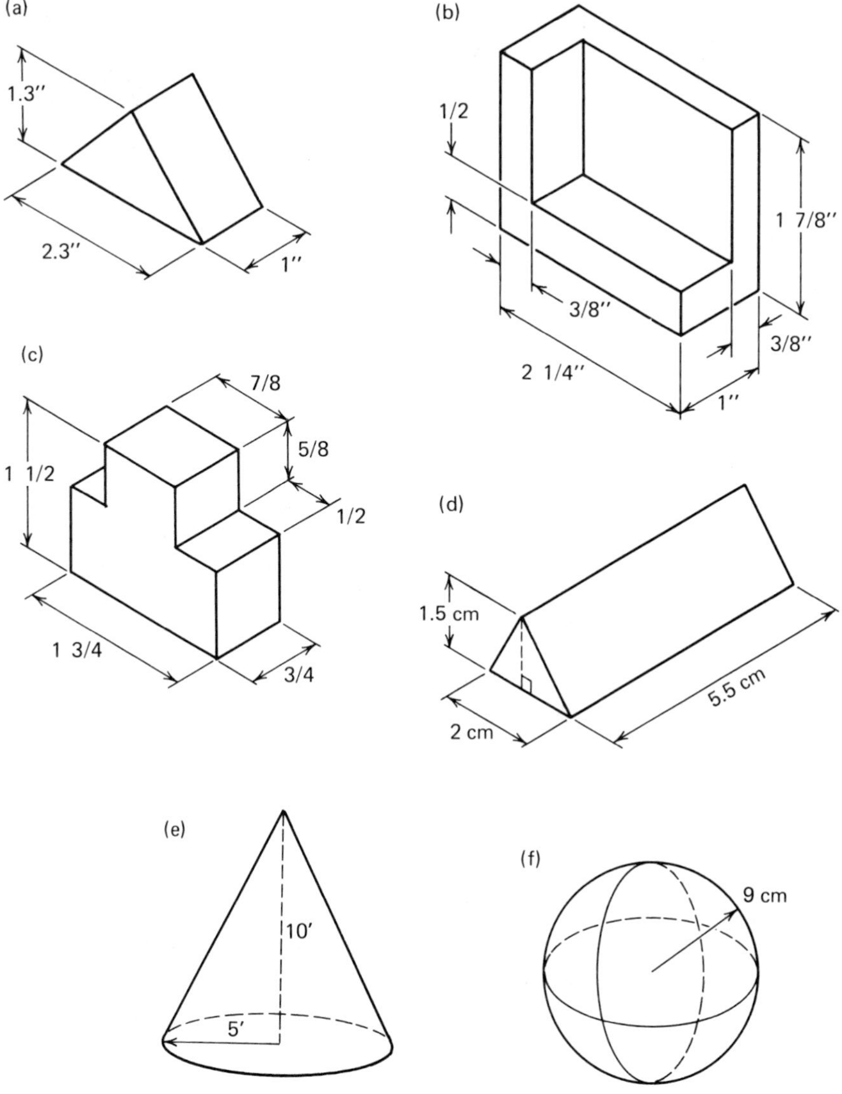

Figure 12-48

4. Find the lateral surface area and the total surface area of the objects shown in Fig. 12-49 (use $\pi \doteq 3.14$).

5. How much paint would it take to paint a room that is 25 ft by 17 ft and 8 1/2 ft high? It takes 1 qt of paint to cover 200 ft^2; the room will need two coats.

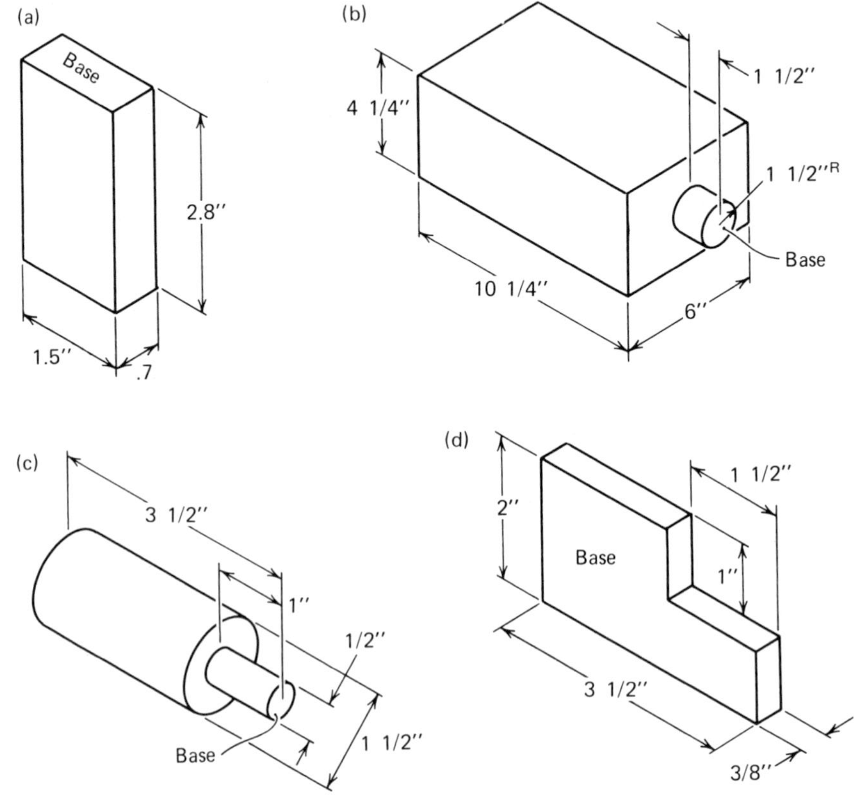

Figure 12-49

Part B

6. How many cubic yards of earth will be removed in digging a hole that is 30 ft long, 25 ft wide, and 7 ft deep?
7. What is the lateral surface area of a pipe with a 4-in. outside diameter and 24-ft length? Compute the answer to the nearest hundredth of a square foot.
8. How many 9 in. by 9 in. asphalt tiles would be required to cover the floor of a room 24 ft by 36 ft?
9. If 1 gal of liquid occupies 231 in.^{3}, how much water could be stored in a cylindrical tank 75 ft high with a 20-ft diameter?
10. What is the displacement of a six-cylinder engine with a bore of 3 in. and a stroke of 4 in.?

Part C

11. How many quarts of paint should you buy to paint a swimming pool with a flat bottom that is 100 ft by 60 ft and 8 ft deep? We shall give it two coats

of paint. One quart of paint will cover 150 ft^2 of surface, and you cannot buy part of a quart. How many gallons of water will you need to fill it after the paint dries?

Exercises 12 and 13 refer to the house plan pictured in Fig. 12-50.

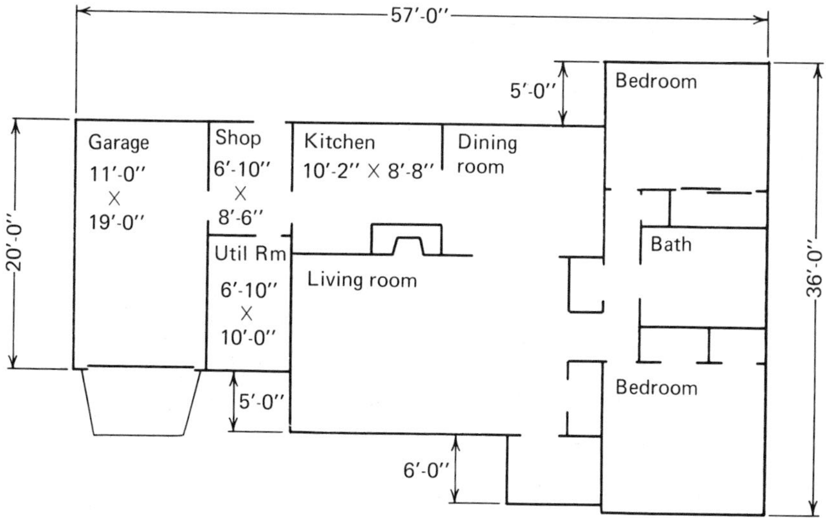

Figure 12-50

12. The builder of the house plans to put 9-in. square tiles costing 30¢ each in the utility room, shop, and kitchen. What will be the cost to do this?

13. The garage, shop, and utility room have 6-in. thick concrete floors. If the concrete costs \$2.16 per ft^3, what was the cost of constructing these floors?

14. Find the total surface area and volume of each object in Fig. 12-51 (use $\pi \doteq 3.14$).

15. An asphalt racetrack has two parallel straightaways, connected on both ends by semicircles that are not banked. If the track should be 2 mi around on the inside and it has a width of 35 yd, show that the area of the surface of the track is the same no matter how long or short the straightaways are. (See Exercise 12 of Exercise Set 12-1.) Suggested procedure:

    (a) Make a scale drawing of the track.
    (b) Let L be the length of one straight side.
    (c) Let r be the inside radius of the semicircles.
    (d) Break the 2-mi length up as the sum of four pieces of track. (This should give you an equation involving L and r; solve for L in terms of r.)
    (e) Find the area of the track by adding the areas of the four simple pieces.
    (f) Substitute the result of Step (d) into the result of Step (e).

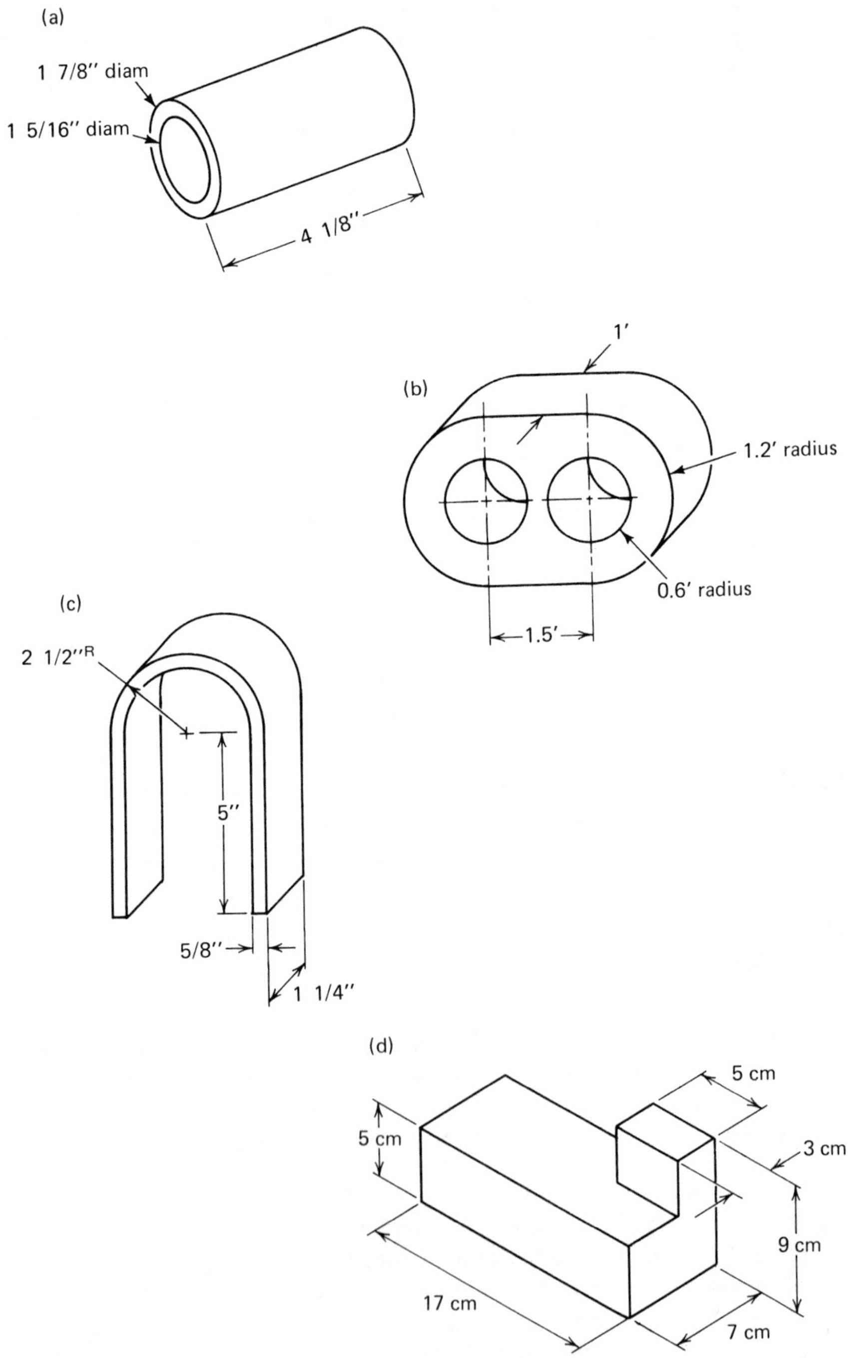

(a)

1 7/8″ diam

1 5/16″ diam

4 1/8″

(b)

1′

1.2′ radius

0.6′ radius

1.5′

(c)

2 1/2″R

5″

5/8″

1 1/4″

(d)

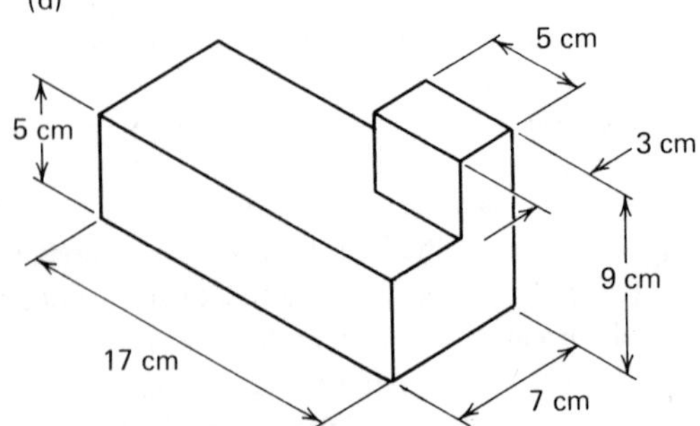

Figure 12-51

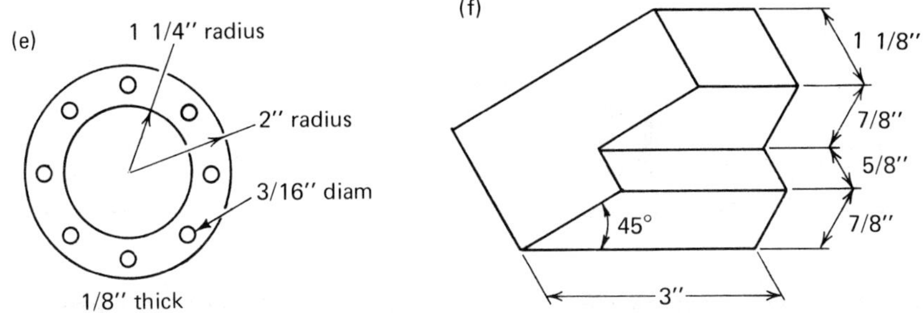

Figure 12-51. continued.

Answers

2. (a) 2,808 ft^2; 228 ft (b) 2,106 ft^2; 210 ft (c) 283.5 ft^2; 69 ft
4. (a) S_L = 12.32 in.^{2}; S_T = 14.42 in.^{2}
   (b) S_L = 242.66 in.^{2}; S_T = 276.26 in.^{2}
   (c) S_L = 13.35 in.^{2}; S_T = 16.88 in.^{2}
   (d) S_L = 4.13 in.^{2}; S_T = 15.13 in.^{2}
6. 194 4/9 yd^3
8. 1,536 tiles
10. 169.56 in.^{3}
12. Buy 382 tiles for $114.60
14. (a) S = 44.11 in.^{2}; V = 5.80 in.^{3}
    (b) S = 29.79 ft^2; V = 6.99 ft^3
    (c) S = 48.02 in.^{2}; V = 13.17 in.^{3}
    (d) S = 542 cm^2; V = 655 cc
    (e) S = 19.87 in.^{2}; V = 0.93 in.^{3}
    (f) S = 21.50 in.^{2}; V = 5.58 in.^{3}

Chapter 13

Using and Manipulating Formulas

Since ancient times, mathematicians have used formulas to express basic relationships in mathematics. These formulas range in complexity from the simple statement $A = l \cdot w$, relating the area of a rectangle to its length and width, to the complex proposition $E = mc^2$, Einstein's basis for relativity theory. A formula is really just an equation relating two or more quantities, but its tremendous power lies in the fact that the quantities can take on arbitrary numerical values. For instance $A = l \cdot w$ expresses the area of a rectangle in terms of its length and width, so this one single formula represents each of the following statements.

The area of a 2 in. by 3 in. rectangle is 2 in. · 3 in. = 6 in.^{2}

The area of a 4 cm by 15 cm rectangle is 4 cm · 15 cm = 60 cm^2.

The area of 1.2 in. by 2.0 in. rectangle is 1.2 in. · 2.0 in. = 2.4 in.^{2}

It also represents *all* the infinitely many possibilities for rectangles.

13-1 SPECIAL FORMULAS

There are many special formulas that are useful in different areas of work. We shall list a few here, but you will probably notice that some with which you are familiar are not included.

1. $d = rt$

   d = distance traveled
   r = rate of speed
   t = time traveled

2. $I = P \times i$

I = interest earned
P = principal sum
i = rate of interest

3. $A = \pi r^2$

A = area of circle
r = radius of circle

4. $\text{mph} = \dfrac{\text{rpm} \times W}{R \times 168}$

mph = speed in miles per hour
rpm = crankshaft speed
W = tire radius in inches
R = tire axle ratio

5. $V = 0.7854 \times L \times B^2$

V = displacement volume
L = stroke length
B = bore

6. $AR = 0.0025 \times (\text{mph})^2 \times FA \times DC$

AR = air resistance in pounds
mph = speed in miles per hour
FA = frontal area in square feet
DC = drag coefficient

7. $p = \dfrac{1}{n}$

p = pitch of a screw
n = number of threads per inch

8. $cs = \dfrac{\pi d \times \text{rpm}}{12}$

cs = cutting speed in feet per minute
d = diameter in inches
rpm = rotational speed of the stock

9. $L = \pi D + 2C$

L = length of pulley belt
D = diameter of pulley
C = center-to-center distance

10. $D = \dfrac{W}{17.28 \times T}$

D = density in grams per cubic centimeter
W = basis weight in pounds
T = thickness in thousandths

11. $N = \dfrac{60Q}{V}$

N = air changes per hour
Q = airflow in cubic feet per minute
V = room volume in cubic feet

12. $F = KS$

F = stress
S = strain
K = elasticity coefficient

13. $E = IR$

E = voltage in volts
I = current in amperes
R = resistance in ohms

14. $\frac{1}{R} = \frac{1}{R_1} + \frac{1}{R_2}$

R = total resistance (parallel)
R_1 = branch No. 1 resistance
R_2 = branch No. 2 resistance

Actually, Formulas 4 through 14 are technical formulas useful in special areas, and you should not be disturbed if they are all new to you. You are not expected to know these formulas.

13-2 SOLUTIONS INVOLVING ONE UNKNOWN

The most frequent use of a formula is in finding a value for one unknown (variable), given a specific numerical value for each of the other variables. For instance, if you know that a rectangle has length 8 in. and width 5.5 in. you would say

$$A = l \cdot w$$

$$A = (8 \text{ in.})(5.5 \text{ in.})$$

$$A = 44 \text{ in.}^2$$

thus finding the appropriate value for the variable representing area, given values for the length variable and the width variable.

In order to find the value for a variable, what we do is "plug in" the numerical value given for each of the other variables and proceed algebraically to solve the resulting equation for the variable we are after.

Example 13-1: In Formula 2, suppose we are told that a principal sum of \$400 earns \$18 interest and we are asked to find the rate of interest.

$$I = P \times i$$

$$\$18 = \$400 \times i$$

$$\frac{\$18}{\$400} = i$$

$$0.045 = i$$

$$4.5\% = i$$

The interest rate is 4.5%.

Example 13-2: In Formula 9, suppose we are told that a 4-ft belt fits two pulleys 18 in. apart (center-to-center) and we are asked to find the diameter of the pulleys.

$$L = \pi D + 2C$$

$$48 \text{ in.} = \pi D + 2(18 \text{ in.})$$

$$48 \text{ in.} = \pi D + 36 \text{ in.}$$

$$12 \text{ in.} = \pi D$$

$$\frac{12 \text{ in.}}{\pi} = D$$

$$D \doteq 3.82 \text{ in.}$$

The diameter of the pulleys is approximately 3.82 in.

Example 13-3: In Formula 5, suppose we want a displacement of 45 in.^3 and the bore is 3 in. What should the stroke length be?

$$V = 0.7854 \times L \times B^2$$

$$45 = 0.7854 \times L \times (9)$$

$$45 = 7.0686 \times L$$

$$\frac{45}{7.0686} = L$$

$$L \doteq 6.36$$

The stroke length should be about 6.35 in.

13-3 DERIVATION OF NEW FORMULAS FROM GIVEN FORMULAS

The purpose of this section is to illustrate the algebraic techniques for transforming a given formula into a new formula. Suppose we have the formula $d = rt$ (where d = distance traveled, r = rate of speed, and t = time traveled) and want to find the amount of time required to travel 220 mi at 50 mph. It is a simple matter to solve

$$d = rt$$

$$220 = 50t$$

$$\frac{220}{50} = t$$

$$4\frac{2}{5} = t$$

So t = 4 hr 24 min.

Suppose we wanted to do several similar calculations or we wanted to solve this problem on a computer or calculator. In such a case, a much faster method is to perform the algebraic manipulations before using numbers. That is, we solve for t before d and r take on numerical values.

$$d = rt$$

$$\frac{d}{r} = \frac{r}{r}t$$

$$\frac{d}{r} = t \quad \text{or} \quad t = \frac{d}{r}$$

Now we have a "new" formula giving travel time in terms of distance traveled and rate of speed. You might want to discuss whether or not this equation is new, but it can make solutions easier to obtain.

Review Exercises

Numbered formulas refer to the list of formulas in Sec. 13-1.

Part A

1. Use Formula 7 to find the pitch of a screw that has 16 threads per inch.
2. Use Formula 2 to find the interest earned in 1 year at 9% interest on a principal sum of $5,000.
3. Use Formula 5 to find the cylinder volume displacement of a piston with a 3-in.-bore diameter and a 4-in. stroke length.
4. Use Formula 11 to find the number of air changes per hour in a 12 ft × 10 ft × 8 ft room caused by an air conditioner with an airflow of 80 ft^3 per minute.
5. Use Formula 13 (Ohm's law) to find the voltage needed to produce 0.8 amp of current through a resistance of 150 ohms.

Part B

6. Use Formula 13 to find the current that will flow through a resistance of 45 ohms when 231 volts is applied.
7. Formula 14 gives the joint resistance when two resistances are connected in parallel. Find the joint resistance of a 10-ohm resistor and a 12.5-ohm resistor in parallel.
8. Use Formula 8 to find the cutting speed on a lathe when a metal rod 1.5 in. in diameter is rotating at 500 rpm.
9. Use Formula 1 to find the time required to travel 140 km at 40 km per hour.

10. Use Formula 4 to find the speed of a car whose wheels have a 13-in. radius if the rear axle ratio is 4.5 and the crankshaft is turning at 4,000 rpm.

Part C

11. Solve Formula 14 for R_2 and find the resistance needed parallel to a 20-ohm resistor to obtain a joint resistance of 15 ohms.
12. Solve Formula 11 for Q and find the airflow of a forced-air heating system needed to change the air in a 20-ft by 15-ft by 8-ft room once every 10 min.
13. Solve Formula 3 for r and find the radius of a circle that has an area of 707 cm^2.
14. Use Formula 6 to find the air resistance (in pounds) of a car with a 30-ft^2 frontal area and a drag coefficient of 0.4 when it is traveling at 60 mph.
15. Solve Formula 9 for D and find the diameter of the pulleys needed for a 30-in belt if the pulleys must be set 9 in. apart.

Answers

2. \$450
4. N = 5
6. 5 2/15 amp
8. 196 1/4 ft/min
10. 68.78 mph
12. $Q = (N \times V)/60$; 240 ft^3/min
14. 108 lb

Chapter 14

Graphs

Although the ability to understand and apply mathematical methods is necessary for the technician, there are times when a quick visual presentation will suffice. Any visual (nonverbal) presentation of the relationship between two quantities is known as a *graph*.

Graphs are frequently used in newspapers, magazines, pamphlets, and books to present information concisely. There are several types of graphs which one should be able to read and to construct. The basic types of graphs that we will be concerned with are the bar graph, picture graph, circle graph, and line graph.

14-1 BAR GRAPHS

> The *bar graph* is made up of vertical or horizontal bars that compare related values.

Example 14-1: The owners of an auto parts shop requested that the results of a survey be graphed. The survey was conducted to determine the average number of customers that came into the store on a daily basis over a 4-week period. The results of this survey will help to determine which days of the week would be best for each of the nine people involved in direct customer sales to get a day off since the parts shop will be open for business Monday through Saturday. Figure 14-1 shows the bar graph which was constructed from the data of this survey.

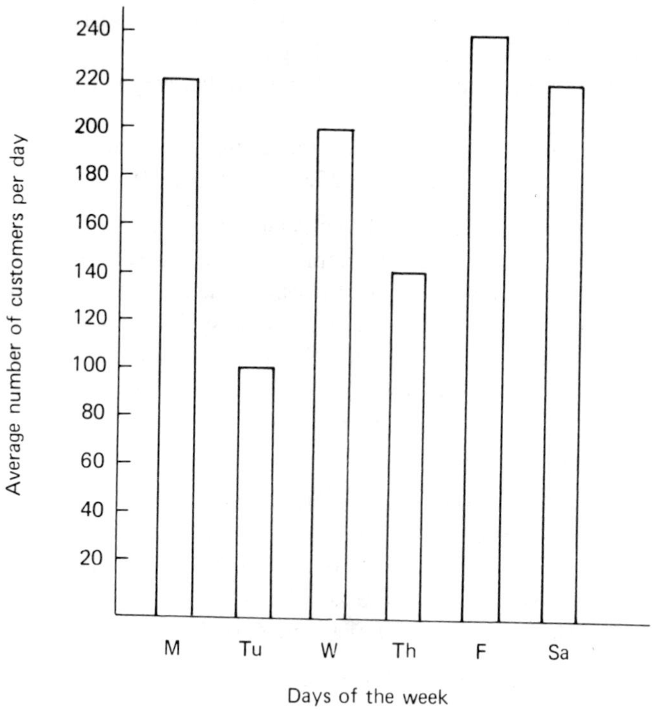

Figure 14-1. Auto parts shop customers over a 4-week period.

The horizontal scale indicates the days of the week that were used in the survey. The vertical scale indicates the average number of people that came into the shop for the recorded days of the week during a 4-week period. From the graph it is easy to tell which days of the week were slack in business activity. In the parts shop there are nine people involved in direct customer sales; therefore, the manager decided that 5 were to have Tuesday off and the other 4 were to be off work on Thursday. No persons involved in direct customer sales shall be off on other days. Do you think that the manager made the correct decision based on the graph?

Example 14-2: Construct a bar graph for the following data. Sales of five different types of wood heaters are listed in the following chart.

Name	Units Sold
Slow-flame	400,000
Hot-box	255,000
Five-side	340,000
Hot-breeze	280,000
Top-fan	390,000

When constructing a bar graph, you must determine a scale that will be large enough to take care of the largest value to be graphed.

Suggested steps to be used in constructing this bar graph are:

Step 1. Determine the range, i.e., the difference between the largest and smallest value (400,000 − 255,000 = 145,000).

Step 2. Assume that there are 10 units lined off on the graph paper. Take the range and divide by the number of units available for use (145,000 ÷ 10 = 14,500). Therefore, the least value that each division may represent is 14,500. *Note:* It may be desirable to round 14,500 up to 20,000 to be used for each division on the graph, since 14,500 would not be convenient to work with as the division. You may not round the number 14,500 down, because there aren't enough divisions on the graph paper.

Step 3. The vertical scale may be started with any number that is smaller than or equal to the lower bound. With 20,000 for each vertical unit, one could start with 240,000 on the vertical scale and label each of the divisions as 260,000, 280,000, 300,000, 320,000, 340,000, 360,000, 380,000, 400,000, and 420,000.

Step 4. Label the horizontal scale as Slow-flame, Hot-box, Five-side, Hot-breeze, and Top-fan.

Step 5. Shade in a bar to represent the number of heaters of each type.

Step 6. Name the graph [see Fig. 14-2(a)].

If you are not careful in reading the bar graph in Fig. 14-2(a), you may think that twice as many Five-side heaters were sold than Hot-breeze heaters because the

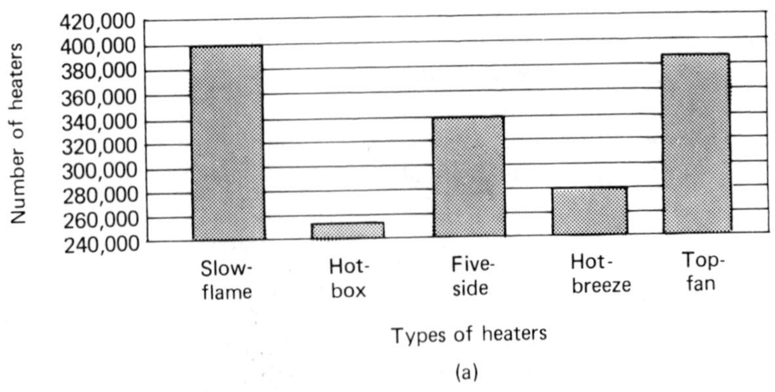

Figure 14-2. Wood heater sales.

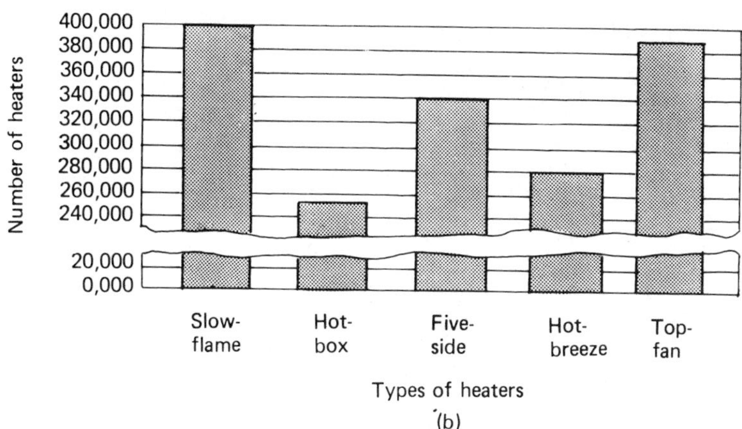

(b)

Figure 14-2. continued.

bar representing Five-side heaters sold is more than twice as high as that representing Hot-breeze heaters sold. However, you must remember that the vertical scale did not start at zero. So, you must compare the number rather than the actual height of the bars. An alternate way to show the previous graph and to draw attention to the numbers on the vertical scale is to make use of a broken space between 20,000 and 240,000. This indicates that the graph could be started with zero and continued up to two hundred forty thousand. The broken space in the bar graph represents 220,000 [see Fig. 14-2(b)].

14-2 PICTURE GRAPHS

> The *picture graph* makes use of appropriate symbols (pictures) that represent specific numbers of things that are being compared.

Example 14-3: Suppose that you took the results of several surveys that were taken during the period of years 1960 to 1980 and constructed a picture graph. For the surveys, the data represent the number of persons buying a small car or the number of persons buying a large car. This data are represented in the picture graph in Fig. 14-3.

Use the picture graph in Fig. 14-3 to answer the following questions. Answers are given in parentheses.

1. Each car represents what percent of the car buyers? (Each car represents 20% of the car buyers.)
2. In 1975 about what percent of the car buyers purchased large cars? (About 80% of the car buyers purchased large cars in 1975.)

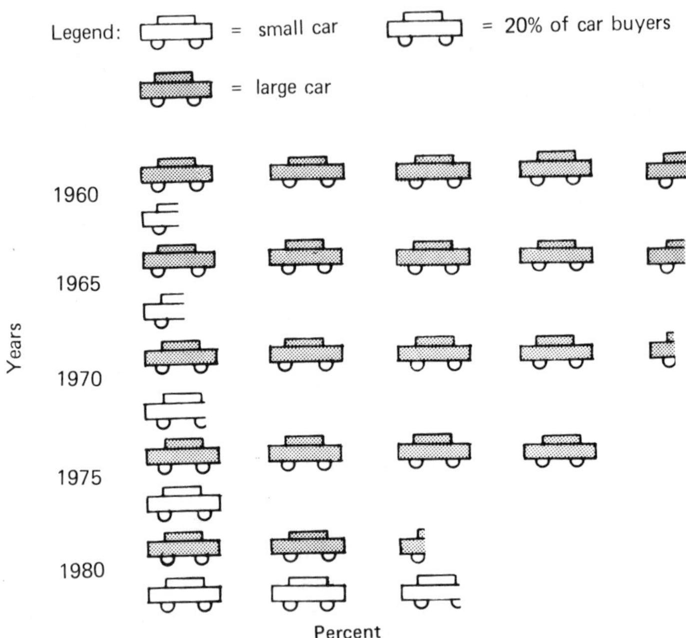

Figure 14-3. Percentage of car buyers, 1960 to 1980: small car versus large car.

3. In 1960 about what percent of the car buyers purchased small cars? (About 10% of the car buyers purchased small cars in 1960.)
4. In 1970 can you tell how many small cars were bought? (No. You can tell only the percentage of car buyers.)
5. In 1980 what was the approximate percent of car buyers who purchased small cars? (About 55% of the car buyers purchased small cars in 1980.)
6. Using the graph, can you explain why more people bought small cars in 1980 than in 1960? (No.)

Example 14-4: In a southeastern city, the percent of cars having air conditioning was about 20% in 1965, about 30% in 1970, about 40% in 1975, and about 70% in 1980. Construct a picture graph to demonstrate this data.

When constructing a picture graph, we must decide on how much of the total data that each picture will represent. The problem above involves percent of cars with air conditioning; therefore, we must decide on what percent each picture will represent. All of the percents in the given data are multiples of 10. Hence, we may let each picture represent 10% of the total. For example, let each car represent 10%; white cars have air conditioning, black cars do not have air conditioning. (*Note:* In deciding on the kind of picture that should represent the data in a problem for a picture graph, you should decide on a picture that relates to the given data.)

Step 1. Label the vertical axis at 10%, 20%, 30%, 40%, 50%, 60%, 70%, 80%, 90%, and 100%.

Step 2. Label the horizontal axis 1965, 1970, 1975, and 1980.

Step 3. (a) Since 20% of the cars had air conditioning in 1965, shade all the cars above 20% for the year 1965.
(b) Since 30% of the cars had air conditioning in 1970, shade all the cars above 30% for the year 1970.
(c) Since 40% of the cars had air conditioning in 1975, shade all the cars above 40% for the year 1975.
(d) Since 70% of the cars had air conditioning in 1980, shade all the cars above 70% for the year 1980.

Step 4. Name the graph (see Fig. 14-4).

Note: Although Fig. 14-3 and Fig. 14-4 are drawn differently, they both are picture graphs and either form could be used to represent data. In many instances there is more than one correct way to construct a graph. You should try to follow

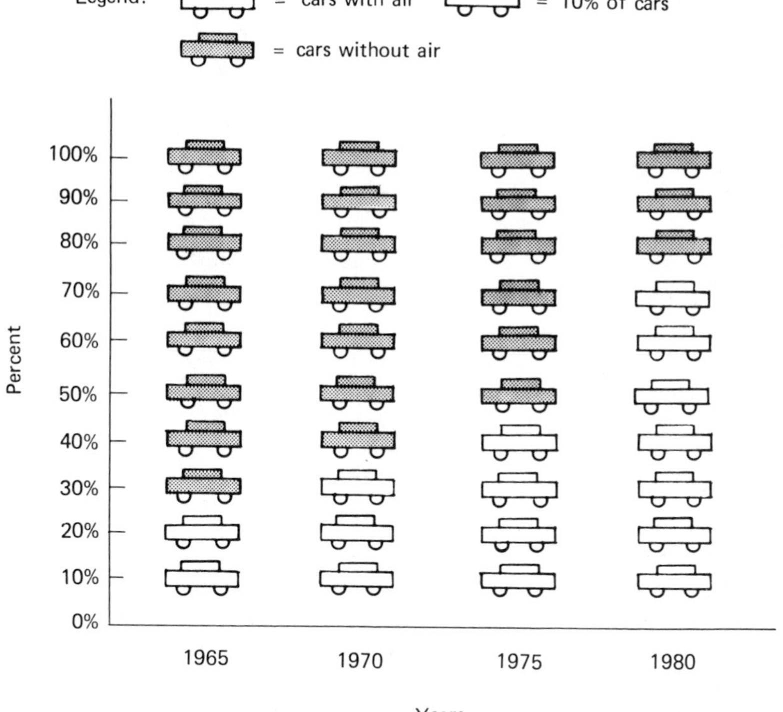

Figure 14-4. Percent of cars with and without air conditioning from 1965 to 1980.

the instructions given to you, and if you are not directed as how to construct the graph, you should construct it in the best manner you can.

14-3 CIRCLE GRAPHS

The *circle graph* presents the breakdown of the whole quantity into its parts so that the parts may be compared or any number of parts may be compared with the whole quantity.

Example 14-5: The cost of owning a new car over a 6-year period is $14,400.

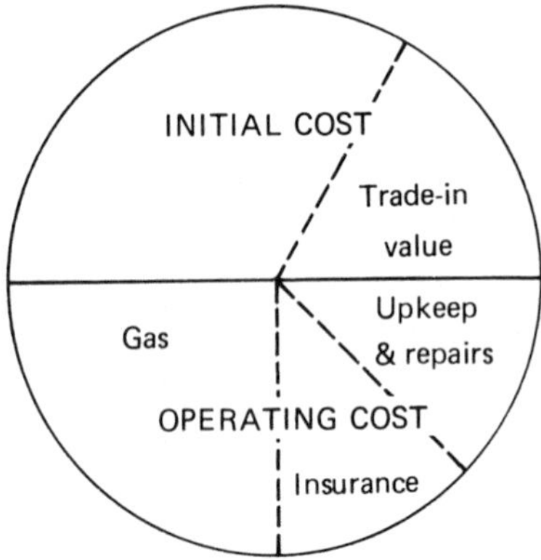

Figure 14-5. Cost of owning a new car for 6-year period.

Use the circle graph in Fig. 14-5 to answer the following questions. Answers are given in parentheses.

1. What was the original (initial) cost of the car? (The original cost was one-half the total cost: 1/2 of $14,400 = $7,200.)
2. What was the cost of gas for the 6-year period? (The cost of gas was one-half the operating cost or one-fourth of total cost: 1/2 of $7,200 = 1/4 of $14,400 = $3,600.)
3. What was the total cost of insurance? (The cost of insurance was one-fourth of the operating cost: 1/4 of $7,200 = $1,800.)
4. What was the total cost of upkeep and repairs? (The cost of upkeep and repairs was one-fourth of operating cost: 1/4 of $7,200 = $1,800.)
5. What was the trade-in value of the car at the end of 6 years? (The trade-in value after 6 years was one-third of initial cost: 1/3 of $7,200 = $2,400.)
6. What was the cost of owning the car for 6 years less the trade-in value? (The cost of owning the car was the initial cost less the trade-in value: $14,400 – $2,400 = $12,000.)

7. What was the average cost per year? (The average cost per year was 1/6 of \$12,000 or \$2,000.)

Example 14-6: Construct a circle graph of the projected operating expenses of an auto parts distributor for the year 1990 for the following data.

Office supplies	\$5,000
Up-keep	\$5,000
Miscellaneous	\$5,000
Payment on loan	\$10,000
Utilities	\$10,000
Salaries	\$65,000
Delivery cost	\$20,000
Total cost	\$120,000

When constructing a circle graph, convert each item (part) to be graphed into a ratio. Determine the amount of the circle to be used for each item by multiplying the ratio for each item by 360°. Use a compass to draw the circle, and then use a protractor to measure the angles needed for each item.

$$\frac{5{,}000}{120{,}000} = \frac{5}{120} \qquad \frac{5}{\cancel{120}_{1}} \times \frac{\cancel{360}^{3}}{1} = 15^\circ \text{ represents office supplies}$$

$$\frac{5{,}000}{120{,}000} = \frac{5}{120} \qquad \frac{5}{\cancel{120}_{1}} \times \frac{\cancel{360}^{3}}{1} = 15^\circ \text{ represents up-keep and repairs}$$

$$\frac{5{,}000}{120{,}000} = \frac{5}{120} \qquad \frac{5}{\cancel{120}_{1}} \times \frac{\cancel{360}^{3}}{1} = 15^\circ \text{ represents miscellaneous}$$

$$\frac{10{,}000}{120{,}000} = \frac{1}{12} \qquad \frac{1}{\cancel{12}_{1}} \times \frac{\cancel{360}^{30}}{1} = 30^\circ \text{ represents payment on loan}$$

$$\frac{10{,}000}{120{,}000} = \frac{1}{12} \qquad \frac{1}{\cancel{12}_{1}} \times \frac{\cancel{360}^{30}}{1} = 30^\circ \text{ represents utilities}$$

$$\frac{65{,}000}{120{,}000} = \frac{13}{24} \qquad \frac{13}{\cancel{24}_{1}} \times \frac{\cancel{360}^{15}}{1} = 195^\circ \text{ represents salaries}$$

$$\frac{20{,}000}{120{,}000} = \frac{1}{6} \qquad \frac{1}{\cancel{6}_{1}} \times \frac{\cancel{360}^{60}}{1} = 60^\circ \text{ represents delivery cost}$$

Now represent the above information on a circle graph (see Fig. 14-6).

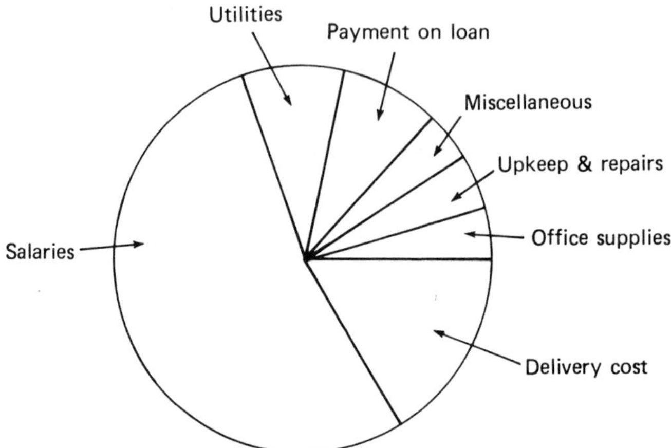

Figure 14-6. Projected operating expenses of an auto parts distributor in 1990.

14-4 LINE GRAPHS

> *Line graphs* are used to record continuous change over a specific time period. A line graph makes use of two axes, the vertical and the horizontal.

The line graph makes use of two axes, which are perpendicular lines labeled X (horizontal) and Y (vertical), to represent the two quantities being related. The point of intersection of the two axes is called the *origin* and is usually given the value of 0 (zero) on the X axis and the value of 0 (zero) on the Y axis.

Points to the right of the origin have a positive X coordinate.

Points to the left of the origin have a negative X coordinate.

Points above the origin have a positive Y coordinate.

Points below the origin have a negative Y coordinate.

For instance, a point P on the graph, which is 2 units to the right of the origin and 3 units below the origin, has an X coordinate of +2 ($x = +2$) and a Y coordinate of −3 ($y = -3$). (See Fig. 14-7.)

The units along the axes are labeled as a scale drawing, with some convenient distance on the graph representing the units of the quantity being discussed. The two axes are usually referred to as the X axis and the Y axis; but when they are labeled with the physical units, they can be referred to as the time axis or distance axis or speed axis or income axis or whatever is appropriate.

Figure 14-8 is a graph of gas mileage (on the Y axis) versus speed (on the X

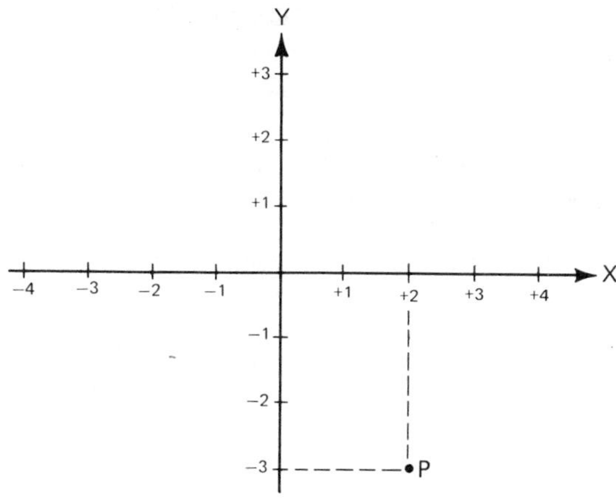

Figure 14-7

axis). Notice that only the positive values of speed and gas mileage make sense, so we omit all but the upper-right portion of the coordinate system. This situation is typical of most physical situations.

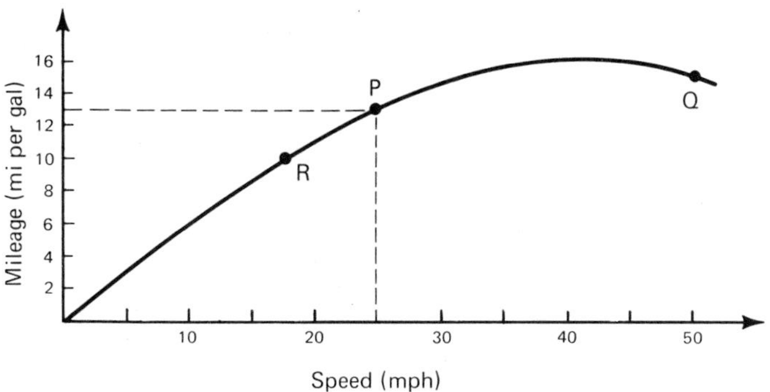

Figure 14-8. Gas mileage versus speed.

In order to read a graph such as Fig. 14-8, we identify a point on the curve and locate its coordinates. As shown by the vertical dashed line in Fig. 14-8, point P is directly above 25 mph on the speed axis, so we can say that point P represents 25 mph. This same point is to the right of a value of the mileage axis, which we can say is about 13 mi per gallon, as shown by the horizontal dashed line. So point P tells us that this car gets about 13 mi per gallon when traveling at 25 mph.

Point Q tells us that this car gets about 15 mi per gallon when traveling at 50 mph, and point R tells us that this car can get 10 mi per gallon by traveling at approximately 18 mph. To obtain practice in reading graphs, you should hold a ruler in the vertical and horizontal positions at points Q and R and verify our results given in the preceding sentence.

In reading a graph, great care should be taken to get the dashed lines or ruler positions as close to vertical and horizontal as possible. The information read from a graph is approximate at best, so we must limit as many of the possible errors as we can. If exact information is needed, you should try to locate it in a table or a listing rather than from a graph.

When constructing line graphs, there are standard guidelines for determining which quantity should go on which axis:

1. If one quantity is part of the other quantity, put the partial quantity on the Y axis.

Example 14-7: In a graph of food expenditures versus family income (see Fig. 14-9), put food expenditures on the Y axis and family income on the X axis.

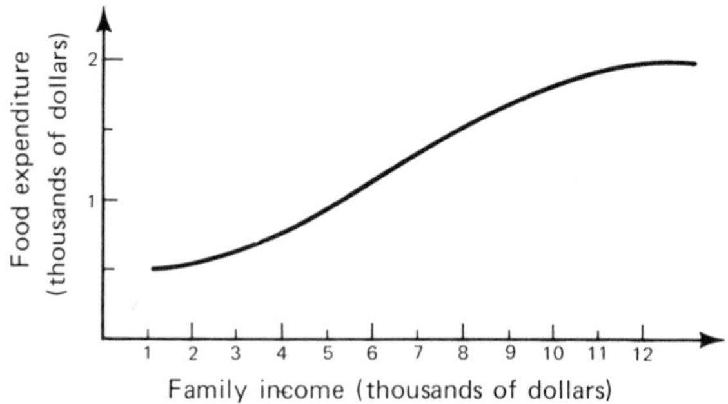

Figure 14-9. Food expenditure versus family income.

2. If one quantity is caused by the other quantity, put the caused quantity on the Y axis.

Example 14-8: In a graph of brake horsepower versus rpm of an engine (see Fig. 14-10), put brake horsepower on the Y axis and rpm on the X axis.

3. If time is one of the quantities, it is usually put on the X axis.

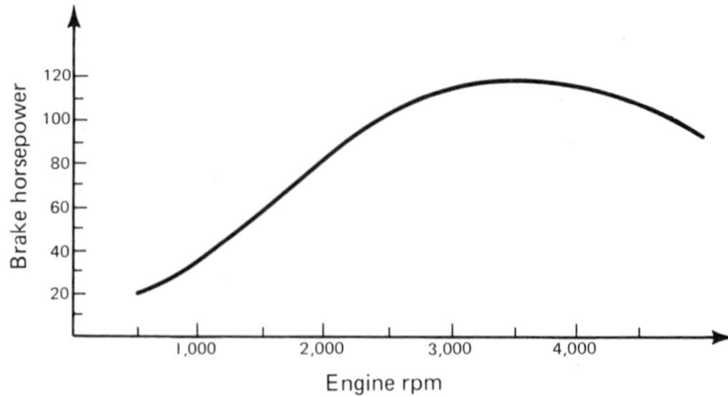

Figure 14-10. Brake horsepower versus rpm.

Example 14-9: In a graph of the standard fire test time-temperature curve (see Fig. 14-11), time is put on the X axis and temperature is put on the Y axis.

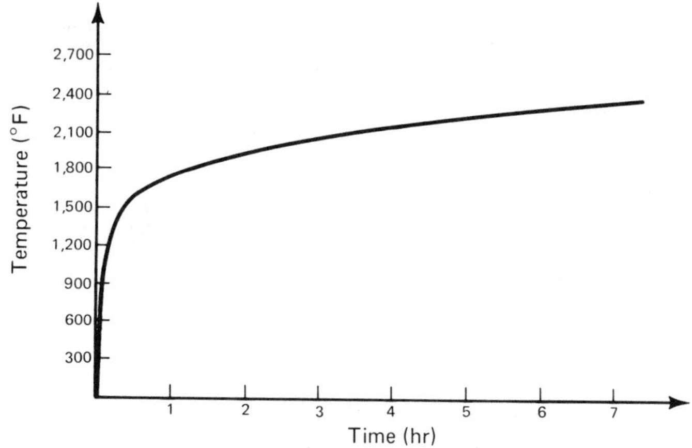

Figure 14-11. Standard fire test.

After we have decided which quantity goes where, we need to label the axes with numerical values of the quantities. Here is a general suggestion:

> If you have a size in mind for the graph, choose your units so that the graph will come close to that size, but remember that multiples of 2, 5, or 10 are easiest to work with.

Example 14-10: See Fig. 14-12. If you want the horizontal axis to be 4 in. long and the water flow ranges from 250 to 400 liters per hour, you need to cover a 150-unit range within 4 in. on the graph. Since 150/4 = 37.5, a choice of 40 might be appropriate. Thus, the inch divisions could now be labeled 250, 290, 330, 370, and 410, with the notation *liters per hour* below them.

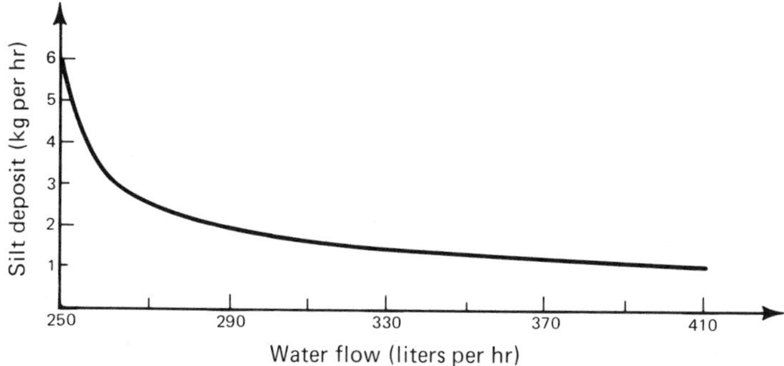

Figure 14-12. Silt deposit versus water flow.

If you chose 50 units, your graph would cover 3 in. along the horizontal axis.

If you chose 30 units, your graph would cover 5 in. along the horizontal axis.

It should be pointed out here that the intersection of the two axes does not have to be labeled 0. In the example above, the horizontal axis starts at 250 ℓ per hour.

After the axes are labeled, we are ready to plot points and sketch a curve through them. Suppose that you have the following information about the yield point of a cast-iron part at different temperatures.

At 50°F it will withstand up to 80,000 lb per square inch (psi).

At 75°F it will withstand up to 70,000 psi.

At 100°F it will withstand up to 45,000 psi.

At 150°F it will withstand up to 31,000 psi.

At 225°F it will withstand up to 26,000 psi.

Since temperature influences yield point, the yield point should go on the Y axis and temperature should go on the X axis. The axes are labeled with appropriate values of temperature and psi, and we are now ready to plot the points.

To plot a point reflecting the 80,000 psi at 50°F, start at 50 on the X axis and move directly above it to a height across from 80,000 on the Y axis. Mark this point with a heavy dot or a small x. Repeat this procedure for each point, and you should have a set of points similar to those in Fig. 14-13.

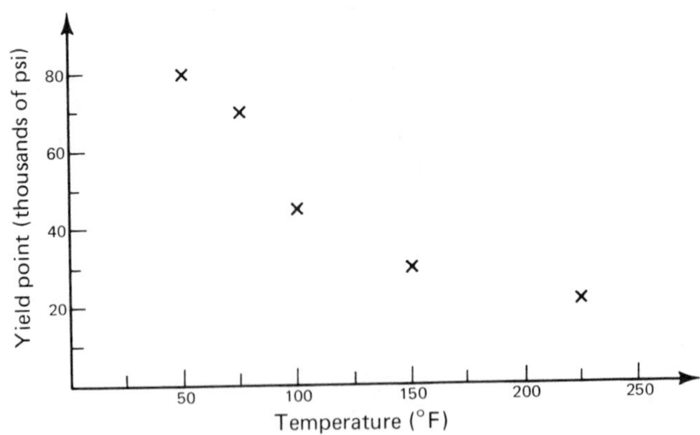

Figure 14-13. Yield point versus temperature.

The next thing we want to do is sketch a curve through these points in order to make reasonable estimates of information for which we have no exact data. If someone asks us the yield point for this cast-iron part at a temperature of 125°F, we want to make an "educated guess" by reading a value from our curve. So we sketch the curve (not necessarily a straight line) which passes through the plotted points and seems to reflect the general relationship between temperature and yield point. Such a curve might look like the curve in Fig. 14-14.

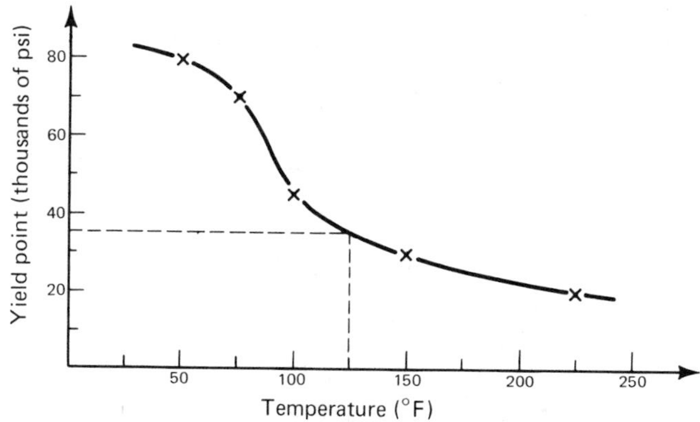

Figure 14-14. Yield point versus temperature.

Now that we have our curve, we can estimate a yield point of about 36,000 psi at a temperature of 125°F, as indicated by the dashed lines in Fig. 14-14. Remember that this value is just an educated guess because there was no set procedure for sketching the curve. If we had more points to plot, we could improve the curve and thus make our answers more accurate.

Review Exercises

Part A

1. Refer to the graph in Fig. 14-15 and answer the following questions.
   (a) What percent of the houses had a value of under $50,00 in 1965?
   (b) For the year 1970, most houses had a sale value of ____________.
   (c) The dark-shaded part of the graph represents houses that have a value of ____________.
   (d) In 1955, most houses were valued at ____________. In 1975, most houses were valued at ____________. In 1980, most houses were valued at ____________.
   (e) From the information given in the graph, can you explain why 98% of the houses were valued over $20,000?
   (f) In 1955, the median price house was valued at ____________.
   (g) In 1985, the median price house will be ____________.

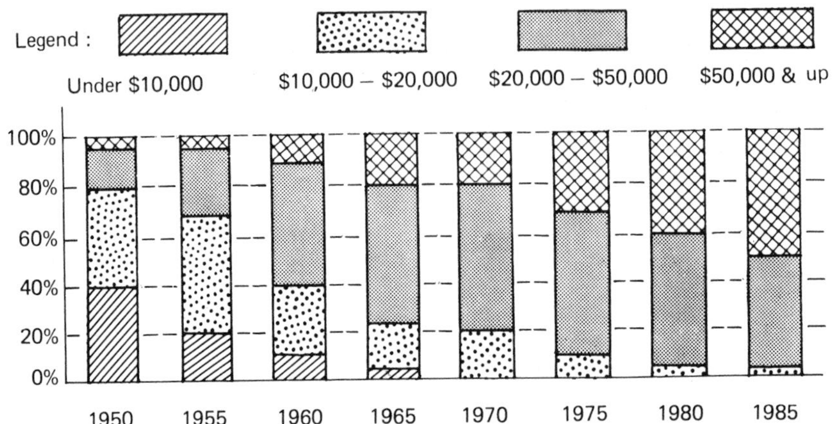

Figure 14-15. Percentage of single-family houses in estimated value ranges from 1950 to 1985.

2. Read the graph in Fig. 14-16 and answer the following questions.

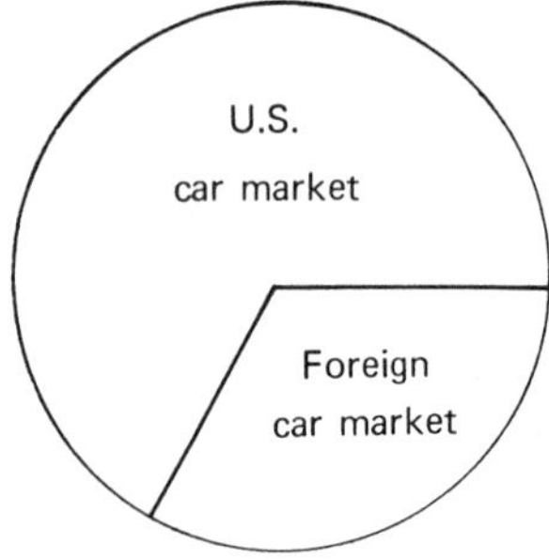

Figure 14-16. Estimated U.S. versus foreign car market for the year 1980.

(a) What was the approximate share of the total car market taken by foreign car makers?
(b) What share of the market did U.S. car makers get of the total?
(c) Can you explain from the information in the graph why foreign cars took such a large share of the market?
(d) Do you know from other sources why foreign cars took such a large share of the market?

3. Read the graph in Fig. 14-17 and answer the following questions.
(a) What was the high temperature for the 24-hour period?
_______°F; _______°C
(b) What was the low temperature for the 24-hour period?
_______°F; _______°C
(c) What does the graph suggest has happened during the 24-hour period of time?

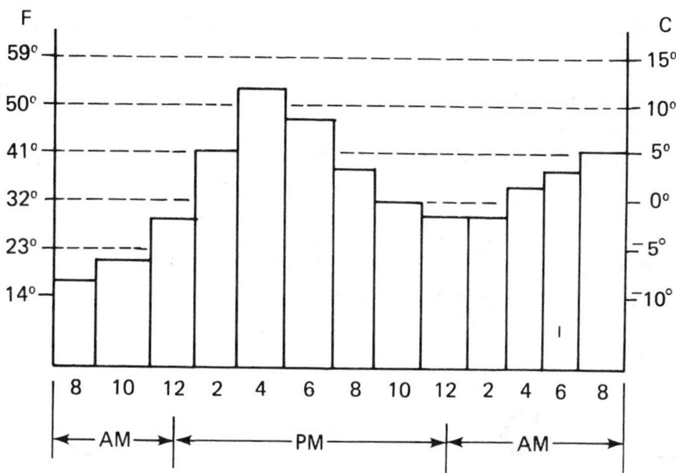

Figure 14-17. Temperatures for a day in January 1979.

4. Answer the following questions using the information in the graphs in Fig. 14-18(a) and (b) which compare the distribution of family expenses in 1950 to those in 1980 for an equivalent occupation.

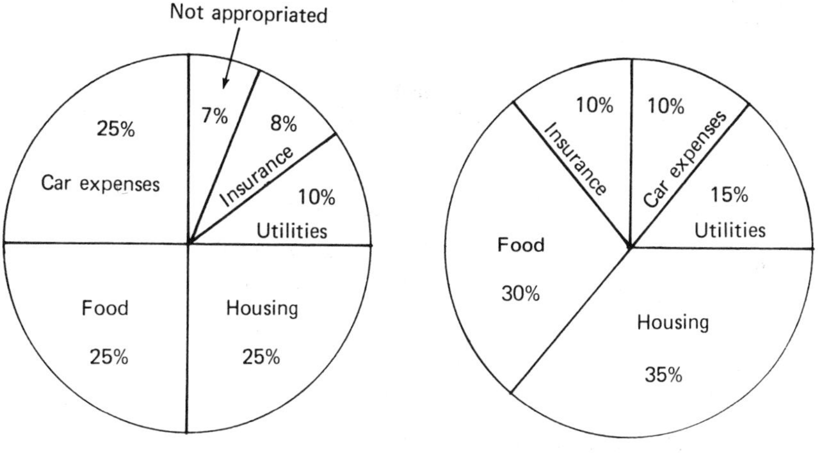

Figure 14-18. Expense distribution.

(a) How much does the family of 1950 have to pay for housing?
(b) How much does the family of 1980 have to pay for housing?
(c) How much does the family of 1950 spend for food?
(d) How much does the family of 1980 spend for food?
(e) How much does the family of 1980 spend for outside entertainment?

(f) How much can the family of 1950 spend for outside entertainment if they desire?
(g) Does the graphed information indicate which family was better off financially?
(h) What is the maximum that the family of 1950 can save each year? Compare this to the savings of the family of 1980.

5. Answer the following questions based on the graph in Fig. 14-19.
   (a) Determine the number of tires sold each month.
   (b) Determine the gross sales for the month of July.
   (c) Compare the June sales with the October sales.

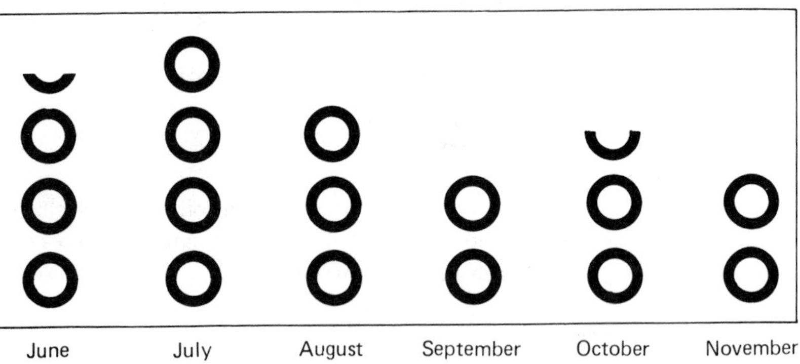

Figure 14-19. Monthly sales of tires. O each tire represents 1,000 tires sold; the average cost of a tire was $60.

6. In Fig. 14-8, estimate the gas mileage this car gets at 10 mph.
7. In Fig. 14-8, estimate the best value of gas mileage this car can get, and give an approximate speed needed to obtain this mileage.
8. In Fig. 14-9, estimate the average food expenditure of families with an income of $7,000. What about families with an income of $7,200? How much confidence do you place in the accuracy of your readings?
9. In Fig. 14-10, estimate the engine rpm needed to produce a brake horsepower of 110. (Be careful!) At what rpm is a maximum brake horsepower achieved?
10. In Fig. 14-11, estimate the temperature required 30 min after the start of the test.

Part B

11. In Fig. 14-12, estimate the silt deposit produced by a water flow of 430 ℓ per hour. Estimates like this one that are beyond the ends of a curve are called *extrapolated data*.

12. Plot points and sketch a curve for the following information:

Charging Characteristics of Generator X

Generator rpm	Amperes
800	0
1,600	22
2,400	28
3,200	31
4,000	29
5,600	23
7,200	13

13. Plot points and sketch a curve for the following information:

Engine rpm	Torque (lb-ft)
800	215
1,200	232
2,000	245
3,200	212
4,400	141

14. Given the graph in Fig. 14-20 of projected national income and consumption over the next 5 years, estimate present consumption and projected consumption in 3 years. Will the ratio of income to consumption be larger or smaller in 3 years than it is now?

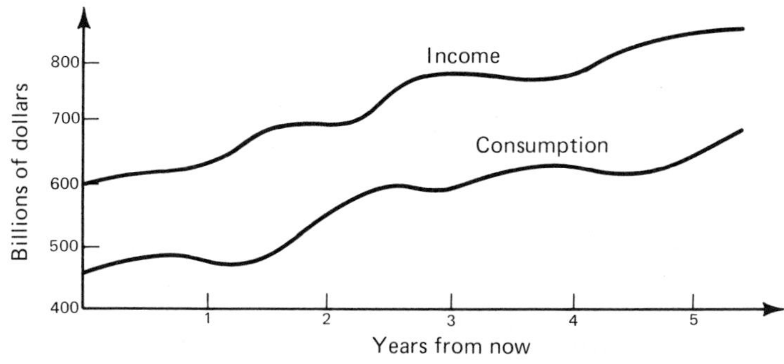

Figure 14-20

15. By studying the graph in Fig. 14-21 of unit profit compared with weekly production, determine the most profitable level of weekly production.

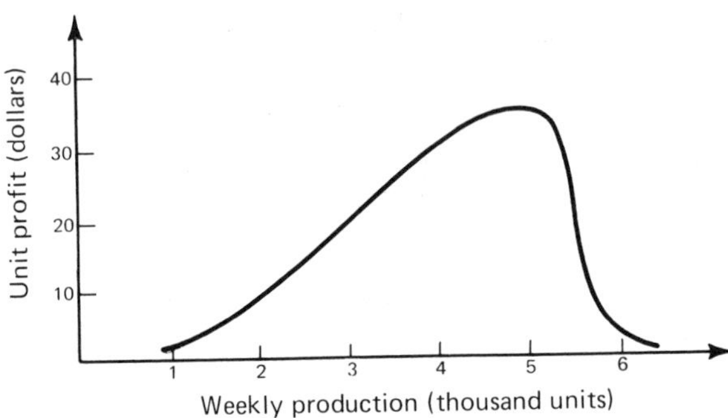

Figure 14-21

Part C

16. Take the enrollment at your school by curriculum area and make a bar graph showing how they compare.
17. Make a bar graph that compares the time that you spend in class, the time that you spend studying outside of class, the time that you waste, and the time that you spend sleeping.
18. Make a circle graph showing your personal budget outlays for 1 month.

Answers

2. (a) About 1/3 (b) About 2/3 (c) No (d) Probably because of fuel mileage
4. (a) \$1,200 (b) \$5,600 (c) \$1,200 (d) \$4,800
   (e) None (f) \$336 (g) No (h) \$336
6. 6 mi/gal
8. \$1,300; \$1,350; very low accuracy
10. 1650°F
12.

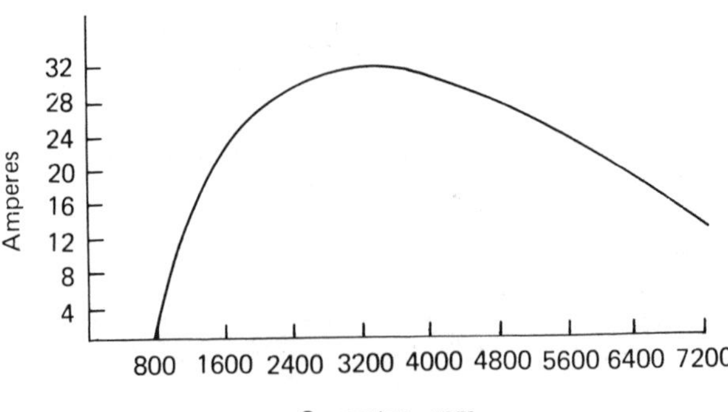

14. Present \$460 billion; projected \$590 billion. Since 780/590 is larger than 600/460, the ratio of income to consumption will be larger in 3 years than it is now.

Chapter 15

Practical Trigonometry and the Right Triangle

This chapter is designed as an introduction to basic trigonometry. Let us define *trigonometry* as *the study of the measure of angles and sides of triangles and the relationships between the angles and sides*. The trigonometry of triangles will be presented in this chapter. We intend to define the relationship between an acute angle and the various parts of a right triangle.

15-1 RIGHT TRIANGLE

As you saw in previous chapters, a triangle is a figure formed by joing three line segments. In particular, a right triangle is a triangle that includes a right (90°) angle. Figure 15-1 shows a right triangle with its parts labeled.

Angles will be denoted by capital letters. Sides will be denoted by lowercase letters. Angle C (denoted ∠C) will always be the right angle unless specified otherwise. Triangle ABC will be denoted as △ABC. You may wish to review Sec. 6-5 concerning angles and lines. As mentioned there, the measure of angle A is denoted as m∠A. In the remainder of this chapter, we shall use a shorter form that

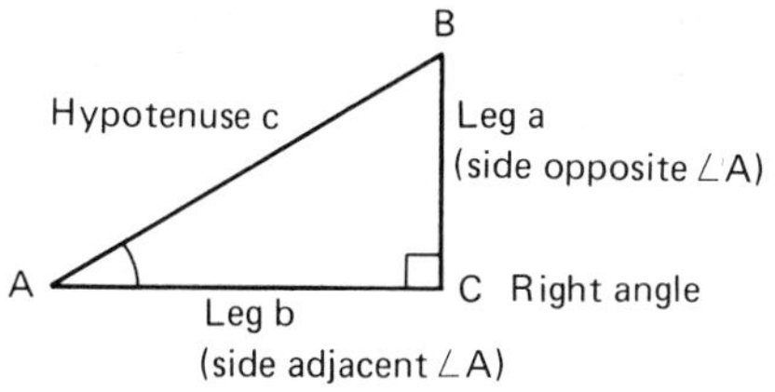

Figure 15-1. Right triangle.

is commonly used in trigonometry. Saying that A = 45° will be the same as saying that m∠A = 45° when we are talking about ∠A. It should be clear from each problem whether such letters as A, B, and C represent angles.

Applications of trigonometry occur frequently in everyday life. Typical of these applications is the problem of determining a distance that cannot be directly measured. In order to determine the cost of constructing a tunnel underneath a river, the contractor must know the distance across the river. She might decide to use her basic knowledge of right-triangle trigonometry to determine this distance. The illustration in Fig. 15-2 indicates her approach to solving this problem; a solution will be given in Sec. 15-5.

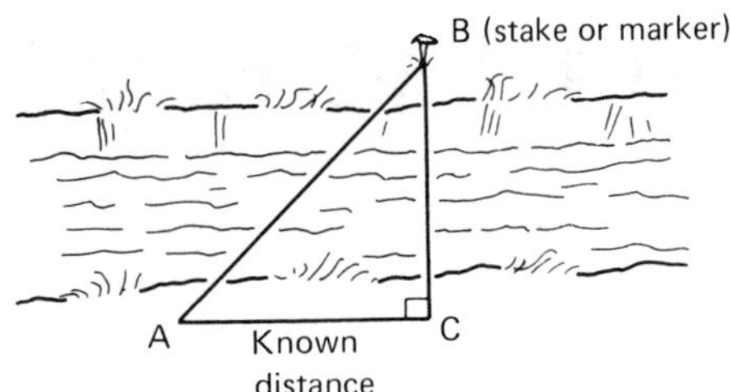

Figure 15-2. Use of right-triangle trigonometry to determine distance.

15-2 PYTHAGOREAN THEOREM

The famous and important *Pythagorean theorem* has been discussed in Sec. 8-2 as a good example of applications of powers and roots. Its primary usefulness, however, comes in the study of trigonometry, and we shall review it again at this point in our study.

The Pythagorean theorem gives a relationship between the lengths of the two sides and the length of the hypotenuse in a right triangle. You will recall that the hypotenuse is the longest side and is always located opposite the right angle.

> The Pythagorean theorem says that
>
> $$a^2 + b^2 = c^2$$
>
> where a and b are the lengths of the sides and c is the length of the hypotenuse.

A very important point that must be stressed is that this theorem applies only for *right triangles*, which are those triangles with a 90° angle (a right angle).

Example 15-1: Find the length of the hypotenuse of a right triangle that has legs 2 in. and 3 in. long.

$$2^2 + 3^2 = c^2$$
$$4 + 9 = c^2$$
$$13 = c^2$$
$$\sqrt{13} = c$$

So the hypotenuse is $\sqrt{13}$ in. long, which is about 3.6 in.

Example 15-2: What is the distance from home plate directly to second base on a baseball diamond where the length between two bases is 90 ft?

Since there is a right angle at first base, we can use the Pythagorean theorem with a = b = 90.

$$c^2 = a^2 + b^2$$
$$c^2 = 90^2 + 90^2$$
$$c^2 = 8{,}100 + 8{,}100$$
$$c^2 = 16{,}200$$
$$c = \sqrt{16{,}200}$$
$$= 10\sqrt{162}$$
$$\doteq 127.28 \text{ ft}$$
$$\doteq 127 \text{ ft } 3\frac{1}{2} \text{ in.}$$

Exercise Set 15-2

Part A

1. Find the third side of a right triangle with the given two sides (a and b are sides, c is the hypotenuse):
   (a) a = 3 b = 4 c = ?
   (b) a = 8 b = ? c = 10
   (c) a = ? b = 12 c = 13
   (d) a = 2 b = 2 c = ?
   (e) a = 1 b = ? c = 2
2. Find the length of the cut to be made (indicated by a dashed line) in the metal piece shown in Fig. 15-3.

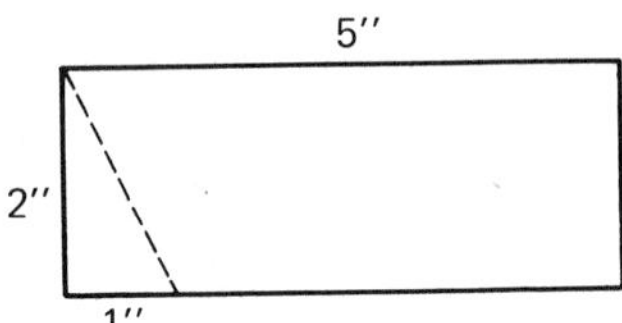

Figure 15-3

3. What is the distance from first base to third base on a softball diamond with 60-ft base paths?

Answer

2. $\sqrt{5}$ in. $\doteq$ 2.23 in.

15-3 THE 30°-60°-90° AND 45°-45°-90° TRIANGLES

There are two standard triangles that occur frequently in applications of trigonometry: the 30°-60°-90° triangle and the 45°-45°-90° triangle.

An equilateral triangle has three 60° angles and three legs of equal length. If we bisect one of these angles, cutting the original triangle in half, we now have two 30°-60°-90° triangles as shown in Fig. 15-4.

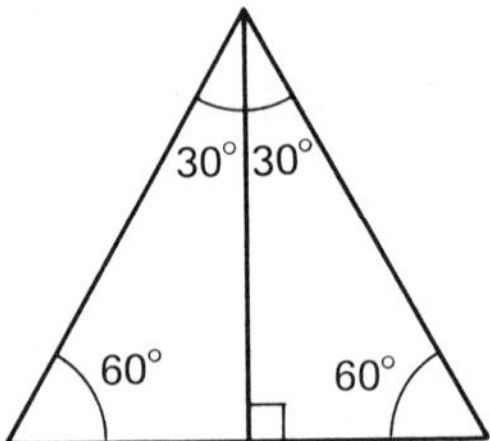

Figure 15-4. Equilateral triangle.

Look at the equilateral triangle in Fig. 15-5(a) that has 2-in. legs. Since the base is cut in half, the new 30°-60°-90° triangle has a 2 in. hypotenuse and a 1 in. leg opposite the 30° angle. The side opposite the 60° angle (the bisecting line) must have length $\sqrt{3}$ in. because

$$a^2 + b^2 = c^2$$

$$(1)^2 + b^2 = (2)^2$$

$$1 + b^2 = 4$$

$$b^2 = 3$$

$$b = \sqrt{3} \text{ in.}$$

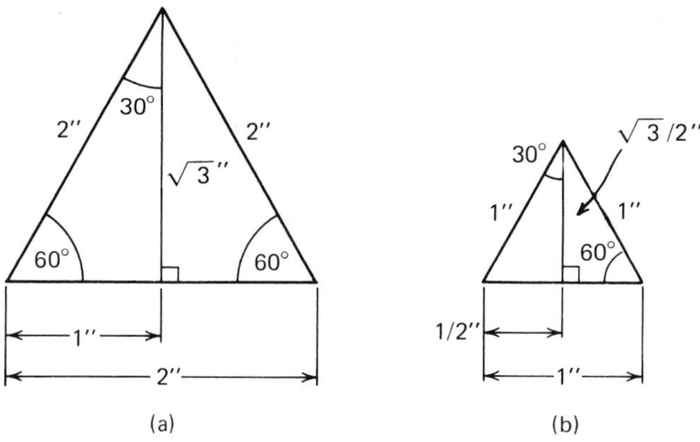

Figure 15-5. Similar 30°-60°-90° triangles.

Now look at the equilateral triangle in Fig. 15-5(b) that has 1-in. legs. Since the bisector cuts the base in half, the new triangle has hypotenuse of length 1 in. and one side of length 1/2 in. (opposite the 30° angle). Thus, the other side (opposite the 60° angle) must have length b where

$$\left(\frac{1}{2}\right)^2 + b^2 = 1^2$$

$$\frac{1}{4} + b^2 = 1$$

$$b^2 = \frac{3}{4}$$

$$b = \sqrt{\frac{3}{4}}$$

$$b = \frac{\sqrt{3}}{2} \text{ or about 0.866 in.}$$

If we had started with lengths other than 1-in. or 2-in. on the equilateral triangle, we would still observe the same relative lengths on the three sides of the 30°-60°-90° triangle. The ratios of the lengths would remain unchanged: *The hypotenuse is twice as long as the shorter leg, and the other leg is* $\sqrt{3}$ ($\doteq$ 1.732) *times the length of the shorter leg*. (See Example 15-3.)

Example 15-3: If you are 20 ft from a tree and you can see its top by looking up at a 60° angle, how tall is the tree? (Assume your eyes are at ground level.)

[You should draw a sketch of this situation, and remember that a sketch is often the best way to start working on a trigonometry (trig) problem.]

The 20-ft length is the length of the side adjacent to the 60° angle so it is the shorter leg. Thus, we have

$$\begin{aligned} \text{hypotenuse} &= 2 \times 20 \\ &= 40 \text{ ft} \\ \text{tree} &= 20 \times \sqrt{3} \\ &\doteq 20 \times 1.732 \\ &= 34.64 \text{ ft} \end{aligned}$$

So the tree is almost 35 ft tall.

An easy way to visualize a 45°-45°-90° triangle is to start with a square with 1-unit sides and draw one of its diagonals (see Fig. 15-6). The length of the hypotenuse must be $\sqrt{2}$ because

$$\begin{aligned} 1^2 + 1^2 &= c^2 \\ 1 + 1 &= c^2 \\ 2 &= c^2 \\ c &= \sqrt{2} \text{ or about } 1.4142 \end{aligned}$$

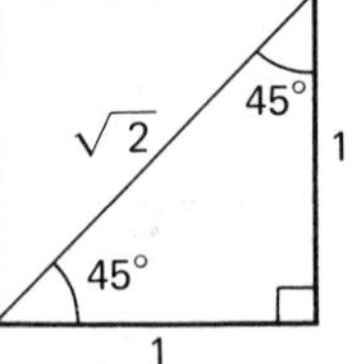

Figure 15-6. A square with 1-unit sides contains two 45°-45°-90° triangles that share a hypotenuse of $\sqrt{2}$.

Again, we did not have to start with 1 unit on each side because we shall always have the same ratio on the lengths of the sides. That is, *the length of the hypotenuse of a* 45°-45°-90° *triangle will always be* $\sqrt{2}$ *times the common length of the two equal legs.*

Example 15-4: What is the center-to-center distance between the two holes on the template shown in Fig. 15-7?

Identify the triangle that is needed to find the center-to-center distance. Note that it has been indicated with dashed lines in the figure. Since the two legs

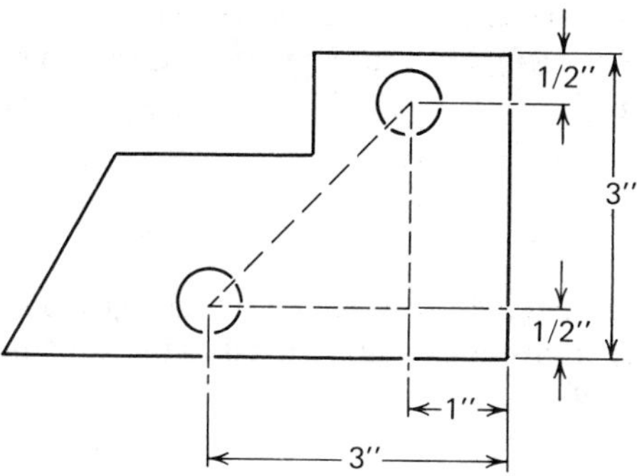

Figure 15-7

of this triangle are equal, we are dealing with a 45°-45°-90° triangle, and so the length of the hypotenuse (which is the distance we are trying to find) is approximately 1.4142 times the common length of the legs. That would make it 2.828 in. to the nearest thousandth.

Example 15-5: If you have a 3-ft wide tabletop (with no overhang) that you want to support with crossed legs to a height of 36 in., how long should the legs be? (Draw a picture!)

We are again dealing with a 45°-45°-90° triangle with a = b = 3 ft, so the legs of the table have length c, where

$$c \doteq (1.4142)(3 \text{ ft})$$
$$= 4.2426 \text{ ft}$$

which is about 4 ft 3 in.

Exercise Set 15-3

Part A

1. Find the questioned lengths in the following 30°-60°-90° triangles, where a is the length of the short side opposite the 30° angle, b is the length of the longer side opposite the 60° angle, and c is the length of the hypotenuse:
   (a) a = 1 b = ? c = 2
   (b) a = 3 b = ? c = ?
   (c) a = ? b = 8.66 c = ?
   (d) a = 40 b = ? c = ?
   (e) a = ? b = ? c = 1/2
   (f) a = ? b = 6.5 c = ?

2. Find the questioned lengths in the following 45°-45°-90° triangles, where a and b are the lengths of the sides and c is the length of the hypotenuse:

(a) a = 3	b = ?	c = 4.24
(b) a = ?	b = ?	c = 1
(c) a = 16.5	b = ?	c = ?
(d) a = ?	b = 31	c = ?

3. If you can stand 60 ft away from a building and see its top by looking along a 30° angle from horizontal, how tall is the building? (Assume your eyes are at ground level.)

Answers

2. (a) 3 (b) 0.707; 0.707 (c) 16.5; 23.34 (d) 31; 43.847

15-4 THE TRIGONOMETRIC RATIOS

It is our intention at this time to develop some intuitive notions about a right triangle and use these notions to define the trigonometric ratios.

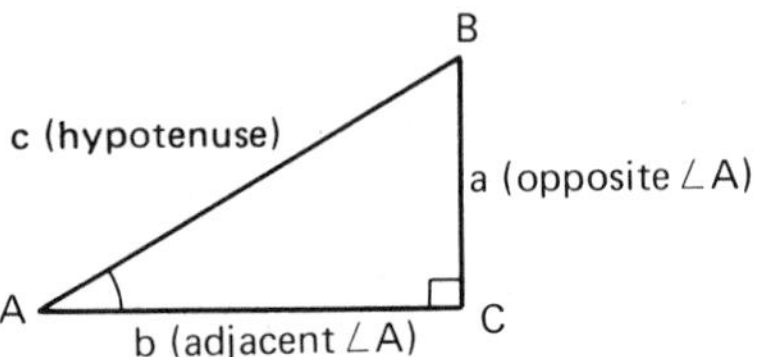

Figure 15-8. Right triangle.

In any right triangle, such as the triangle in Fig. 15-8, if we fix the length of sides a and b, does this fix the other parts of the right triangle? Since we could use the Pythagorean theorem to find the length of the third side, it seems reasonable that the other parts are fixed in length and in size.

Now, let us form as many ratios as possible using the sides a and b and the hypotenuse c, indicating where these are located in relationship to ∠A.

1. $\dfrac{\text{a (side opposite } \angle\text{A)}}{\text{b (side adjacent } \angle\text{A)}}$

2. $\dfrac{\text{b (side adjacent } \angle\text{A)}}{\text{a (side opposite } \angle\text{A)}}$

3. $\dfrac{\text{a (side opposite } \angle\text{A)}}{\text{c (hypotenuse)}}$

4. $\dfrac{\text{c (hypotenuse)}}{\text{a (side opposite } \angle\text{A)}}$

5. $\dfrac{\text{b (side adjacent } \angle \text{A)}}{\text{c (hypotenuse)}}$

6. $\dfrac{\text{c (hypotenuse)}}{\text{b (side adjacent } \angle \text{A)}}$

Notice that we have only six of these ratios that we can form.

In this section we shall limit our discussion to the ratio a (side opposite ∠A)/b (side adjacent ∠A), which we shall give the name of *tangent* ratio. Now let us fix side a to be 3 units and side b to be 4 units in Fig. 15-9; in Fig. 15-10, fix side a to be 6 units and side b to be 8 units. Now, suppose that Fig. 15-11 is a result of moving the triangle ABC (△ABC) of Fig. 15-9 and placing it on a part of the triangle A′B′C′ (△A′B′C′) of Fig. 15-10, as shown in Fig. 15-11. Notice that ∠A of the first triangle appears to be the same size as ∠A′ of the second triangle.

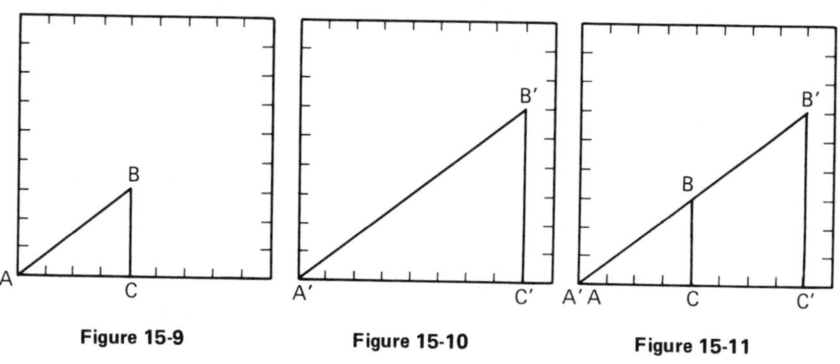

Figure 15-9 **Figure 15-10** **Figure 15-11**

It's not enough to say that ∠A of the first triangle appears to be the same size as ∠A′ of the second triangle; there is a branch of mathematics called *geometry* in which angles like these are proved to be equal.

Using the triangles in Fig. 15-9 and 15-10, we see that the tangent ratios exist. Thus we may think of the correspondence between an angle and its tangent ratio as a relation that can be expressed as follows:

The tangent ratio in △ABC is (side opposite ∠A)/(side adjacent ∠A) or 3/4.

The tangent ratio in △A′B′C′ is (side opposite ∠A′)/(side adjacent ∠A′) or 6/8.

Is the ratio 6/8 equal to the ratio 3/4? Yes, 6/8 = 3/4, and hence we can make the following definition.

> *Definition:* The tangent ratio of ∠A = (side opposite ∠A)/(side adjacent ∠A).
>
> *Abbreviation:* tan A = opposite/adjacent.

This new definition does not depend in any way on the size of the triangle we are working with. It only depends on $\angle A$.

Suppose we are interested in the tangent ratio of a 45° angle (abbreviated tan 45°). Does it matter how large we make the triangle? Look at the three possibilities in Fig. 15-12.

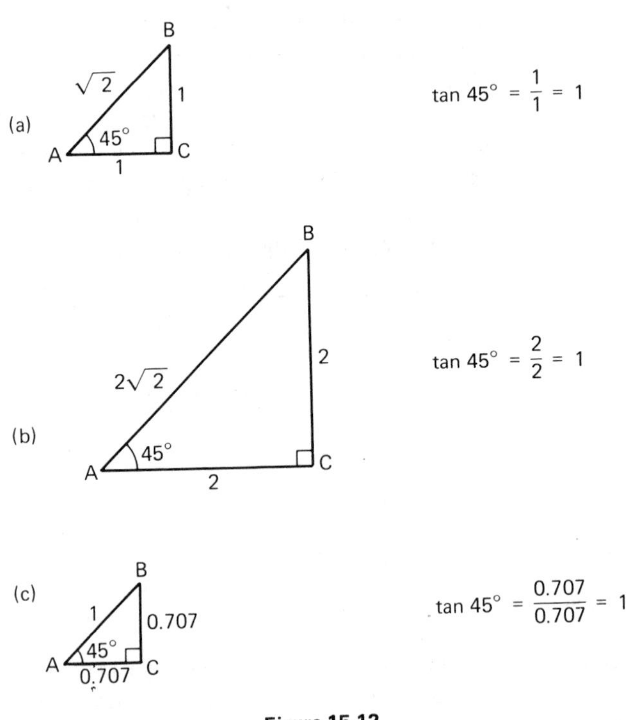

Figure 15-12

Based on the illustrations in Fig. 15-12, we can conclude that the tangent of a 45° angle (with no reference to any particular triangle) is 1.

Example 15-6: Consider the 45°-45°-90° triangle shown in Fig. 15-13 and find the side b.

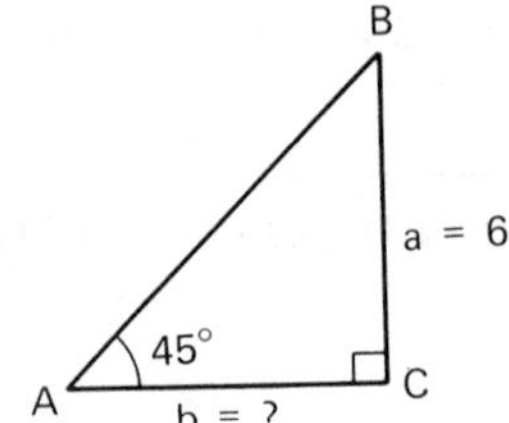

Figure 15-13

Solution

$$\tan \angle A = \frac{\text{side opposite } \angle A}{\text{side adjacent } \angle A}$$

$$\tan 45° = \frac{6}{b}$$

$$1 = \frac{6}{b}$$

$$b(1) = b\left(\frac{6}{b}\right)$$

$$b = 6$$

Example 15-7: If, as in Fig. 15-14, $\angle A = 30°$, $\angle B = 60°$, and $\angle C = 90°$, write the tangent of $\angle A$.

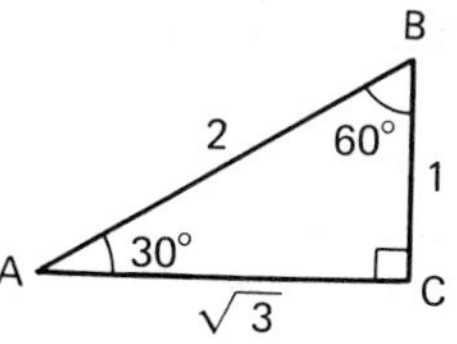

Figure 15-14

Solution

$$\text{tangent } \angle A = \text{tangent } 30° = \frac{1}{\sqrt{3}}$$

Using the six different ratios that we expressed at the beginning of this section, we may now give five new definitions similar to the definition of the tangent of an angle:

$$\text{sine } \angle A = \frac{\text{side opposite } \angle A}{\text{hypotenuse}}$$

$$\text{cosine } \angle A = \frac{\text{side adjacent } \angle A}{\text{hypotenuse}}$$

$$\text{cotangent } \angle A = \frac{\text{side adjacent } \angle A}{\text{side opposite } \angle A}$$

$$\text{cosecant } \angle A = \frac{\text{hypotenuse}}{\text{side opposite } \angle A}$$

$$\text{secant } \angle A = \frac{\text{hypotenuse}}{\text{side adjacent } \angle A}$$

The five ratios above are usually abbreviated as below.

$$\sin \angle A = \frac{\text{opposite}}{\text{hypotenuse}}$$

$$\cos \angle A = \frac{\text{adjacent}}{\text{hypotenuse}}$$

$$\cot \angle A = \frac{\text{adjacent}}{\text{opposite}}$$

$$\csc \angle A = \frac{\text{hypotenuse}}{\text{opposite}}$$

$$\sec \angle A = \frac{\text{hypotenuse}}{\text{adjacent}}$$

Again, notice that none of these definitions depends on any particular triangle.

Example 15-8: Determine the value (see Fig. 15-15) of tan ∠A, cot ∠A, sin ∠A, csc ∠A, cos ∠A, and sec ∠A, where ∠A is an acute angle of the right △ABC, if a = 3 and b = 4.

Solution

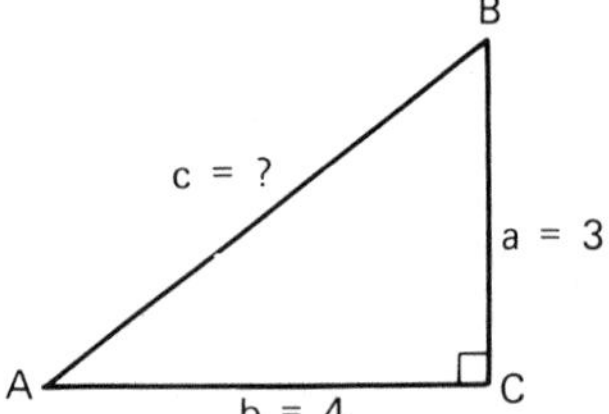

Figure 15-15

First, determine the length of side c.

$c^2 = a^2 + b^2$ by the Pythagorean theorem

$c^2 = (3)^2 + (4)^2$

$c^2 = 9 + 16$

$c^2 = 25$

$c = 5$

$\tan \angle A = \frac{3}{4}$ $\sin \angle A = \frac{3}{5}$ $\cos \angle A = \frac{4}{5}$

by definition

$\cot \angle A = \frac{4}{3}$ $\csc \angle A = \frac{5}{3}$ $\sec \angle A = \frac{5}{4}$

Example 15-9: Using the triangle shown in Fig. 15-16, we can form the following trig ratio table for a 30° angle.

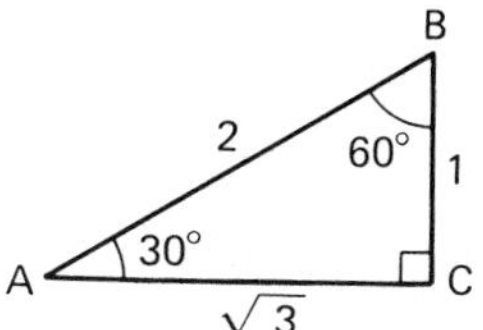

Figure 15-16

Angle	Tan	Cot	Sin	Cos	Csc	Sec
30°	$\frac{1}{\sqrt{3}}$	$\frac{\sqrt{3}}{1}$	$\frac{1}{2}$	$\frac{\sqrt{3}}{2}$	$\frac{2}{1}$	$\frac{2}{\sqrt{3}}$

Example 15-10: In right $\triangle ABC$ in Fig. 15-17, $\angle C = 90°$, $a = 10$, and $\angle A = 30°$. Find side c using the table in Example 15-9.

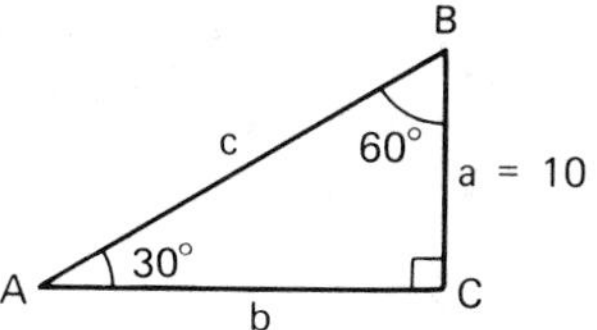

Figure 15-17

Solution

$$\sin \angle A = \frac{a}{c}$$

$$\sin 30° = \frac{10}{c}$$

$$\frac{1}{2} = \frac{10}{c}$$

$$2\left(\frac{1}{2}\right) = 2\left(\frac{10}{c}\right)$$

$$1 = \frac{20}{c}$$

$$c = 20$$

Exercise Set 15-4

Part A

1. Determine the value of tan ∠A, cot ∠A, sin ∠A, csc ∠A, cos ∠A, and sec ∠A, where ∠A is an acute angle of the right △ABC, if a = 5 and b = 12.

2. In a right $\triangle ABC$, $\angle A = 45°$, $\angle B = 45°$, $a = 10$, and $b = 10$. Find c and then determine $\sin \angle A$ and $\tan \angle A$.

3. In right $\triangle ABC$, $\angle C = 90°$, $\angle A = 30°$, and $a = 15.6$. Find side c using the table in Example 15-9.

4. Given right $\triangle ABC$ with hypotenuse AC and $\tan A = 5/12$, find the length of AC and then determine $\sin \angle A$ and $\cos \angle A$.

5. Given right $\triangle ABC$, with $\angle C = 90°$, $AC = 4$, and $BC = 3$, find the six trigonometric ratios of $\angle B$.

6. Find the distance d across the pond between points B and C using the information given in Fig. 15-18.

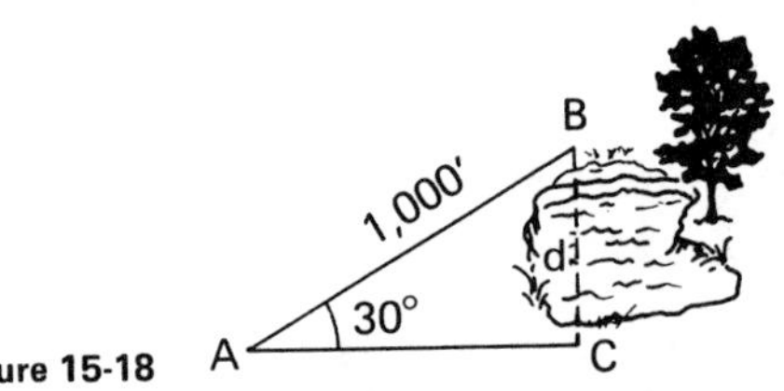

Figure 15-18

Answers

2. $c = 10\sqrt{2}$; $\sin A = \sqrt{2}/2 \doteq 0.707$; $\tan A = 1$
4. $AC = 13$; $\sin A = 5/13$; $\cos A = 12/13$
6. $d = 500$ ft

15-5 USE OF TRIGONOMETRIC TABLES

We have already discussed the six trigonometric ratios and their values for some special angles. In this section, we shall discuss the use of trigonometric tables, that is, how to find values of the trigonometric ratios for any angle. If you have a calculator with the trig ratios and your instructor permits use of calculators, you might just verify the answers to the examples in this section.

Appendix C shows each of the six trigonometric ratios in decimal form for any angle in degrees from 0 to 90. Figure 15-19 shows a portion of Appendix C.

For angles between 0° and 45°, we find the number of degrees in the *Degrees* column on the *left* side and then go across to the column containing the desired trig ratio as indicated at the *top* of the table.

Degrees	Sin	Cos	Tan	Cot	Sec	Csc	
0							90
1		0.9998					89
2			0.0349				88
⋮							⋮
43				1.0724	1.3673	1.4663	47
44	0.6947					1.4396	46
45						1.4142	45
	Cos	Sin	Cot	Tan	Csc	Sec	Degrees

Figure 15-19

Example 15-11: Find the value for sin 44°.

Across from 44° in the *sin* column, we see that sin 44° = 0.6947.

Example 15-12: What is the value for tan 2°?

In the *tan* column across from the 2, we find 0.0349. So tan 2° = 0.0349.

For angles between 45° and 90°, we find the number of degrees in the *Degrees* column on the *right* side and go across to the column containing the desired trig ratio as indicated at the *bottom* of the table.

Example 15-13: Find sin 89°.

Find 89° in the rightmost column of the table. From the bottom, find the *sin* column. Then sin 89° = 0.9998.

Example 15-14: Find the value for csc 47°.

Across from 47°, we find, reading from the bottom in the *csc* column, 1.3673. Therefore, csc 47° = 1.3673.

Example 15-15: As another use of Appendix C, suppose we know that the cotangent of some angle is 1.0724, that is, cot A = 1.0724. What is the value for A in degrees?

Solution: First look through the *cot* column until we find 1.0724. Since this value was found as we came down (from top to bottom) the cotangent column, we look across to the left and find 43°. Hence A = 43° or cot 43° = 1.0724.

Example 15-16: If sec A = 1.4300, find A to the nearest whole degree.

Solution: As we compare our value with those in the *sec* columns, we find that our value is between 1.4142 and 1.4396. We see that the value closest to 1.4300 is 1.4396. Since the value 1.4300 is not in the table, we choose 1.4396, which is the closest value to 1.4300. Because we read from the bottom of the column, we must go across to the right and find 46°. To the nearest degree, A = 46° or sec 46° ≐ 1.4300.

Example 15-17: A solution to the problem stated in Sec. 15-1: Find the distance across the river. That is, find BC.

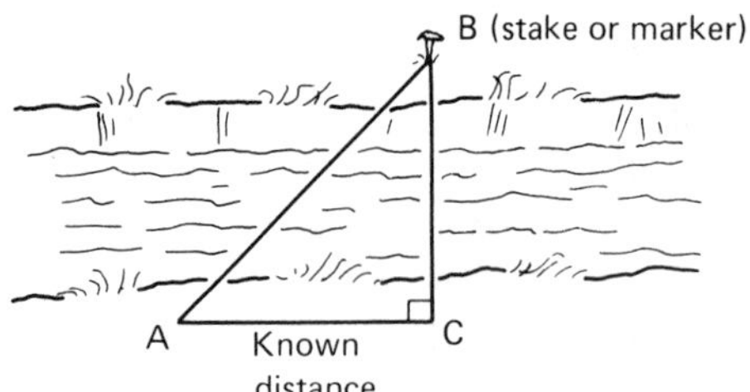

Figure 15-20

Suppose the known distance AC is 150 ft and ∠A = 43°. (See Fig. 15-21.) We know the size of ∠A and the length of the side adjacent to ∠A. We want to find the length of the side opposite ∠A. Recall from Sec. 15-4 that the tangent ratio of ∠A = (side opposite ∠A)/(side adjacent ∠A). Then tan A = BC/AC or tan 43° = BC/150. Solving for BC, we find

$$BC = 150(\tan 43^\circ)$$

$$BC = 150\,(0.9325) \qquad \text{(from Appendix C)}$$

$$BC = 139.875$$

Therefore the river is about 140 ft wide at that point.

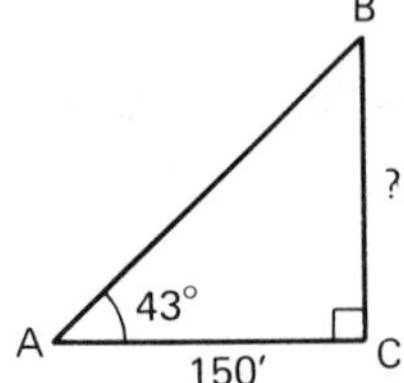

Figure 15-21

Exercise Set 15-5

Part A

1. Find the value for each of the following, using Appendix C.

(a) sin 42°	(b) cos 61°	(c) csc 38°	(d) sec 75°
(e) cot 83°	(f) tan 59°	(g) cos 12°	(h) sin 81°
(i) sec 80°	(j) cot 19°	(k) tan 70°	(l) csc 28°

2. Find the number of degrees for ∠A (to the nearest degree).

(a) tan A = 0.9325	(b) sin A = 0.3584	(c) cot A = 0.3640
(d) cos A = 0.6820	(e) sec A = 1.2868	(f) csc A = 1.6243
(g) sin A = 0.8290	(h) tan A = 1.3764	(i) cos A = 0.5446
(j) cot A = 1.3764	(k) csc A = 1.1924	(l) sec A = 1.0711

3. In ΔABC, C = 90°, A = 25°, and AC = 4.7 in. Draw the triangle with AC as the base, and find the height of the triangle (i.e., side BC). (Use trig ratios rather than scale drawing.)

Answers

2.

(a) 43°	(b) 21°	(c) 70°	(d) 47°	(e) 39°	(f) 38°
(g) 56°	(h) 54°	(i) 57°	(j) 36°	(k) 57°	(l) 21°

15-6 TRIGONOMETRIC APPLICATIONS

In previous sections of this chapter we placed most of the emphasis on knowing the basic trigonometric ratios, using special right triangles and also using trigonometric tables for computation. Now we are presenting physical problems that can best be solved by use of trigonometry. The following examples are intended to pose several mechanical or physical problems and illustrate the use of trigonometry.

Example 15-18: Determine the minimum length of a ladder needed to reach to the top of a barn 35 ft tall if the maximum recommended angle of elevation for safety is 65°.

Solution: We know the height of the barn, and the size of the angle of elevation is indicated in Fig. 15-22. The sine ratio of ∠A = (side opposite ∠A)/(hypotenuse).

$$\sin 65^\circ = \frac{35}{AB}$$

$$0.9063 \doteq \frac{35}{AB}$$

$$0.9063\ AB \doteq 35$$

$$AB \doteq 38.6 \text{ ft}$$

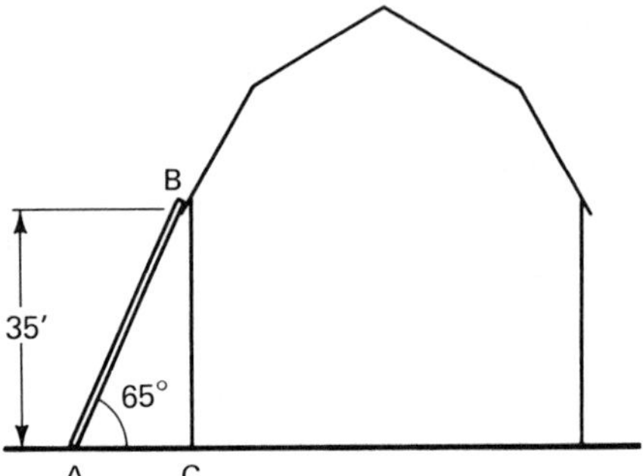

Figure 15-22

Example 15-19: In the circular plate pictured in Fig. 15-23, the radius of the plate is 10 in. and the radius of the circle that holes are to be drilled on is exactly 8 in. If you are required to drill nine holes equally spaced around the circular plate, find the center-to-center distance that two consecutive holes are apart.

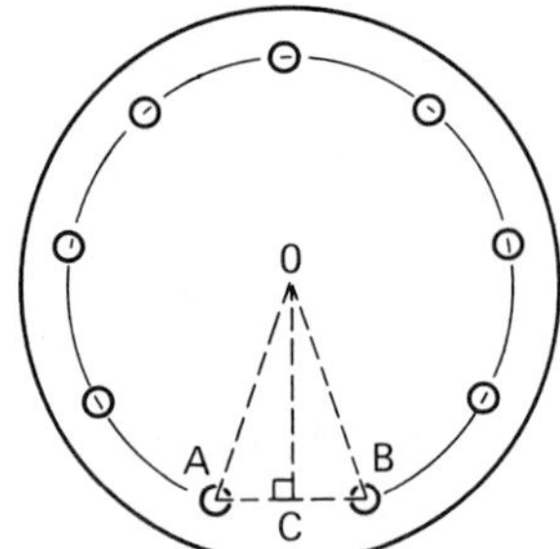

Figure 15-23

Solution: Draw in radii OA and OB. OA equals OB: Radii of the same circles are equal. Draw line AB. In order to produce a right triangle that we can work with, draw OC as an altitude to the base AB.

$$\angle AOC = \frac{1}{2}\angle AOB$$

AC equals 1/2AB: An altitude to the base of a triangle with two equal legs bisects the vertex angle and the base.

$$\angle AOC = \frac{1}{18} \text{ of } 360^\circ \qquad \angle AOC = \frac{1}{2}\angle AOB$$

$$= 20^\circ \qquad \angle AOB = \frac{1}{9} \text{ circle in degrees}$$

OC is perpendicular to AB: definition of an altitude. Now, use the ΔAOC.

$$\sin 20^\circ = \frac{AC}{8}$$

$$0.3420 \doteq \frac{AC}{8}$$

$$2.736 \doteq AC$$

Hence

$$AB \doteq 5.472 \text{ in.}$$

Example 15-20: In Fig. 15-24, $\angle$CAB is the angle of depression, AC = 10 ft, and BC = 1 in. Find the angle of depression to the nearest degree.

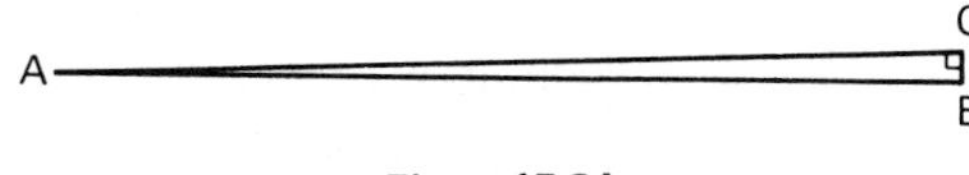

Figure 15-24

Solution

$$AC = 10 \text{ ft} = 120 \text{ in.}$$

$$\tan A = \frac{1}{120}$$

$$\tan A \doteq 0.0083$$

$$\angle A \doteq 1^\circ$$

Example 15-21: If a pentagonal nut is to be constructed to fit a 2-in. bolt with a minimum wall thickness of 1/2 in. (as indicated in Fig. 15-25), find the minimum diameter of stock material that can be used to make this nut.

Solution: In Fig. 15-25, it will be necessary to find the length of OA by using a right triangle. In ΔAOC, $\angle$ACO is a right angle.

$$\angle AOB = 72^\circ$$

because the nut has five equal faces.

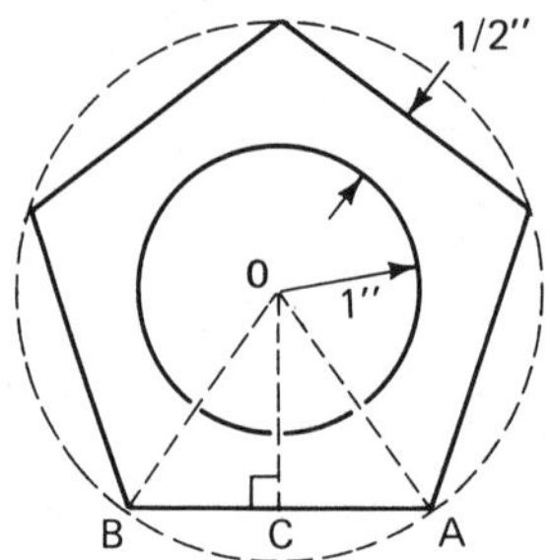

Figure 15-25

$$\angle AOC = 36^\circ \qquad \angle AOC = \frac{1}{2}\angle AOB$$

$$OC = 1.5$$

$$\cos 36^\circ = \frac{1.5}{AO}$$

$$0.8090 \doteq \frac{1.5}{AO}$$

$$0.8090\ AO = 1.5$$

$$AO = 1.85$$

The minimum diameter of stock that is usable would be 3.70 in.

Exercise Set 15-6

1. A hexagonal nut measuring 1.2 in. across (see Fig. 15-26) is to be placed at a minimum distance from the V of a piece of angle iron. Find the minimum distance from O to the angle iron needed to allow the nut to turn. (*Hint:* Consider length OB.)

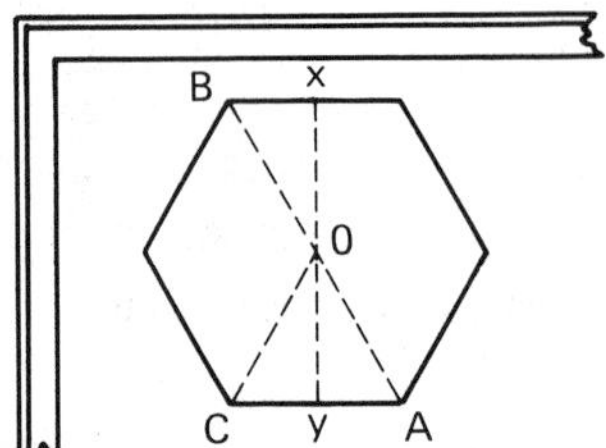

Figure 15-26

2. The design of an engine requires a 2-in. radius for its pistons and a 10-in. connecting rod with which it is connected to the crankshaft. What is the maximum length of the radius of the crankshaft throw if the angle cannot be larger than 20°, as indicated in Fig. 15-27?

3. In Fig. 15-28, the wedge is 4 in. (OP) in length and $\angle MOP = 5^\circ$. Find the minimum width of stock material required to make the wedge.

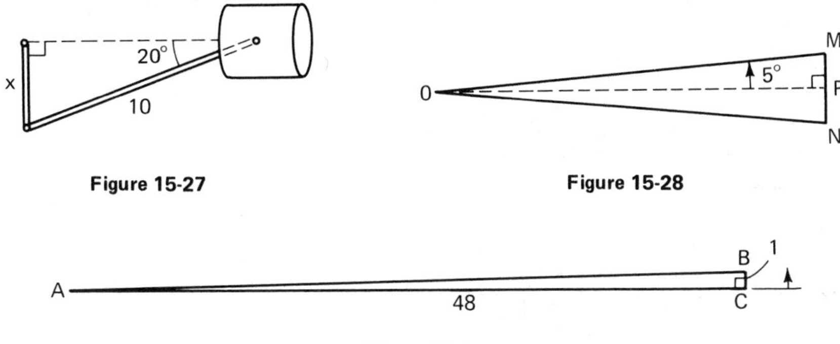

Figure 15-27

Figure 15-28

Figure 15-29

4. In horizontal drainage lines, it is recommended that a ratio 48 to 1 be used for drainage elevation (see Fig. 15-29). Find the angle of elevation ($\angle$A) that produces the ratio of 48 to 1.
5. If the maximum angle of elevation that will be allowed on a 5-mi stretch of new road is 3°, find the maximum difference in feet that one end of the grade is lower than the other.

Answers

2. 3.42 in.
4. 1°11′

Chapter 16

Special Topics

16-1 INDEXING IN THE MACHINE SHOP

The *index head*, or *dividing head*, on a milling machine is a device used to turn a workpiece in a circular motion. The machinist turns the crank on the index head in order to rotate his workpiece through a specified angle or a specified fractional part of a circle. For instance, if he was spacing 20 holes around a circular plate, he would need to rotate the plate 1/20 of a full revolution between two consecutive holes. The basic principle that makes indexing useful is:

40 complete turns on the index crank produces 1 complete rotation of the workpiece

By setting up a direct proportion, the machinist can now determine the number of turns needed to obtain 1/20 of a revolution. He would say that one complete rotation of the workpiece is equivalent to 20 of his spaces, so

$$\frac{\text{40 turns on index crank}}{\text{20 spaces}} = \frac{\text{N turns}}{\text{1 space}}$$

Solving for N, he finds N = 2. This tells him that two turns of the index crank will rotate the workpiece from one hole to the next.

We can now see that the basic mathematical principle behind indexing is the proportion

$$\boxed{\frac{40 \text{ turns}}{\text{total number of spaces}} = \frac{\text{N turns}}{1 \text{ space}}}$$

Example 16-1: If we needed to locate the eight "points" for an octagonal nut, we would set up the proportion

$$\frac{40 \text{ turns}}{8 \text{ spaces}} = \frac{\text{N turns}}{1 \text{ space}}$$

and find $40 = 8N$ or $N = 5$ turns. So we can go from one point to the next by turning the index crank around exactly five times.

Example 16-2: If we were milling a slot that must have a 90° turn in it, how could we use indexing?

Since 90° is one-fourth of a full circle, we set up the proportion

$$\frac{40 \text{ turns}}{4 \text{ spaces}} = \frac{\text{N turns}}{1 \text{ space}}$$

and find $N = 10$ turns. Notice that we could also have set this up in terms of degrees:

$$\frac{40 \text{ turns}}{360^\circ} = \frac{\text{N turns}}{90^\circ}$$

and we would again see that 10 turns would rotate the workpiece 90°.

Now suppose we needed six holes drilled around a circular plate. We would set up

$$\frac{40 \text{ turns}}{6 \text{ spaces}} = \frac{\text{N turns}}{1 \text{ space}}$$

and find that $N = 6\ 2/3$ turns. How are we supposed to find 2/3 of a turn *accurately*? The device used to measure fractional parts of a turn for the index crank accurately is called the *indexing plate*. This plate has several circular rows of holes on it and attaches behind the crank in such a way that a pin on the crank will fit into the holes. The indexing plate remains stationary as the crank is turned so that the crank pin can be moved from one hole to another.

To find 2/3 of a turn, we could use an index plate with a row of 15 holes. If we start at one hole and move around to the tenth hole beyond it, we have moved exactly 2/3 of a turn because $10 = 2/3(15)$. Of course, we could obtain the same result with 12 holes on an 18-hole row, 14 holes on a 21-hole row, 18 holes on a 27-hole row, 22 holes on a 33-hole row, or many other possibilities.

A typical set of index plates has the following numbers of holes in various rows:

Plate No. 1: 15, 16, 17, 18, 19, 20

Plate No. 2: 21, 23, 27, 29, 31, 33

Plate No. 3: 37, 39, 41, 43, 47, 49

In order to select the proper row for a fractional turn, look for a multiple of the denominator of the fraction. In our "2/3 situation," we indicated several possibilities, each of which was a multiple of 3.

Example 16-3: To space seven points around a circle,

$$\frac{40 \text{ turns}}{7 \text{ spaces}} = \frac{\text{N turns}}{1 \text{ space}}$$

We find N = 5 5/7. Looking for a multiple of 7, we could select the 21-hole row. Since 5 5/7 = 5 15/21, we want to turn the crank five complete turns plus 15 holes on the 21-hole row.

Example 16-4: To rotate a workpiece 25°,

$$\frac{40 \text{ turns}}{360^\circ} = \frac{\text{N turns}}{25^\circ}$$

We find N = 1,000/360 = 2 280/360 = 2 7/9. Looking for a multiple of 9, we could select the 27-hole row. Since 2 7/9 = 2 21/27, we want to turn the crank two complete turns plus 21 holes on the 27-hole row.

Exercise Set 16-1

1. How would you use indexing to space the teeth on a 16-tooth gear?
2. How would you use indexing to rotate a workpiece through an angle of 55°?
3. How would you use indexing to space 52 holes around a circular plate?
4. How would you use indexing to produce an index plate with an 85-hole circle?
5. How many positions would you be marking around a circle if, between two consecutive positions, you move forward 60 holes on the 21-hole circle and then backward 60 holes on the 27-hole circle? (This is called *compound* indexing.)

Answers

2. 6 turns plus 3 holes on the 27-hole row
4. 8 holes on the 17-hole row

16-2 LAWS OF SINES AND COSINES

Up to this point, all the problems involving practical application of trig ratios have dealt with right triangles. Many physical problems that commonly occur (in many vocations) involve triangles that do not have a right angle. Mathematically, the laws of sines and cosines are formulated to handle situations like this with a minimum of known information.

For example, if a side and any two angles of a triangle are known, then the remaining parts can be determined. A second situation involves two sides and the included angle (the angle between them) in a triangular plot, without having access to the other dimensions of the plot. By using one or both of these special laws, the missing side can be determined and, if needed, the measure of the other angles can also be determined. A third situation occurs when two sides and an angle opposite one of them are known. In this last situation, we plan to present examples that have a unique solution in that they will relate to practical problems with specific limitations. (Note that in theory this last situation does not necessarily have a unique solution. It might have one, two, or no solutions for the missing side, based on the lengths of the two sides that are given as they relate to the known angle.)

Law of Sines: In any triangle, such as the triangle in Fig. 16-1, the law of sines states that the length of any side of a triangle is proportional to the sine of the angle opposite that side.

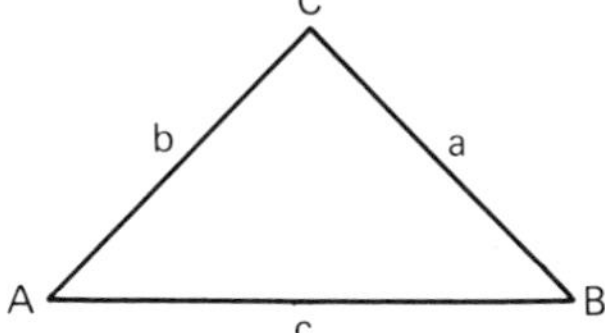

Figure 16-1

The *law of sines* can be stated mathematically as follows:

$$\frac{a}{\sin A} = \frac{b}{\sin B} = \frac{c}{\sin C}$$

Example 16-5: Find the missing parts of the triangular lake site by using the information as provided on the sketch in Fig. 16-2.

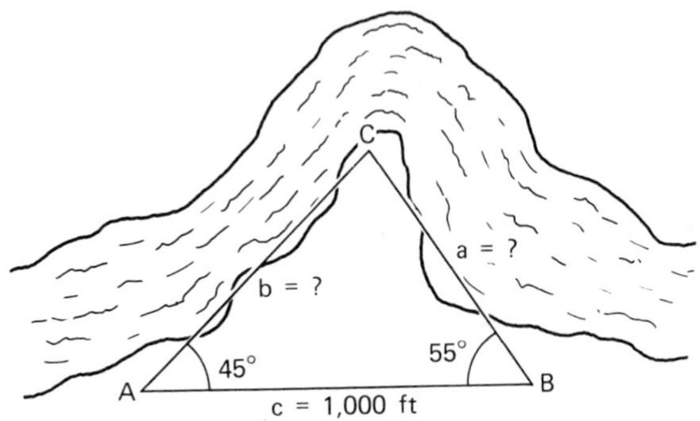

Figure 16-2

Solution

$$A + B + C = 180°$$

$$45° + 55° + C = 180° \qquad \text{by substituting}$$

$$C = 80°$$

$$\frac{a}{\sin 45°} = \frac{b}{\sin 55°} = \frac{1{,}000}{\sin 80°} \qquad \text{by substituting}$$

Now to find **a** we select two ratios; one of the ratios contains **a**, and the other is the ratio with no missing parts.

$$\frac{a}{\sin 45°} = \frac{1{,}000}{\sin 80°}$$

$$\frac{a}{0.7071} \doteq \frac{1{,}000}{0.9848}$$

$$(0.9848)(a) \doteq (1{,}000)(0.7071)$$

$$0.9848a \doteq 707.1$$

$$a \doteq 718 \text{ ft}$$

Next, select the ratio containing **b** and the ratio with no missing parts.

$$\frac{b}{\sin B} = \frac{c}{\sin C}$$

$$\frac{b}{\sin 55°} = \frac{1{,}000}{\sin 80°}$$

$$\frac{b}{0.8192} \doteq \frac{1{,}000}{0.9848}$$

$$(0.9848)(b) \doteq (1,000)(0.8192)$$

$$0.9848b \doteq 819.2$$

$$b \doteq 832 \text{ ft}$$

Law of Cosines: The law of cosines can be used to find a missing side of a triangle (for example, side a in Fig. 16-3) if the other two sides and the included angle are known.

The law of cosines can be stated as follows: The square of the side opposite any angle is equal to the sum of the squares of the other two sides minus twice the product of these sides and the cosine of that angle.

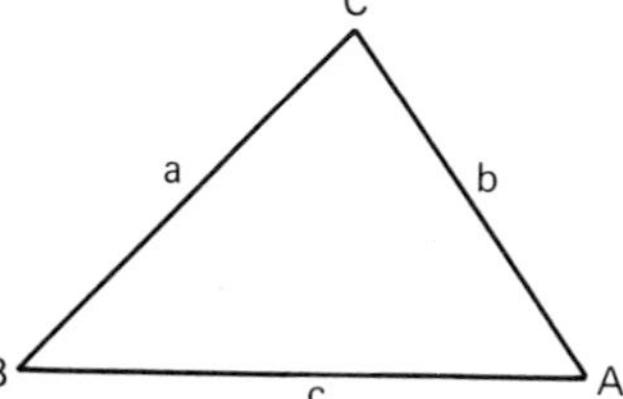

Figure 16-3

The *law of cosines* can be stated mathematically as follows:

$$a^2 = b^2 + c^2 - 2bc \cos A$$

Note that the method of lettering a triangle is completely arbitrary, so you may wish to use the following illustrations for later reference in working problems.

Illustration 1: $y^2 = x^2 + z^2 - 2xz \cos Y$. (See Fig. 16-4.)

Illustration 2: $b^2 = a^2 + c^2 - 2ac \cos B$. (See Fig. 16-5.)

Illustration 3: $c^2 = a^2 + b^2 - 2ab \cos C$. (See Fig. 16-6.)

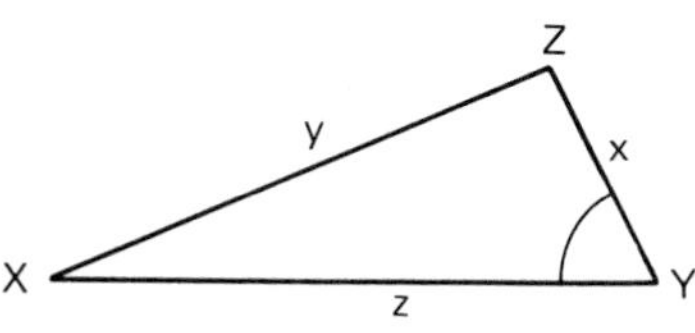

Figure 16-4

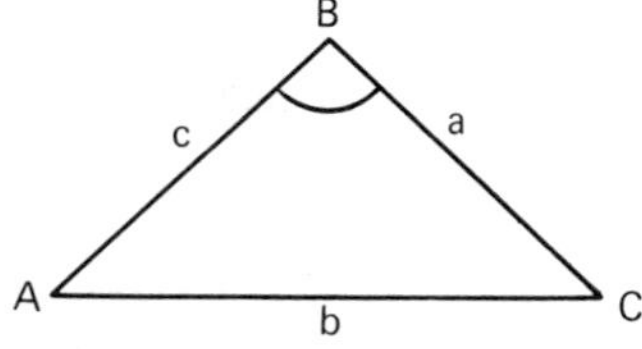

Figure 16-5

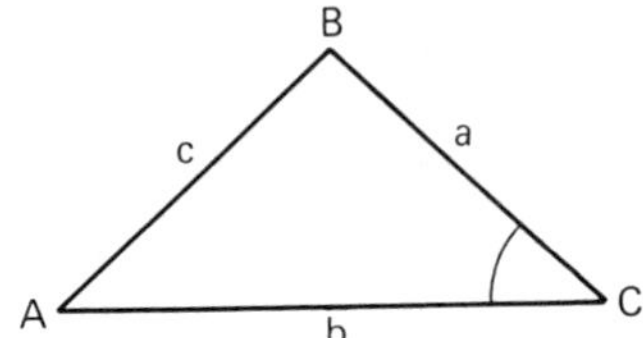

Figure 16-6

Example 16-6: Find the missing side in Fig. 16-7 if a = 6, b = 7, and C = 42°.

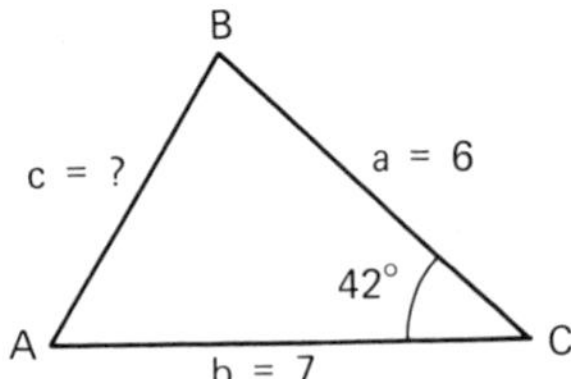

Figure 16-7

Solution

$$c^2 = a^2 + b^2 - 2ab \cos C$$

$$c^2 = 6^2 + 7^2 - 2(6)(7) \cos 42^\circ$$

$$c^2 \doteq 36 + 49 - 84(0.7431)$$

$$c^2 \doteq 85 - 62.4$$

$$c^2 \doteq 22.6$$

$$c \doteq \sqrt{22.6}$$

$$c \doteq 4.8$$

Obtuse Angles: In many application problems, triangles having an obtuse angle (larger than 90°) are involved. Finding the missing parts of triangles with an obtuse angle quite often necessitates the use of trig ratios and tables. This should not pose a difficult problem in that the tables can still be used to determine the decimal approximation for any trig ratio needed for each obtuse angle. In order to find a trig ratio or value for an obtuse angle, follow these steps:

Step 1. Subtract the given obtuse angle from 180°. This gives us an acute angle (smaller than 90°) that can be found in the trig table.

Step 2. Find the desired trig ratio of the acute angle obtained in Step 1.

Step 3. For a sine ratio, prefix the value from the table with a plus sign. For a cosine ratio, prefix the value from the table with a minus sign.

Example 16-7: In Fig. 16-8, the sine of 150° can be found by subtracting 150° from 180°, looking up the sine of that difference (30°), and placing a plus sign in front of the value.

$$\sin 150^\circ = +\sin 30^\circ$$

$$= +0.5000 \quad \text{since} \quad \sin 30^\circ = 0.5000$$

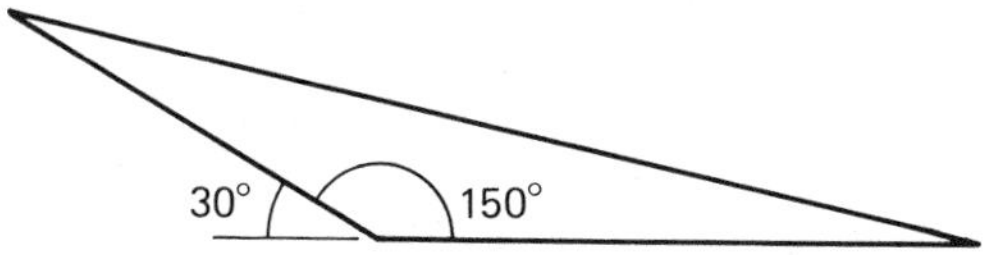

Figure 16-8

Example 16-8: In Fig. 16-9, the sine of 135° can be found by subtracting 135° from 180°, looking up the sine of the difference (45°), and placing a plus sign in front of the value. That is,

$$\sin 135^\circ = +\sin 45^\circ$$

$$\doteq -(0.7071) \quad \text{since} \quad \cos 45^\circ \doteq 0.7071$$

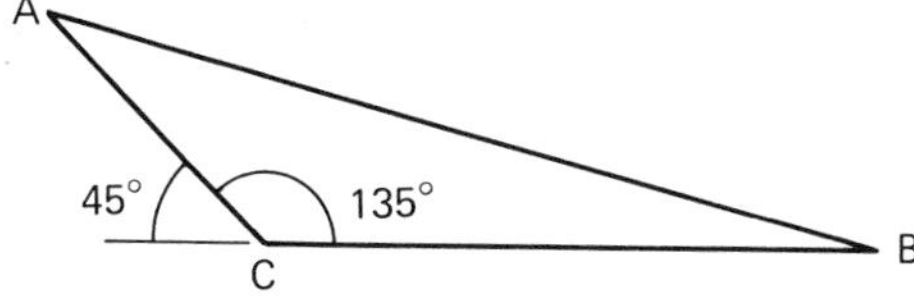

Figure 16-9

Example 16-9: In Fig. 16-10, the cosine of 150° can be found by subtracting 150° from 180°, taking the cosine of that difference (30°), and placing a minus sign in front of the value.

$$\cos 150^\circ = -\cos 30^\circ$$

$$\doteq -(0.8660) \quad \text{since} \quad \cos 30^\circ \doteq 0.8660$$

$$\doteq -0.8660$$

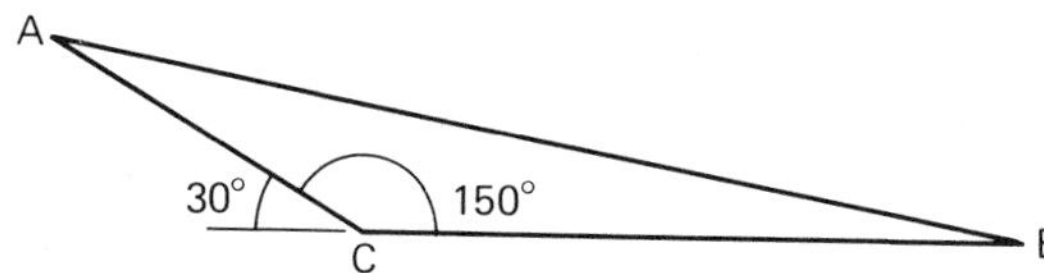

Figure 16-10

Example 16-10: In Fig. 16-11, the cosine of 135° can be found by subtracting 135° from 180°, taking the cosine of that difference (45°), and placing a minus sign in front of the value.

$$\cos 135^\circ = -\cos 45^\circ$$

$$\doteq -(0.7071) \quad \text{since} \quad \cos 45^\circ \doteq 0.7071$$

$$\doteq -0.7071$$

A
45° 135°
C B

Figure 16-11

Exercise Set 16-2

Part A

Use the trig tables to find the indicated values.

1. sin 95°
2. cos 100°
3. sin 120°
4. sin 130°
5. cos 155°
6. cos 175°

Part B

7. In the parallelogram shown in Fig. 16-12, use the given information to find the length of the indicated diagonal.
8. How long would the upper cable in Fig. 16-13 have to be in order for it to be attached to the top of the pole if a 40-ft cable is attached 12 ft from the top? The angle between the lower cable and the pole is 50°.

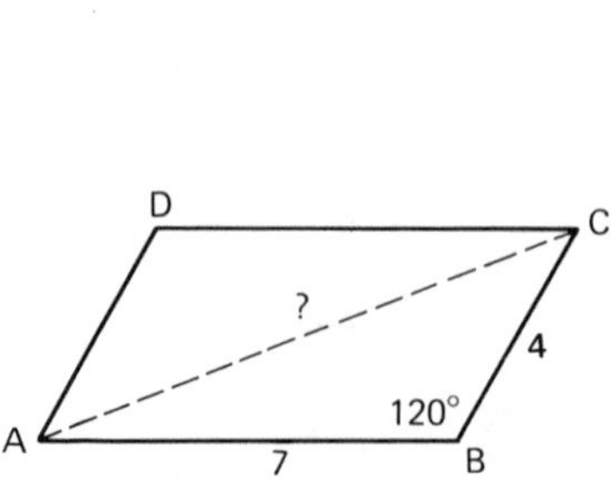

Figure 16-12

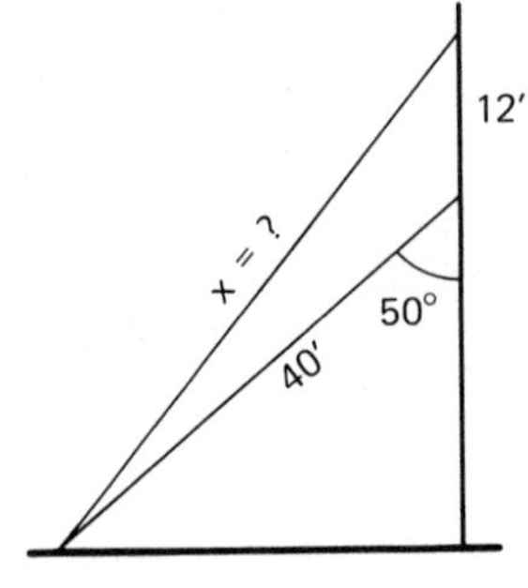

Figure 16-13

9. In Fig. 16-14, find the slant height of the hill if the angle of elevation of the hill is known and another angle of elevation is measured to be 40° at a point 30 ft from the foot of the hill.

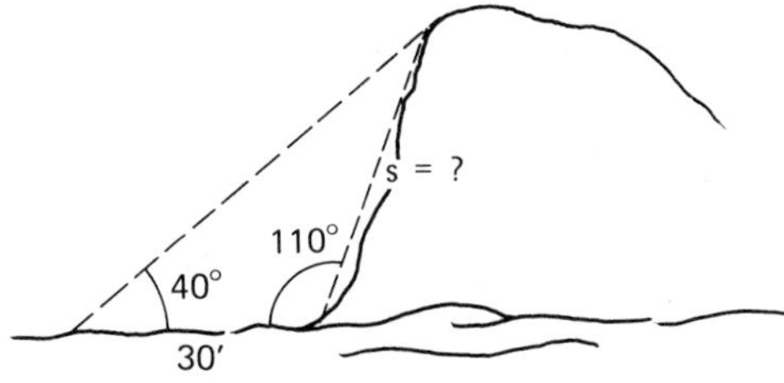

Figure 16-14

Part C

10. In physics, a vector is a quantity that has both magnitude and direction. In the parallelogram of vectors shown in Fig. 16-15, use the given vector magnitudes and the given angle (between vector AB and vector AD) to find the magnitude of the diagonal (resultant) vector. The following are implied by the fact that the figure is a parallelogram:

$$B = 120^\circ \qquad D = 120^\circ \qquad CD = 4 \qquad BC = 6$$

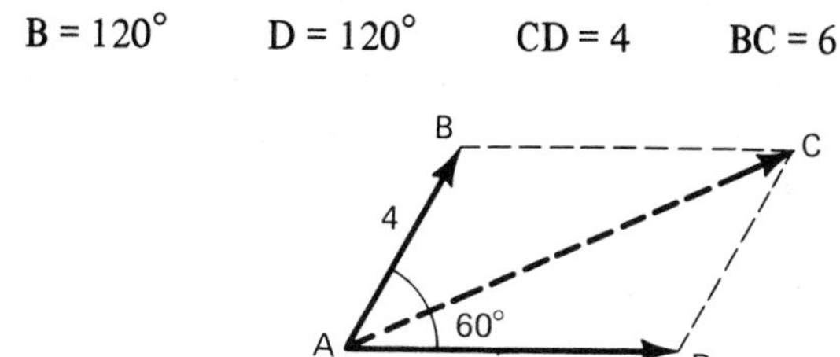

Figure 16-15

11. In Fig. 16-16, find the distance AB between two rocks that are projected above the water level of a lake if $m\angle AMC = 120^\circ$, $m\angle BMC = 25^\circ$, $m\angle MCB = 85^\circ$, and $m\angle MCA = 20^\circ$ with distance MC measured to be 100 ft.

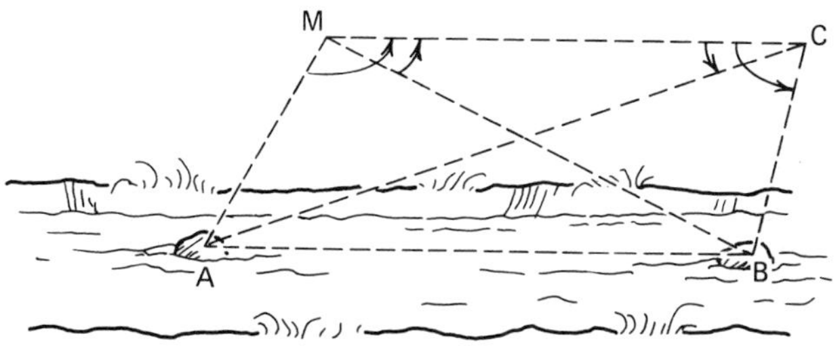

Figure 16-16

12. Refer back to Exercise 9 and find the height of hill above the base line.

Answers

2. −0.1736
4. 0.7660
6. −0.9962
8. 48.6′
10. 8.72
12. 36.27′

16-3 MATHEMATICS OF SURVEYING

In surveying we are concerned with direction, distance, and elevation. A property line is of no value to you or anyone else unless it has the proper direction and the correct length in relation to the other lines that make up the property. If property has correct lines, then a scale drawing can be made of the property and also the area of the property can be determined.

Important to the surveyor are instruments for measuring angles where a high degree of accuracy is required. One such instrument is the *transit*. A transit can be used (1) to measure desired angles to a high degree of accuracy and (2) to carry or construct a straight line.

Azimuth Line (Azimuth Angles or AZ Angles) and Bearing: It is usual to assume that the true meridian line (magnetic north line) is the basic line of reference, although the south end of a coordinate axis is taken as the 0° in common surveying. *Bearing* indicates the direction (compared to the reference line) in which one point lies, as viewed from another.

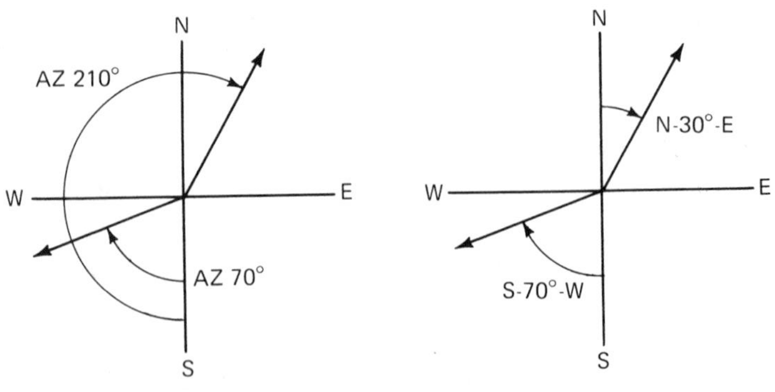

Figure 16-17

Figure 16-18

In Figs. 16-17 and 16-18 note that

1. AZ 70° = S-70°00′-W.
2. AZ 210° = N-30°00′-E.

The polygon in Figure 16-19 is an example of a convex polygon.

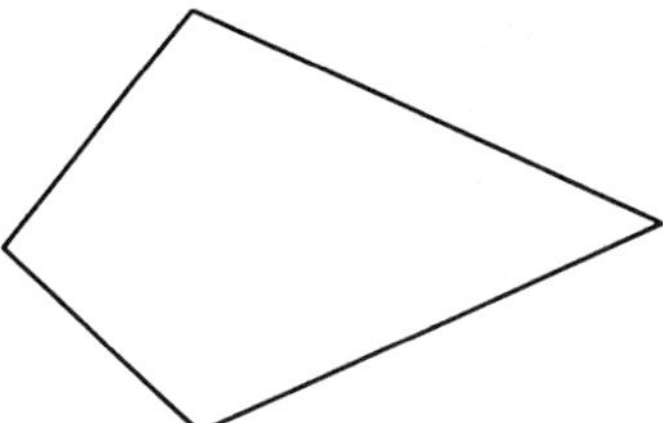

Figure 16-19. Convex polygon.

> *Definitions:* A *convex polygon* is a polygon in which no interior angle is larger than 180°.
>
> The *sum of the interior angles* is (n − 2)180°, where n is the number of sides.

Illustration 1: A triangle is a polygon of three sides. The sum of the interior angles of a triangle is equal to 180°. By the definition above, we can show that (n − 2)180° = 180° for a triangle.

$$(3 - 2)180^\circ = 1(180^\circ) = 180^\circ$$

Illustration 2: Now, if the polygon has four sides, use the definition above to find the sums of the interior angles. Sum of angles is

$$(n - 2)180^\circ = (4 - 2)(180^\circ) = 2(180^\circ) = 360^\circ$$

Example 16-11: The surveyor started at the left corner of a tract of land and traversed it in a clockwise direction.

Station	Line (course)	Direction (bearing)	Distance
A	AB	N-10°10′-E	410 ft
B	BC	N-60°30′-E	250 ft
C	CD	S-40°00′-E	200 ft
D	DA	S-48°10′-W	560 ft

Plot the traverses to a scale of 1 mm = 10 ft. See Fig. 16-20.

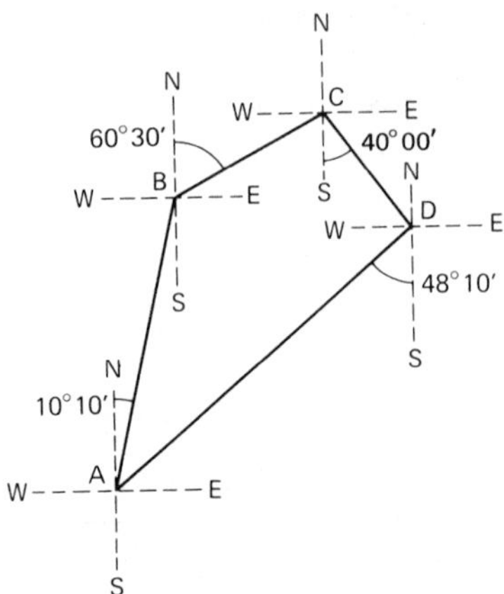

Figure 16-20

Example 16-12: Use the following field observations:

Station	Line (course)	Direction (observed bearing)	Needed Angles	Distance
A	AB	S-83°-15′-W	77°30′	500 ft
A	AC	S-5°45′-W		570 ft
B	BA	N-83°15′-E	55°40′	
B	BC	S-41°05′-E		?
C	CA	N-5°40′-E	46°45′	
C	CB	N-41°05′-W		

1. Plot the traverses to a scale of 1 mm = 10 ft. See Fig. 16-21.

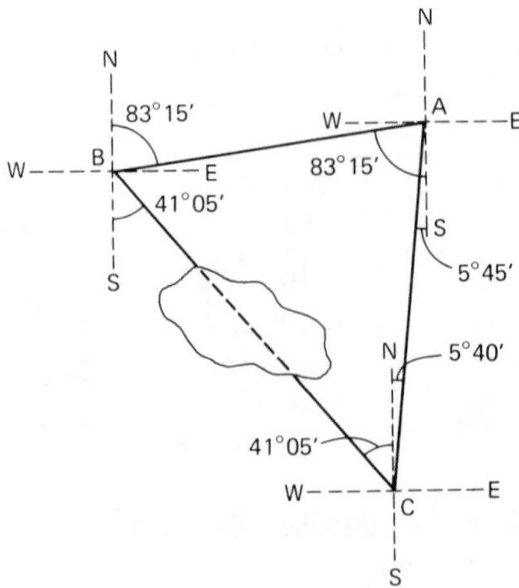

Figure 16-21

2. Find the sum of the interior angles of the tract above based on the surveyors notes and compare it with the sum obtained by using (n − 2)180° for a check. Calculate the error made in the field observation.

Solution

$$\angle BAC = 83°15' - 5°45' = 77°30'$$

$$\angle ABC = (180°) - (41°05' + 83°15')$$

$$= 179°60' - 124°20'$$

$$= 55°40'$$

$$\angle ACB = 41°05' + 5°40'$$

$$= 46°45'$$

Sum of the interior angles of this triangle measured from the bearings are listed below:

$$\angle BAC + \angle ABC + \angle ACB = 77°30' + 55°40' + 46°45'$$

$$= 178°115'$$

$$= 179°55'$$

To check, compare the sum above for the interior angles of the triangular plot with the sum obtained from (n − 2)180°.

$$(3 - 2)180° = 1(180°) = 180°$$

The comparison indicates an error of only 05′, which is within an acceptable range.

3. Compute the length of BC.

$$\frac{BC}{\sin \angle BAC} = \frac{AB}{\sin \angle ACB}$$

$$\frac{BC}{\sin 77°30'} = \frac{500}{\sin 46°45'}$$

$$\frac{BC}{0.9781} = \frac{500}{0.7314}$$

$$0.7314BC = 500(0.9781)$$

$$0.7314BC = 489.05$$

$$BC = 668.65 \text{ ft}$$

Definitions: Latitude is the distance a second point is measured to be either north or south of a first point.

Departure (*longitude*) is defined to be the distance a second point is east or west of a first point.

Example 16-13: Compare the departure of line AC with the sum of the departures AB and BC in Fig. 16-22.

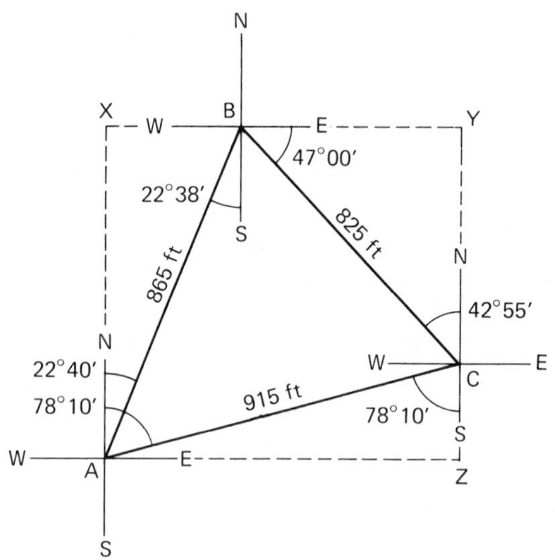

Figure 16-22

Find departure XB, BY, and AZ.

$$\frac{XB}{865} = \sin 22°40' \qquad \frac{BY}{825} = \sin 42°55' \qquad \frac{AZ}{915} = \sin 78°10'$$

$$XB = 865(\sin 22°40') \qquad BY = 825(\sin 42°55') \qquad AZ = 915(\sin 78°10')$$

$$XB = 865(0.3854) \qquad BY = 825(0.6810) \qquad AZ = 915(0.9787)$$

$$XB = 333.371 \qquad BY = 561.825 \qquad AZ = 895.5 \text{ ft}$$

$$XB = 333.4 \text{ ft} \qquad BY = 561.8 \text{ ft}$$

Compare XB + BY with AZ.

$$\begin{aligned} AZ &= 895.5 \\ XB + BY &= 333.4 + 561.8 = \underline{895.2} \\ & 0.3 \text{ ft difference} \end{aligned}$$

Example 16-14: Compute the difference in elevation of points A and D in Fig. 16-23.

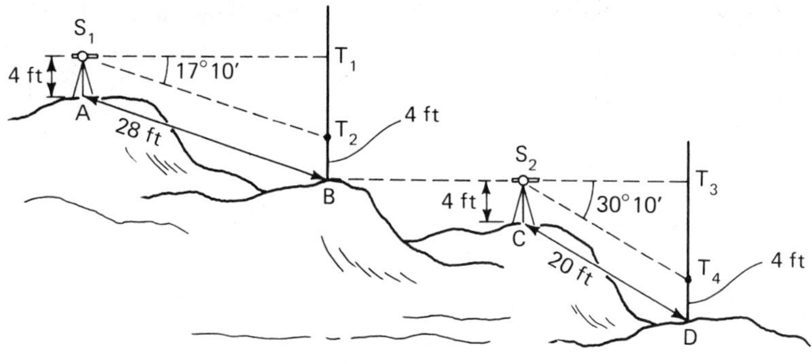

Figure 16-23

Solution

1. In right $\Delta S_1T_1T_2$,

$$\frac{T_1T_2}{28 \text{ ft}} = \sin 17°10'$$

$$T_1T_2 = 28 \text{ ft}(0.2954)$$

$$T_1T_2 = 8.3 \text{ ft} = \text{difference in elevation of points A and B}$$

2. In right $\Delta S_2T_3T_4$,

$$\frac{T_3T_4}{20 \text{ ft}} = \sin 30°10'$$

$$T_3T_4 = 20 \text{ ft}(0.5265)$$

$$T_3T_4 = 10.5 \text{ ft}$$

$$T_3D = 14.5 \text{ ft} = \text{difference in elevation of points B and D}$$

$$T_1T_2 + T_3D = \text{total difference in elevation}$$

$$8.3 \text{ ft} + 14.5 \text{ ft} = 22.8 \text{ ft}$$

Calculation of the Area of a Triangle: In Fig. 16-24,

$$\boxed{\text{area } \Delta ABC = \frac{1}{2} bc \sin A}$$

Figure 16-24

Example 16-15: Find the area of the tract of land shown in Fig. 16-25.

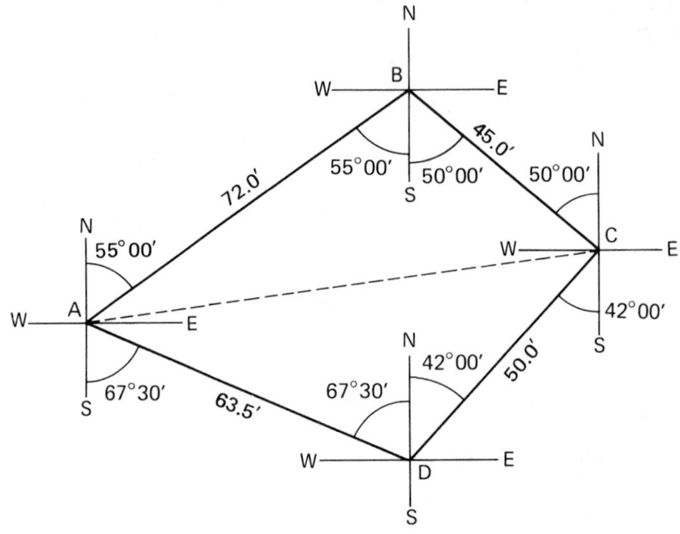

Figure 16-25

Solution: First divide the plot into triangles by drawing line AC and then calculating the area.

$$\angle ABC = 55^\circ + 50^\circ = 105^\circ \text{ and } \angle ADC = 109^\circ 30'$$

$$\text{area } \Delta ABC = \frac{1}{2}(AB)(BC) \sin \angle B$$

$$= \frac{1}{2}(72.0)(45.0)(\sin 105^\circ)$$

$$= \frac{1}{2}(72.0)(45.0)(0.9659)$$

$$\doteq 1{,}565 \text{ ft}^2$$

$$\text{area } \Delta ACD = \frac{1}{2}(AD)(DC)(\sin 109^\circ 30')$$

$$= \frac{1}{2}(63.5)(50.0)(0.9397)$$

$$\doteq 1{,}492 \text{ ft}^2$$

$$\text{total area} \doteq 3{,}057 \text{ ft}^2$$

Exercise Set 16-3

Part A

1. Field observations:

Station	Line (course)	Direction (observed bearing)	Distance	Angles
A	AF	S-83°50′-E		?
A	AB	N-7°50′-E	260 ft	
B	BA	S-7°45′-W		?
B	BC	N-44°55′-E	365 ft	
C	CB	S-45°00′-W		?
C	CD	N-5°20′-E	360 ft	
D	DC	S-5°20′-W		?
D	DE	N-80°00′-E	355 ft	
E	ED	S-79°55′-W		?
E	EF	S-7°10′-W	995 ft	
F	FE	N-7°10′-E		?
F	FA	N-83°55′-W	545 ft	
A	AF	S-83°50′-E		

(a) Plot the traverses to a scale of 1/16 in. = 10 ft.

(b) Find the measure of the following interior angles from the field observations.

(i) ∠FAB = (ii) ∠ABC = (iii) ∠BCD =

(iv) ∠CDE = (v) ∠DEF = (vi) ∠EFA =

2. Use the observed bearing in Exercise 1 and express the bearing of the following lines (courses) in terms of azimuth angles.

Line	Azimuth Angle
(a) AF	
(b) AB	
(c) BA	
(d) BC	
(e) CB	

	Line	Azimuth Angle (*cont'd*)
(f)	CD	
(g)	DC	
(h)	DE	
(i)	ED	
(j)	EF	
(k)	FE	
(l)	FA	

3. Compute the difference in elevation of A and C in Fig. 16-26.
4. Find the area of the plot shown in Fig. 16-20.

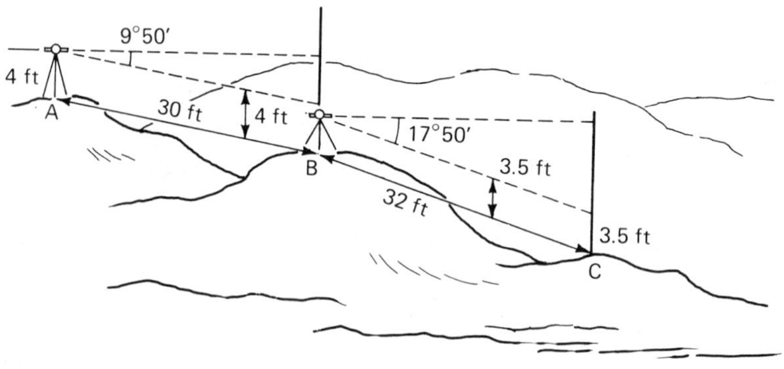

Figure 16-26

5. Find the area of the tract of land shown in Fig. 16-21.
6. Find the area of the triangular tract shown in Fig. 16-22.

Part B

7. Use all information in Exercise 1 and find the area. (*Hint:* Use the scale drawing and measure any angles that may be required that are not given.)

Answers

2. (a) 276°10′ (b) 187°50′ (c) 7°45′
   (d) 224°55′ (e) 45°00′ (f) 185°20′
   (g) 5°20′ (h) 260°00′ (i) 79°55′
   (j) 7°10′ (k) 187°10′ (l) 96°05′
4. 95,435 ft^2 or 95,262 ft^2 depending on triangles used
6. 328,066 ft^2

16-4 NUMERICAL CONTROL

The main effect of the computer age in the area of machine tools has been *numerical control*. Use of this concept and its techniques has allowed greater efficiency and precision in production machining. Just as a computer uses electronic impulses to do computations with tremendous speed, a machine tool equipped for numerical control uses electronic impulses to perform machining operations with tremendous speed.

The basic concept of numerical control for machine tools is the use of numbers to regulate electronic impulses that control the movements of the machine. This type of automation allows rapid completion of repetitive tasks. It is not a threat to the competent machinist, who is needed to set up and supervise the procedure; but it will spare the machinist from the tedious repetitive work that can be so boring.

Suppose that we have a milling machine set up with numerical control. The table can move to the right or left, which we arbitrarily call the X direction, and it can also move in or out, which we arbitrarily call the Y direction. If a single impulse produces a motion of 0.001 in. in the specified direction, then various combinations of impulses can produce any motion that is composed of so many thousandths in the X direction or so many thousandths in the Y direction.

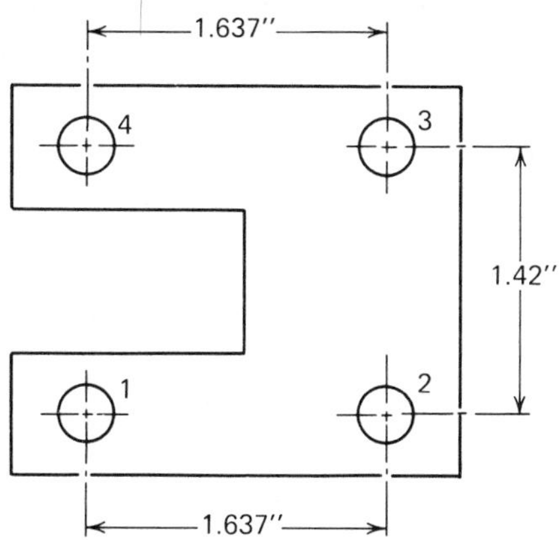

Figure 16-27

In Fig. 16-27, there are four holes to be drilled on a milling machine. We would start with the workpiece positioned under the drill bit in position to drill hole no. 1 and identify this as the initial *reference point*. The drill would then be brought down to do its job and let up again to prepare to move to hole no. 2. Now, since hole no. 2 is 1.637 in. directly to the right of hole no. 1, we would notice that we need 1,637 impulses in the positive X direction. This change in the X direction, which we call ΔX ("delta X"), can be obtained by telling the machine with numbers to let ΔX = 1,637 impulses. When the machine arrives at that point (almost immediately), the drill is brought down to produce hole no. 2 and let up again to prepare to move to hole no. 3. To reach hole no. 3, which is 1.42 in. directly in from hole no. 2, we need 1,420 impulses in the positive Y direction. In the language of numerical control, $\Delta Y = 1{,}420$, which moves the machine to hole no. 3. The drill is again brought down and let up, drilling hole no. 3. To reach hole no. 4, we need to go 1.637 in. to the left, which is the negative X direction. So we use $\Delta X = -1{,}637$, bring the tool down, let it back up, and the job is complete.

To summarize this procedure, we can set up the kind of information table that is the first step in numerical control programming.

ΔX	ΔY	Tool	Comments
			Reference point at hole no. 1
		Down	Drill hole no. 1
		Up	
+1,637			Move 1.637 in. to right
		Down	Drill hole no. 2
		Up	
	+1,420		Move 1.42 in. in
		Down	Drill hole no. 3
		Up	
-1,637			Move 1.637 in. to left
		Down	Drill hole no. 4
		Up	Finished

The only unusual procedure we need to clarify is combined motion in both directions. On most machines the ΔX and ΔY motions would proceed at the same rate until one of them was completed, and then the remainder of the larger change would be finished. For example, if we requested $\Delta X = 1{,}320$ and $\Delta Y = -1{,}055$, the table would move at a 45° angle to $\Delta X = 1{,}055$ and $\Delta Y = -1{,}055$, and then the table would move to the right for the additional 265 X-direction impulses.

Example 16-16: To drill the holes in the workpiece sketched in Fig. 16-28, we could start at the hole on the lower left corner and follow the dotted line. Note the information table.

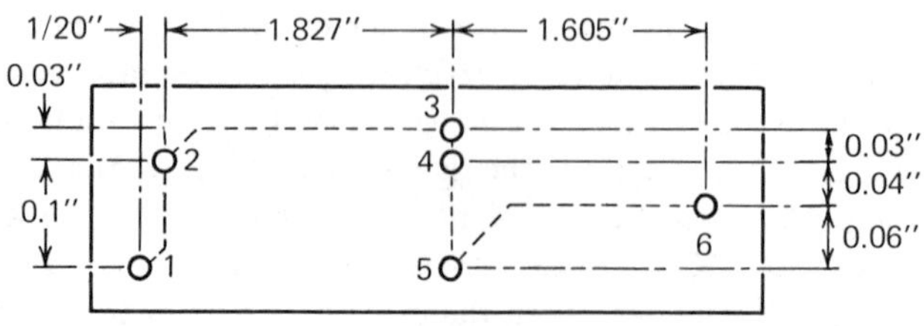

Figure 16-28

ΔX	ΔY	Tool	Comments
			Start lower left
		Down	Drill
		Up	
0050	0100		Move to second
		Down	Drill
		Up	
1,827	0030		Move to third
		Down	Drill
		Up	
	-0030		Move to fourth
		Down	Drill
		Up	
	-0100		Move to fifth
		Down	Drill
		Up	
1,605	0060		Move to sixth
		Down	Drill
		Up	Finished
-3,482	-0060		Back to starting point; ready for next piece

In Example 16-16, could you find a more efficient order for drilling the holes?

Example 16-17: Suppose we needed to cut a groove in a workpiece as indicated in Fig. 16-29. Starting at the lower left, we can let the machine do a joint movement of 400 impulses in the X direction along with 400 impulses in the Y direction. Then, to do the horizontal section of the groove, we need 600 impulses in the X direction. For the last section we have a slight problem. If we asked for $\Delta X = 1{,}100$ and $\Delta Y = 400$, the machine would use 400 impulses at the same time in each direction and then give us the remaining 700 impulses in the X direction. This obviously will not put the groove where we want it, so we must give the machine more information. Suppose we asked for $\Delta X = 275$ and $\Delta Y = 100$ a total

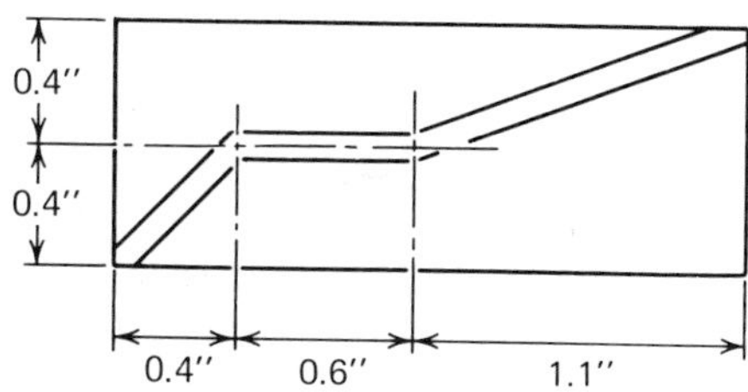

Figure 16-29

of four times (see Fig. 16-30). That would give us the correct total changes and would come closer to the straight groove we want. What about $\Delta X = 20$ and $\Delta Y = 55$ a total of 20 times (see Fig. 16-31)? This would be better, but it would require more separate inputs to the machine. To go to the extreme case, $\Delta X = 4$ and $\Delta Y = 11$ a total of 100 times would produce a groove almost indistinguishable from the straight groove requested, but it would require 100 lines of input in the information table. Unless you are lucky enough to be working on a machine that will do *different* changes in X and Y in the same amount of time, you must make a decision comparing accuracy of the work with amount of input needed.

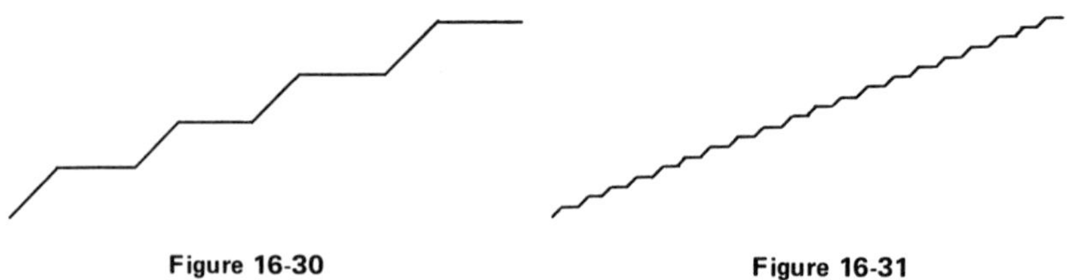

Figure 16-30 Figure 16-31

Exercise Set 16-4

1. Graph the location of the holes and grooves drilled by the following numerical control program:

ΔX	ΔY	Tool	Comments
			Starting point
		Down	
		Up	
+100	-100		
		Down	
		Up	
+250			
		Down	
	-200		
		Up	
-350	-200		
		Down	
		Up	
-150	+600		
		Down	
+150	+300		
		Up	Finishing point

2. Graph the locations of the holes and grooves drilled by the following numerical control program.

ΔX	ΔY	Tool	Comments
			Starting point
		Down	
		Up	
600			
		Down	
400	400		
		Up	
-300	200		
		Down	
-500	-400		
		Up	
-200			
		Down	
	-200		
		Up	Finishing point

3. Develop a numerical control program that would drill the template illustrated in Fig. 16-32.

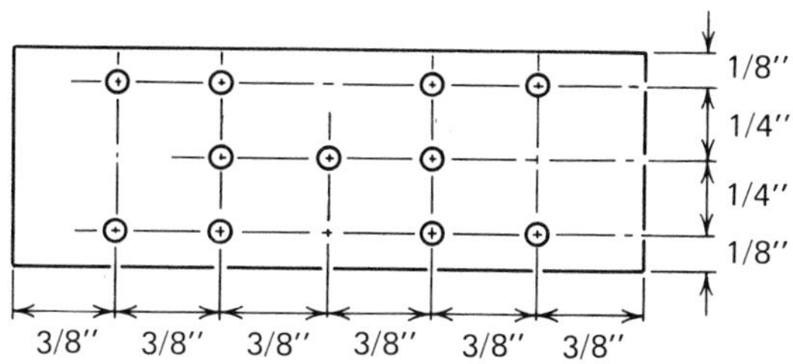

Figure 16-32

4. Develop a numerical control program that would drill a workpiece to the indicated specifications. (See Fig. 16-33.)

Figure 16-33

5. Use values of sine and cosine to develop a numerical control program that will cut a circle with a 1-in. radius.

Answers

2.

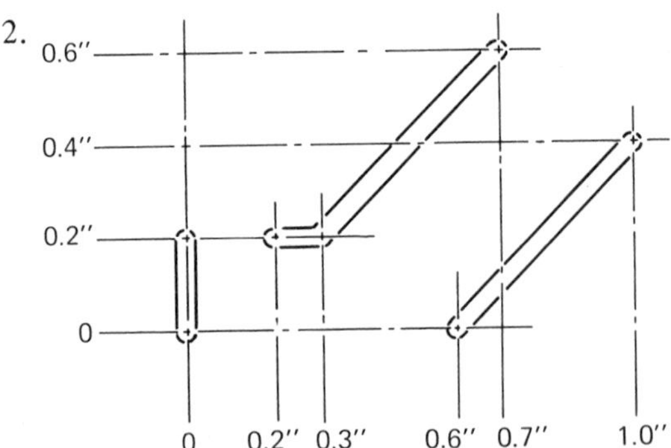

4.

ΔX	ΔY	Tool	Comments
		Down	Start lower left on M
	+813		5/8 + 3/16 = 0.625 + 0.1875 = 0.8125
+125			
+188	-188		
+313	+188		
	-813		
		Up	M is complete
+813	+688		Upper right on G
		Down	

ΔX	ΔY	Tool	Comments (*cont'd*)
-313	+125		
-125	-688		
+313	-125		
+125	+250		
-250	+125		
		Up	G is complete
+625			Bottom bar on P
		Down	
+250			
+125	+313		
-375	+125		
	-813		
		Up	P is complete
-1,814			Back to starting point

16-5 MANAGEMENT APPLICATIONS

Management applications is discussed as a special topic; however, it is a topic of importance to all businesses, professions, and vocations. In this section, we shall briefly discuss taxes, depreciation, and payroll deductions. It should be pointed out that this discussion is intended to give only an introduction to these three areas.

There are no tables given in this book to be used in computing taxes, depreciation, or payroll deductions. You should be able to find current tables in your library or at an Internal Revenue Service office for use as needed in this section.

Taxes: We shall discuss sales tax and property tax, although there are many other types of taxes. Income tax will be discussed in connection with payroll deductions.

Sales tax and property tax are usually calculated on a percentage basis. Whenever sales tax is figured from a chart, there is no calculation involved for you except adding the amount of the tax to the appropriate figure. Thus, we shall be concerned with finding tax based on a percentage.

Example 16-18: Smith's Retail Store must charge a 4% sales tax on all items sold.

1. If an item sells for $38.73, how much tax must be charged?

Sales tax is equal to 4% of the selling price of the item. (Sales tax should always be rounded to the nearest cent.)

$$\text{tax} = 4\% \text{ of price}$$

$$\text{tax} = 0.04 \times \$38.73$$

$$\text{tax} = \$1.55$$

So $1.55 tax should be charged for this item.

2. A sales tax of $2.70 was charged on another item sold. What was the selling price of the item before the tax was added?

   Again, we have the equation

$$\text{tax} = 4\% \text{ of price}$$

$$\$2.70 = 0.04 \times \text{price}$$

$$\$2.70 \div 0.04 = \text{price}$$

$$\$67.50 = \text{price}$$

So we see that this item sold for about $67.50 before tax was added.

Example 16-19: Jones' Construction Company owns a shop and a warehouse. The shop is valued at $4,720.00, and the warehouse is valued at $2,400.00. The property tax is charged on 60% of the value of the property. If the tax rate is 2.5%, how much is the tax on each building and what is the total property tax for the two buildings?

Solution: First, we find the taxable value for each building.

Shop

$$\text{taxable value} = 60\% \text{ of } \$4{,}720$$
$$= 0.60(4{,}720)$$
$$= \$2{,}832$$

Warehouse

$$\text{taxable value} = 60\% \text{ of } \$2{,}400$$
$$= 0.60(2{,}400)$$
$$= \$1{,}440$$

Second, we find the amount of tax for each building.

Shop

$$\text{tax} = 2.5\% \text{ of } \$2{,}832$$
$$= 0.025(\$2{,}832)$$
$$= \$70.80$$

Warehouse

$$\text{tax} = 2.5\% \text{ of } \$1{,}440$$
$$= 0.025(\$1{,}440)$$
$$= \$36.00$$

The property tax is \$70.80 for the shop and \$36.00 for the warehouse. The total tax is \$106.80.

Alternate Solution: Since we were not interested in the taxable value except to find the amount of tax, we could have multiplied 60%, 2.5%, and the original value of the property to find the amount of tax.

Shop	*Warehouse*
tax = 60% of 2.5% of \$4,720	tax = 60% of 2.5% of \$2,400
= 0.6 × 0.025 × \$4,720	= 0.6 × 0.025 × \$2,400
= \$70.80	= \$36.00

Thus, we obtain the same answers.

Depreciation: As a building, an automobile, a machine, or other item increases in age, it usually decreases in value. This decrease in value is called *depreciation*. There are various methods that may be used to calculate depreciation. We shall discuss the *straight-line method* and the *sum-of-the-years'-digits method*, which are two frequently used methods for calculating depreciation.

Before computing depreciation, we must consider some terms that will be used. In order to depreciate an item, a *useful life* (in years) must be determined; that is, how many years will the item be used before being traded or junked. The *trade-in value* or *scrap value* of a piece of equipment is the value of the equipment at the time it is traded or junked. The difference between the original value and the scrap value of an item is called the *wearing value*. Finally, a *depreciation schedule* should be set up showing the years of depreciation, the book value of the item, the annual depreciation, and the accumulated depreciation. (*Note:* The accumulated depreciation should never be larger than the wearing value.)

The *straight-line method* allows the same annual depreciation each year for the useful life of an item. The wearing value is divided by the number of years of the useful life.

Example 16-20: The Precision Machine Shop purchased a new machine for \$4,000. This machine is expected to be used for 5 years and then sold for \$500. Use the straight-line method to prepare a depreciation schedule.

$$\begin{aligned}\text{wearing value} &= \text{original value} - \text{scrap value}\\ &= \$4{,}000 - \$500\\ &= \$3{,}500\end{aligned}$$

$$\text{annual depreciation} = \frac{\text{wearing value}}{\text{no. of years used}}$$

$$= \frac{\$3{,}500}{5}$$

$$= \$700$$

Depreciation Schedule for Straight-Line Method

Year	Book Value	Annual Depreciation	Accumulated Depreciation
0 (new)	$4,000	–	–
1	3,300	$700	$ 700
2	2,600	700	1,400
3	1,900	700	2,100
4	1,200	700	2,800
5	500	700	3,500

The *sum-of-the-years'-digits method* allows more depreciation the first year than the second, more the second than the third, and so forth. Thus, as the equipment gets older and is worth less, it depreciates less.

Example 16-21: Using the previous example and the sum-of-the-years'-digits method, prepare a depreciation schedule. (*Remember:* Original value is $4,000; scrap value is $500; useful life is 5 years.)

Step 1. Add the digits representing the years.

$$1 + 2 + 3 + 4 + 5 = 15$$

Step 2. Form fractions using the years' digits as numerators and the sum of these digits (15) as the denominator.

Step 3. Beginning with the largest fraction in the first year, multiply the wearing value ($3,500) by each corresponding fraction. This gives each year's depreciation.

Year	Annual Depreciation
1	5/15 × $3,500 = $1,166.67
2	4/15 × 3,500 = 933.33
3	3/15 × 3,500 = 700.00
4	2/15 × 3,500 = 466.67
5	1/15 × 3,500 = 233.33

Depreciation Schedule for Sum-of-the-Years'-Digits Method

Year	Book Value	Annual Depreciation	Accumulated Depreciation
0	$4,000.00	–	–
1	2,833.33	$1,166.67	$1,166.67
2	1,900.00	933.33	2,100.00
3	1,200.00	700.00	2,800.00
4	733.33	466.67	3,266.67
5	500.00	233.33	3,500.00

Payroll Deductions: As an employee or an employer, there are some things with which you should be familiar concerning payroll deductions. Federal income tax and social security tax (F.I.C.A., Federal Insurance Contributions Act) are two compulsory payroll deductions required by the federal government. Other deductions may be made also. Some of these may include state income tax, retirement insurance, health or hospital insurance, savings plan, and charitable contributions. After all payroll deductions are made, the amount remaining is called *net wages*.

Social security tax is withheld on a percentage basis. The amount withheld from an employee's wages must be equally matched by the employer. For instance, if you pay 6.65% of your wages for F.I.C.A., your employer must also pay 6.65% to match the amount paid by you. This means that an amount equal to 13.3% of your wages (up to the maximum set by law) goes for F.I.C.A. Some past and projected rates for F.I.C.A. are listed below.

Social Security (F.I.C.A.) Withholding Rates

Year	Rate	Maximum Wage Taxable
1973	5.85	$10,800
1974	5.85	13,200
1975	5.85	14,100
1976	5.85	15,300
1977	5.85	16,500
1978	6.05	17,700
1979	6.13	22,900
1980	6.13	25,900
1981	6.65	29,700
1982	6.70	To be determined using a formula provided by law.
1983	6.70	
1984	6.70	
1985	7.05	
1986–1989	7.15	
1990 and later	7.65	

Example 16-22: Stephanie Miller's gross earnings per week are $283.52 in 1981.

1. How much is withheld per week for F.I.C.A.?

To find this amount, find 6.65% of $283.52.

F.I.C.A. = 6.65% of $283.52

F.I.C.A. = 0.0665 ($283.52)

F.I.C.A. = $18.85 per week

2. How much does Stephanie contribute to F.I.C.A. during a year (52 weeks)?

Since Stephanie Miller earns less than $29,700 a year, she pays F.I.C.A. on all of her earnings. So Stephanie pays $980.20 for F.I.C.A. during the year.

Federal income tax withholdings may be found in a tax table, or they may be computed on a percentage basis. However, we shall discuss only the use of a tax table. Tables 16-1 and 16-2 are only parts of the actual tables.

Example 16-23: Mr. Hardy (who is single) earns $235.42 per week. If he claims one exemption for himself, how much federal tax should be withheld from his weekly check?

Since Mr. Hardy is single, his withholdings are found in Table 16-1. In the row "at least 230 but less than 240," go across to the column with one exemption. Here you should find his withholdings of $36.40.

Example 16-24: John Dolittle is married and has four children. If he earns $450 per week and claims only five exemptions, how much are his federal tax withholdings per week?

Table 16-2 shows withholdings for married employees. So look in the "at least 450 but less than 460" row and go across the the column with five exemptions. You should find $60.50 to be Mr. Dolittle's weekly withholdings.

Exercise Set 16-5

Part A

1. Find the sales tax and total cost to buy a set of spark plugs that list for $9.95 if a 4% sales tax is charged.
2. John bought two spring and mattress sets for twin beds. The sets were on sale for 36% off, and a 4% sales tax was charged. If the original price was $238.00 per set, how much did John pay for the two sets including tax?
3. Mr. Smith bought a new truck to use in his business. His truck sold for $7,275.55. Sales tax, property tax for first year, down payment, insurance and licenses totaled 24% of the cost of the truck. If Mr. Smith financed his

TABLE 16-1. Partial Listing of Federal Income Tax Withheld for a Single Person on a Weekly Payroll.

And the wages are—		And the number of withholding allowances claimed is—										
At least	But less than	0	1	2	3	4	5	6	7	8	9	10 or more
		The amount of income tax to be withheld shall be—										
$135	$140	$19.00	$15.30	$11.80	$8.40	$5.00	$2.10	$0	$0	$0	$0	$0
140	145	20.00	16.20	12.70	9.30	5.80	2.90	0	0	0	0	0
145	150	21.10	17.10	13.60	10.20	6.70	3.60	.70	0	0	0	0
150	160	22.60	18.60	15.00	11.50	8.10	4.70	1.80	0	0	0	0
160	170	24.70	20.70	16.80	13.30	9.90	6.40	3.30	.50	0	0	0
170	180	26.80	22.80	18.80	15.10	11.70	8.20	4.80	2.00	0	0	0
180	190	28.90	24.90	20.90	16.90	13.50	10.00	6.50	3.50	.60	0	0
190	200	31.00	27.00	23.00	18.90	15.30	11.80	8.30	5.00	2.10	0	0
200	210	33.60	29.10	25.10	21.00	17.10	13.60	10.10	6.70	3.60	.70	0
210	220	36.20	31.20	27.20	23.10	19.10	15.40	11.90	8.50	5.10	2.20	0
220	230	38.80	33.80	29.30	25.20	21.20	17.20	13.70	10.30	6.80	3.70	.80
230	240	41.40	36.40	31.40	27.30	23.30	19.20	15.50	12.10	8.60	5.20	2.30
240	250	44.00	39.00	34.00	29.40	25.40	21.30	17.30	13.90	10.40	6.90	3.80
250	260	46.60	41.60	36.60	31.60	27.50	23.40	19.40	15.70	12.20	8.70	5.30
260	270	49.20	44.20	39.20	34.20	29.60	25.50	21.50	17.50	14.00	10.50	7.10
270	280	51.80	46.80	41.80	36.80	31.80	27.60	23.60	19.60	15.80	12.30	8.90
280	290	54.80	49.40	44.40	39.40	34.40	29.70	25.70	21.70	17.60	14.10	10.70
290	300	57.80	52.10	47.00	42.00	37.00	32.00	27.80	23.80	19.70	15.90	12.50
300	310	60.80	55.10	49.60	44.60	39.60	34.60	29.90	25.90	21.80	17.80	14.30
310	320	63.80	58.10	52.30	47.20	42.20	37.20	32.20	28.00	23.90	19.90	16.10
320	330	66.80	61.10	55.30	49.80	44.80	39.80	34.80	30.10	26.00	22.00	17.90
330	340	70.00	64.10	58.30	52.50	47.40	42.40	37.40	32.40	28.10	24.10	20.00
340	350	73.40	67.10	61.30	55.50	50.00	45.00	40.00	35.00	30.20	26.20	22.10
350	360	76.80	70.30	64.30	58.50	52.80	47.60	42.60	37.60	32.60	28.30	24.20
360	370	80.20	73.70	67.30	61.50	55.80	50.20	45.20	40.20	35.20	30.40	26.30

TABLE 16-2. Partial Listing of Federal Income Tax Withheld for a Married Person on a Weekly Payroll.

And the wages are—		And the number of withholding allowances claimed is—										
At least	But less than	0	1	2	3	4	5	6	7	8	9	10 or more
		The amount of income tax to be withheld shall be—										
$150	$160	$17.20	$13.70	$10.60	$7.70	$4.80	$1.90	$0	$0	$0	$0	$0
160	170	19.00	15.50	12.10	9.20	6.30	3.40	.50	0	0	0	0
170	180	20.80	17.30	13.80	10.70	7.80	4.90	2.00	0	0	0	0
180	190	22.60	19.10	15.60	12.20	9.30	6.40	3.50	.60	0	0	0
190	200	24.40	20.90	17.40	14.00	10.80	7.90	5.00	2.10	0	0	0
200	210	26.20	22.70	19.20	15.80	12.30	9.40	6.50	3.60	.80	0	0
210	220	28.10	24.50	21.00	17.60	14.10	10.90	8.00	5.10	2.30	0	0
220	230	30.20	26.30	22.80	19.40	15.90	12.50	9.50	6.60	3.80	.90	0
230	240	32.30	28.30	24.60	21.20	17.70	14.30	11.00	8.10	5.30	2.40	0
240	250	34.40	30.40	26.40	23.00	19.50	16.10	12.60	9.60	6.80	3.90	1.00
250	260	36.50	32.50	28.50	24.80	21.30	17.90	14.40	11.10	8.30	5.40	2.50
260	270	38.60	34.60	30.60	26.60	23.10	19.70	16.20	12.70	9.80	6.90	4.00
270	280	40.70	36.70	32.70	28.60	24.90	21.50	18.00	14.50	11.30	8.40	5.50
280	290	42.80	38.80	34.80	30.70	26.70	23.30	19.80	16.30	12.90	9.90	7.00
290	300	45.10	40.90	36.90	32.80	28.80	25.10	21.60	18.10	14.70	11.40	8.50
300	310	47.50	43.00	39.00	34.90	30.90	26.90	23.40	19.90	16.50	13.00	10.00
310	320	49.90	45.30	41.10	37.00	33.00	28.90	25.20	21.70	18.30	14.80	11.50
320	330	52.30	47.70	43.20	39.10	35.10	31.00	27.00	23.50	20.10	16.60	13.20
330	340	54.70	50.10	45.50	41.20	37.20	33.10	29.10	25.30	21.90	18.40	15.00
340	350	57.10	52.50	47.90	43.30	39.30	35.20	31.20	27.20	23.70	20.20	16.80

350	360	59.50	54.90	50.30	45.70	41.40	37.30	33.30	29.30	25.50	22.00	18.60
360	370	61.90	57.30	52.70	48.10	43.50	39.40	35.40	31.40	27.30	23.80	20.40
370	380	64.60	59.70	55.10	50.50	45.90	41.50	37.50	33.50	29.40	25.60	22.20
380	390	67.40	62.10	57.50	52.90	48.30	43.70	39.60	35.60	31.50	27.50	24.00
390	400	70.20	64.80	59.90	55.30	50.70	46.10	41.70	37.70	33.60	29.60	25.80
400	410	73.00	67.60	62.30	57.70	53.10	48.50	43.80	39.80	35.70	31.70	27.60
410	420	75.80	70.40	65.00	60.10	55.50	50.90	46.20	41.90	37.80	33.80	29.70
420	430	78.60	73.20	67.80	62.50	57.90	53.30	48.60	44.00	39.90	35.90	31.80
430	440	81.40	76.00	70.60	65.20	60.30	55.70	51.00	46.40	42.00	38.00	33.90
440	450	84.20	78.80	73.40	68.00	62.70	58.10	53.40	48.80	44.20	40.10	36.00
450	460	87.00	81.60	76.20	70.80	65.40	60.50	55.80	51.20	46.60	42.20	38.10
460	470	90.20	84.40	79.00	73.60	68.20	62.90	58.20	53.60	49.00	44.40	40.20
470	480	93.40	87.30	81.80	76.40	71.00	65.60	60.60	56.00	51.40	46.80	42.30
480	490	96.60	90.50	84.60	79.20	73.80	68.40	63.10	58.40	53.80	49.20	44.60
490	500	99.80	93.70	87.50	82.00	76.60	71.20	65.90	60.80	56.20	51.60	47.00

truck and paid only for the five charges mentioned above, how much did he have to pay before his first monthly payment?

Jane Johnson earns $249.95 per week in 1981. She is single and claims no exemptions. (Use with Exercises 4–8 below.)

4. How much F.I.C.A. does Jane pay weekly?
5. How much federal income tax does Jane pay per week?
6. If Jane has a retirement plan that deducts 5.4% of her wages, how much is that per week?
7. If Jane pays 4.3% for state income tax, how much should be withheld weekly for state tax?
8. Assuming that the amounts in Exercises 4–7 represent Jane's total payroll deductions, find Jane's weekly net wages.

Part B

9. While away on vacation, your boss (who normally figures the payroll) asked you to figure the weekly payroll. All employees are paid on an hourly rate. Each employee pays F.I.C.A., federal income tax, 4.5% for retirement, and insurance at the rate of $10.97 per employee and $1.45 for each additional dependent. Complete the table below.

Employee Number	Marital Status Exemptions	Total hours	Hourly rate	Gross Wages
58	M/3	38	$4.40	
63	S/1	30	4.99	
65	S/2	40	5.55	
76	M/6	40	6.50	

F.I.C.A.	Tax	Ret.	Ins.	Total Deductions	Net Wages

10. The total cost of an electric drill was $62.14 (tax included). If sales tax was at the rate of 4%, find (a) the original price of the drill and (b) the amount of sales tax.

11. Suppose you bought five notebook binders for $19.50 with a 4% sales tax included. (a) Find the list price per notebook binder. (b) Find the sales tax per notebook binder.

12. The cost of an air-powered nail gun, tax included, was $82.95. If a 5% tax was charged, find (a) the list price of the nail gun and (b) the amount of sales tax charged.

Part C

13. Mr. Smith paid $5,275.54 for a used truck for his business. He expects to use it for 6 years and sell it for $550. Set up a depreciation schedule using (a) the straight-line method and (b) the sum-of-the-years'-digits method.

14. The building used by Prospect Motors is valued at $12,000. It is expected to be worth $4,000 after 15 years. (a) Set up a depreciation schedule for the first 5 years using the straight-line method. (b) What is the accumulated depreciation at the end of 12 years?

15. Johnson's Printing Company purchased three new offset printing machines for $6,200 each. These machines are to be used 20 years and sold together for $600. (a) What is the wearing value? (b) Using the sum-of-the-years'-digits method, set up a depreciation schedule for the first 6 years. (c) What is the accumulated depreciation after 15 years?

16. Bibby Construction Contractor bought a ditcher for $25,000. If the scrap value after 30 years is $500, use the sum-of-the-years'-digits method to find the depreciation for each year.

Answers

2. $316.83
4. $16.62
6. $13.50
8. $165.08
10. (a) $59.75 (b) $2.39
12. (a) $79.00 (b) $3.95
14. (a)

Year	Value	Annual Depreciation	Accumulated Depreciation
0	$12,000.00	–	–
1	11,466.67	$533.33	$ 533.33
2	10,933.34	533.33	1,066.66
3	10,400.01	533.33	1,599.99
4	9,866.68	533.33	2,133.32
5	9,333.35	533.33	2,666.65

(b) $6,399.96

16.

Year	Depreciation	Year	Depreciation
1	$1,580.70	16	$790.35
2	1,528.01	17	737.66
3	1,475.32	18	684.97
4	1,422.63	19	632.28
5	1,369.94	20	579.59
6	1,317.25	21	526.90
7	1,264.56	22	474.21
8	1,211.87	23	421.52
9	1,159.18	24	368.83
10	1,106.49	25	316.14
11	1,053.80	26	263.45
12	1,001.11	27	210.76
13	948.42	28	158.07
14	895.73	29	105.38
15	843.04	30	52.69

Appendices

APPENDIX A

Squares and Square Roots*

N	N^2	$\sqrt{N}$	$\sqrt{10N}$	N	N^2	$\sqrt{N}$	$\sqrt{10N}$
1.0	1.00	1.000	3.162	5.5	30.25	2.345	7.416
1.1	1.21	1.049	3.317	5.6	31.36	2.366	7.483
1.2	1.44	1.095	3.464	5.7	32.49	2.387	7.550
1.3	1.69	1.140	3.606	5.8	33.64	2.408	7.616
1.4	1.96	1.183	3.742	5.9	34.81	2.429	7.681
1.5	2.25	1.225	3.873	6.0	36.00	2.449	7.746
1.6	2.56	1.265	4.000	6.1	37.21	2.470	7.810
1.7	2.89	1.304	4.123	6.2	38.44	2.490	7.874
1.8	3.24	1.342	4.243	6.3	39.69	2.510	7.937
1.9	3.61	1.378	4.359	6.4	40.96	2.530	8.000
2.0	4.00	1.414	4.472	6.5	42.25	2.550	8.062
2.1	4.41	1.449	4.583	6.6	43.56	2.569	8.124
2.2	4.84	1.483	4.690	6.7	44.89	2.588	8.185
2.3	5.29	1.517	4.796	6.8	46.24	2.608	8.246
2.4	5.76	1.549	4.899	6.9	47.61	2.627	8.307
2.5	6.25	1.581	5.000	7.0	49.00	2.646	8.367
2.6	6.76	1.612	5.099	7.1	50.41	2.665	8.426
2.7	7.29	1.643	5.196	7.2	51.84	2.683	8.485
2.8	7.84	1.673	5.292	7.3	53.29	2.702	8.544
2.9	8.41	1.703	5.385	7.4	54.76	2.720	8.602
3.0	9.00	1.732	5.477	7.5	56.25	2.739	8.660
3.1	9.61	1.761	5.568	7.6	57.76	2.757	8.718
3.2	10.24	1.789	5.657	7.7	59.29	2.775	8.775
3.3	10.89	1.817	5.745	7.8	60.84	2.793	8.832
3.4	11.56	1.844	5.831	7.9	62.41	2.811	8.888
3.5	12.25	1.871	5.916	8.0	64.00	2.828	8.944
2.6	12.96	1.897	6.000	8.1	65.61	2.846	9.000
3.7	13.69	1.924	6.083	8.2	67.24	2.864	9.055
3.8	14.44	1.949	6.164	8.3	68.89	2.881	9.110
3.9	15.21	1.975	6.245	8.4	70.56	2.898	9.165
4.0	16.00	2.000	6.325	8.5	72.25	2.915	9.220
4.1	16.81	2.025	6.403	8.6	73.96	2.933	9.274
4.2	17.64	2.049	6.481	8.7	75.69	2.950	9.327
4.3	18.49	2.074	6.557	8.8	77.44	2.966	9.381
4.4	19.36	2.098	6.633	8.9	79.21	2.983	9.434
4.5	20.25	2.121	6.708	9.0	81.00	3.000	9.487
4.6	21.16	2.145	6.782	9.1	82.81	3.017	9.539
4.7	22.09	2.168	6.856	9.2	84.64	3.033	9.592
4.8	23.04	2.191	6.928	9.3	86.49	3.050	9.644
4.9	24.01	2.214	7.000	9.4	88.36	3.066	9.695
5.0	25.00	2.236	7.071	9.5	90.25	3.082	9.747
5.1	26.01	2.258	7.141	9.6	92.16	3.098	9.798
5.2	27.04	2.280	7.211	9.7	94.09	3.114	9.849
5.3	28.09	2.302	7.280	9.8	96.04	3.130	9.899
5.4	29.16	2.324	7.348	9.9	98.01	3.146	9.950
5.5	30.25	2.345	7.416	10	100.00	3.162	10.000

*From Mary P. Dolciani, Simon L. Berman, and William Wooton, *Modern Algebra and Trigonometry: Structure and Method-Book 2,* ©1965, pp. 606–607. Reprinted by permission of Houghton Mifflin Co., Boston, Mass.

APPENDIX B

Cubes and Cube Roots*

N	N^3	$\sqrt[3]{N}$	$\sqrt[3]{10N}$	$\sqrt[3]{100N}$	N	N^3	$\sqrt[3]{N}$	$\sqrt[3]{10N}$	$\sqrt[3]{100N}$
1.0	1.000	1.000	2.154	4.642	**5.5**	166.375	1.765	3.803	8.193
1.1	1.331	1.032	2.224	4.791	**5.6**	175.616	1.776	3.826	8.243
1.2	1.728	1.063	2.289	4.932	**5.7**	185.193	1.786	3.849	8.291
1.3	2.197	1.091	2.351	5.066	**5.8**	195.112	1.797	3.871	8.340
1.4	2.744	1.119	2.410	5.192	**5.9**	205.379	1.807	3.893	8.387
1.5	3.375	1.145	2.466	5.313	**6.0**	216.000	1.817	3.915	8.434
1.6	4.096	1.170	2.520	5.429	**6.1**	226.981	1.827	3.936	8.481
1.7	4.913	1.193	2.571	5.540	**6.2**	238.328	1.837	3.958	8.527
1.8	5.832	1.216	2.621	5.646	**6.3**	250.047	1.847	3.979	8.573
1.9	6.859	1.239	2.668	5.749	**6.4**	262.144	1.857	4.000	8.618
2.0	8.000	1.260	2.714	5.848	**6.5**	274.625	1.866	4.021	8.662
2.1	9.261	1.281	2.759	5.944	**6.6**	287.496	1.876	4.041	8.707
2.2	10.648	1.301	2.802	6.037	**6.7**	300.763	1.885	4.062	8.750
2.3	12.167	1.320	2.844	6.127	**6.8**	314.432	1.895	4.082	8.794
2.4	13.824	1.339	2.884	6.214	**6.9**	328.509	1.904	4.102	8.837
2.5	15.625	1.357	2.924	6.300	**7.0**	343.000	1.913	4.121	8.879
2.6	17.576	1.375	2.962	6.383	**7.1**	357.911	1.922	4.141	8.921
2.7	19.683	1.392	3.000	6.463	**7.2**	373.248	1.931	4.160	8.963
2.8	21.952	1.409	3.037	6.542	**7.3**	389.017	1.940	4.179	9.004
2.9	24.389	1.426	3.072	6.619	**7.4**	405.224	1.949	4.198	9.045
3.0	27.000	1.442	3.107	6.694	**7.5**	421.875	1.957	4.217	9.086
3.1	29.791	1.458	3.141	6.768	**7.6**	438.976	1.966	4.236	9.126
3.2	32.768	1.474	3.175	6.840	**7.7**	456.533	1.975	4.254	9.166
3.3	35.937	1.489	3.208	6.910	**7.8**	474.552	1.983	4.273	9.205
3.4	39.304	1.504	3.240	6.980	**7.9**	493.039	1.992	4.291	9.244
3.5	42.875	1.518	3.271	7.047	**8.0**	512.000	2.000	4.309	9.283
3.6	46.656	1.533	3.302	7.114	**8.1**	531.441	2.008	4.327	9.322
3.7	50.653	1.547	3.332	7.179	**8.2**	551.368	2.017	4.344	9.360
3.8	54.872	1.560	3.362	7.243	**8.3**	571.787	2.025	4.362	9.398
3.9	59.319	1.574	3.391	7.306	**8.4**	592.704	2.033	4.380	9.435
4.0	64.000	1.587	3.420	7.368	**8.5**	614.125	2.041	4.397	9.473
4.1	68.921	1.601	3.448	7.429	**8.6**	636.056	2.049	4.414	9.510
4.2	74.088	1.613	3.476	7.489	**8.7**	658.503	2.057	4.431	9.546
4.3	79.507	1.626	3.503	7.548	**8.8**	681.472	2.065	4.448	9.583
4.4	85.184	1.639	3.530	7.606	**8.9**	704.969	2.072	4.465	9.619
4.5	91.125	1.651	3.557	7.663	**9.0**	729.000	2.080	4.481	9.655
4.6	97.336	1.663	3.583	7.719	**9.1**	753.571	2.088	4.498	9.691
4.7	103.823	1.675	3.609	7.775	**9.2**	778.688	2.095	4.514	9.726
4.8	110.592	1.687	3.634	7.830	**9.3**	804.357	2.103	4.531	9.761
4.9	117.649	1.698	3.659	7.884	**9.4**	830.584	2.110	4.547	9.796
5.0	125.000	1.710	3.684	7.937	**9.5**	857.375	2.118	4.563	9.830
5.1	132.651	1.721	3.708	7.990	**9.6**	884.736	2.125	4.579	9.865
5.2	140.608	1.732	3.733	8.041	**9.7**	912.673	2.133	4.595	9.899
5.3	148.877	1.744	3.756	8.093	**9.8**	941.192	2.140	4.610	9.933
5.4	157.464	1.754	3.780	8.143	**9.9**	970.299	2.147	4.626	9.967
5.5	166.375	1.765	3.803	8.193	**10**	1000.000	2.154	4.642	10.000

*From Mary P. Dolciani, Simon L. Berman, and William Wooton, *Modern Algebra and Trigonometry: Structure and Method-Book 2,* ©1965, pp. 606–607. Reprinted by permission of Houghton Mifflin Co., Boston, Mass.

APPENDIX C Trigonometric Functions of Angles*

Deg	Rad	Sin	Cos	Tan	Cot	Sec	Csc		
0	0.0000	0.0000	1.0000	0.0000	. . . . .	1.0000	. . . . .	1.5708	90
1	0.0175	0.0175	0.9998	0.0175	57.2900	1.0002	57.299	1.5533	89
2	0.0349	0.0349	0.9994	0.0349	28.6363	1.0006	28.654	1.5359	88
3	0.0524	0.0523	0.9986	0.0524	19.0811	1.0014	19.107	1.5184	87
4	0.0698	0.0698	0.9976	0.0699	14.3007	1.0024	14.336	1.5010	86
5	0.0873	0.0872	0.9962	0.0875	11.4301	1.0038	11.474	1.4835	85
6	0.1047	0.1045	0.9945	0.1051	9.5144	1.0055	9.5668	1.4661	84
7	0.1222	0.1219	0.9925	0.1228	8.1443	1.0075	8.2055	1.4486	83
8	0.1396	0.1392	0.9903	0.1405	7.1154	1.0098	7.1853	1.4312	82
9	0.1571	0.1564	0.9877	0.1584	6.3138	1.0125	6.3925	1.4137	81
10	0.1745	0.1736	0.9848	0.1763	5.6713	1.0154	5.7588	1.3963	80
11	0.1920	0.1908	0.9816	0.1944	5.1446	1.0187	5.2408	1.3788	79
12	0.2094	0.2079	0.9781	0.2126	4.7046	1.0223	4.8097	1.3614	78
13	0.2269	0.2250	0.9744	0.2309	4.3315	1.0263	4.4454	1.3439	77
14	0.2443	0.2419	0.9703	0.2493	4.0108	1.0306	4.1336	1.3265	76
15	0.2618	0.2588	0.9659	0.2679	3.7321	1.0353	3.8637	1.3090	75
16	0.2793	0.2756	0.9613	0.2867	3.4874	1.0403	3.6280	1.2915	74
17	0.2967	0.2924	0.9563	0.3057	3.2709	1.0457	3.4203	1.2741	73
18	0.3142	0.3090	9.9511	0.3249	3.0777	1.0515	3.2361	1.2566	72
19	0.3316	0.3256	0.9455	0.3443	2.9042	1.0576	3.0716	1.2392	71
20	0.3491	0.3420	0.9397	0.3640	2.7475	1.0642	2.9238	1.2217	70
21	0.3665	0.3584	0.9336	0.3839	2.6051	1.0711	2.7904	1.2043	69
22	0.3840	0.3746	0.9272	0.4040	2.4751	1.0785	2.6695	1.1868	68
23	0.4014	0.3907	0.9205	0.4245	2.3559	1.0864	2.5593	1.1694	67
24	0.4189	0.4067	0.9135	0.4452	2.2460	1.0946	2.4586	1.1519	66
25	0.4363	0.4226	0.9063	0.4663	2.1445	1.1034	2.3662	1.1345	65
26	0.4538	0.4384	0.8988	0.4877	2.0503	1.1126	2.2812	1.1170	64
27	0.4712	0.4540	0.8910	0.5095	1.9626	1.1223	2.2027	1.0996	63
28	0.4887	0.4695	0.8829	0.5317	1.8807	1.1326	2.1301	1.0821	62
29	0.5061	0.4848	0.8746	0.5543	1.8040	1.1434	2.0627	1:0647	61
30	0.5236	0.5000	0.8660	0.5774	1.7321	1.1547	2.0000	1.0472	60
31	0.5411	0.5150	0.8572	0.6009	1.6643	1.1666	1.9416	1.0297	59
32	0.5585	0.5299	0.8480	0.6249	1.6003	1.1792	1.8871	1.0123	58
33	0.5760	0.5446	0.8387	0.6494	1.5399	1.1924	1.8361	0.9948	57
34	0.5934	0.5592	0.8290	0.6745	1.4826	1.2062	1.7883	0.9774	56
35	0.6109	0.5736	0.8192	0.7002	1.4281	1.2208	1.7434	0.9599	55
36	0.6283	0.5878	0.8090	0.7265	1.3764	1.2361	1.7013	0.9425	54
37	0.6458	0.6018	0.7986	0.7536	1.3270	1.2521	1.6616	0.9250	53
38	0.6632	0.6157	0.7880	0.7813	1.2799	1.2690	1.6243	0.9076	52
39	0.6807	0.6293	0.7771	0.8098	1.2349	1.2868	1.5890	0.8901	51
40	0.6981	0.6428	0.7660	0.8391	1.1918	1.3054	1.5557	0.8727	50
41	0.7156	0.6561	0.7547	0.8693	1.1504	1.3250	1.5243	0.8552	49
42	0.7330	0.6691	0.7431	0.9004	1.1106	1.3456	1.4945	0.8378	48
43	0.7505	0.6820	0.7314	0.9325	1.0724	1.3673	1.4663	0.8203	47
44	0.7679	0.6947	0.7193	0.9657	1.0355	1.3902	1.4396	0.8029	46
45	0.7854	0.7071	0.7071	1.0000	1.0000	1.4142	1.4142	0.7854	45
		Cos	Sin	Cot	Tan	Csc	Sec	Rad	Deg

*From Carl B. Allendoerfer and Cletus O. Oakley, *Principles of Mathematics,* 2nd ed., ©1963, p. 473. Reprinted by permission of McGraw-Hill Book Co., New York, N.Y.

APPENDIX D

English Measures

Measure of Length

1 foot (ft) = 12 inches (in.)
1 yard (yd) = 3 feet (ft)
= 36 inches (in.)
1 rod = 16 1/2 feet (ft)
= 5 1/2 yards (yd)
1 mile (mi) = 5,280 feet (ft)
= 1,760 yards (yd)
= 320 rods

Measure of Area

1 square foot (ft^2) = 144 square inches ($in.^2$)
1 square yard (yd^2) = 9 square feet (ft^2)
1 acre = 4,840 square yards (yd^2)
= 43,560 square feet (ft^2)
1 square mile (mi^2) = 640 acres

Measure of Volume

1 cubic foot (ft^3) = 1,728 cubic inches ($in.^3$)
1 cubic yard (yd^3) = 27 cubic feet (ft^3)

Liquid Measure

1 fluid ounce (fl oz) = 2 tablespoons (tbs)
1 tablespoon (tbs) = 3 teaspoons (tsp)
1 cup (c) = 8 fluid ounces (fl oz)
1 pint (pt) = 16 fluid ounces (fl oz)
1 quart (qt) = 2 pints (pt)
1 gallon (gal) = 4 quarts (qt)

Dry Measure

1 quart (qt) = 2 pints (pt)
1 peck (pk) = 8 quarts (qt)
1 bushel (bu) = 4 pecks (pk)

Measure of Weight

1 pound (lb) = 16 ounces (oz)
1 ton = 2,000 pounds (lb)

Volume and Capacity

1 gallon (gal) = 231 cubic inches ($in.^3$)

1 bushel (bu) = 1 1/4 cubic feet (ft^3)

Angular Measure

1 circle = 360 degrees ($^\circ$)

1 degree ($^\circ$) = 60 minutes ($'$)

1 minute ($'$) = 60 seconds ($''$)

APPENDIX E Metric Measures and Conversions

Metric Measures of Length

10 millimeters (mm) = 1 centimeter (cm)
10 centimeters (cm) = 1 decimeter (dm)
10 decimeters (dm) = 1 meter (m)
10 meters (m) = 1 dekameter (dam)
10 dekameters (dam) = 1 hectometer (hm)
10 hectometers (hm) = 1 kilometer (km)

Metric to English

1 meter (m) = 39.37 inch (in.)
= 1.0936 yard (yd)
1 centimeter (cm) = 0.3937 inch (in.)
1 millimeter (mm) = 0.03937 inch (in.)
1 kilometer (km) = 0.6214 mile (mi) (approx. 5/8 mile)
1 liter (ℓ) = 1.0567 liquid quart (qt)
= 0.9081 dry quart (qt)
1 gram (g) = 0.0353 ounce (oz)
1 kilogram (kg) = 2.2 pound (lb)

English to Metric

1 inch (in.) = 2.54 centimeter (cm)
1 foot (ft) = 0.3048 meter (m)
1 yard (yd) = 0.9144 meter (m)
1 mile (mi) = 1.6093 kilometer (km)
1 liquid quart (qt) = 0.9463 liter (ℓ)
1 dry quart (qt) = 1.1012 liter (ℓ)
1 ounce (oz) = 28.35 gram.
1 pound (lb) = 0.4536 kilogram (kg)

Index

I

K

L

M

N

O

P

Q

R

S

T

U

V

W

X

Y